Particle Technology and Engineering

Particle Technology and Engineering

An Engineer's Guide to Particles and Powders: Fundamentals and Computational Approaches

Jonathan Seville

Chuan-Yu Wu

AMSTERDAM • BOSTON • HEIDELBERG • LONDON
NEW YORK • OXFORD • PARIS • SAN DIEGO
SAN FRANCISCO • SINGAPORE • SYDNEY • TOKYO

Butterworth-Heinemann is an imprint of Elsevier

Butterworth-Heinemann is an imprint of Elsevier
The Boulevard, Langford Lane, Kidlington, Oxford OX5 1GB, UK
50 Hampshire Street, 5th Floor, Cambridge, MA 02139, USA

British Library Cataloguing-in-Publication Data
A catalogue record for this book is available from the British Library

Library of Congress Cataloging-in-Publication Data
A catalog record for this book is available from the Library of Congress

ISBN: 978-0-08-098337-0

For information on all Butterworth-Heinemann publications visit our website at https://www.elsevier.com/

Publisher: Joe Hayton
Acquisition Editor: Fiona Geraghty
Editorial Project Manager: Lindsay Lawrence
Production Project Manager: Nicky Carter
Designer: Maria Inês Cruz

Typeset by TNQ Books and Journals

Contents

Preface

Particles—small discrete elements of matter—are all around us and are important as natural phenomena (mist, rain, snow, sand, etc.) and as products (food, building materials, pharmaceuticals, etc.). The study of Particle Technology has become an essential part of the education of many kinds of engineers and scientists. It overlaps significantly with—and draws material from—related subjects such as the sciences of aerosols, colloids, and surfaces in general.

Particle Technology as a discipline must include both a study of the fundamentals of how particulate materials behave and some indications of how the science can be applied in practice. We have both been involved in numerous discussions of the scope and organization of Particle Technology in association with our teaching in UK universities and our work on the Editorial Board of the Elsevier journal *Powder Technology*. In our view, an organization of the subject of Particle Technology which reflects the interests of those who carry out research in it is as shown in the diagram below.

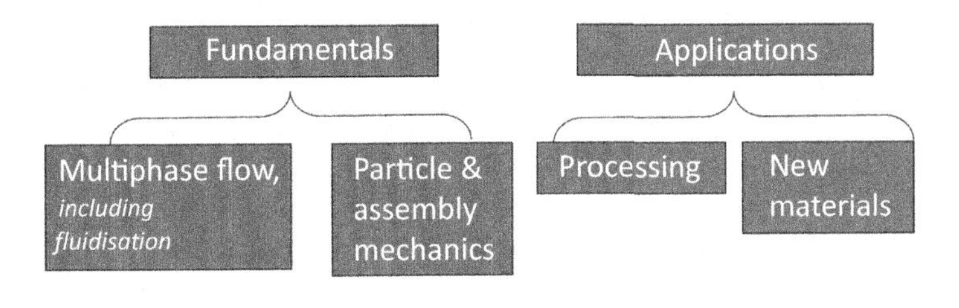

Organization of research effort in particle technology.

In our view, the interaction of particles with fluids in multiphase flow needs always to be accompanied by an appreciation of the mechanics of particle-to-particle contact and the peculiar behavior of assemblies of particles. On the application side, the processing of products containing particles is hugely important, but so are the properties of materials containing particles. Many new materials containing particles emerge every year, and the ways in which their microstructures are formed and are subsequently modified in use are important parts of the emerging discipline of Formulation Engineering.

As we summarize in Chapter 1 of this book, the fundamental science on which Particle Technology draws was established in the nineteenth century by physicists such as Stokes, Smoluchowski, and Hertz, and elaborated in the twentieth century by numerous others, spurred on by the rapid increase in applications of the subject, from particulate catalysts in reactors to composite materials. The number of applications continues to increase, but the scientific fundamentals remain the same. What has changed the subject very significantly in the last two decades is the rise of computational methods, enabled by the development of better computational codes and the affordability of massive computer power.

Built upon an earlier book authored by Seville, Tüzün, and Clift (1997), this book attempts to do two things: to summarize the essential scientific fundamentals and to introduce the basics required to perform computations in Particle Technology. For the former, we start by introducing the fundamental characteristics of powders in bulk form in Chapter 2, explaining the important bulk properties of powders and how they can be determined. As the bulk properties are ultimately determined by the properties of individual particles, we then introduce individual particle properties in Chapter 3—in particular, particle shape and size, how they can be defined, and how they can be appropriately measured. We then introduce the complexity of a surrounding fluid phase, first in interaction with a single particle (Chapter 4), then through considering multiple particles in gases, in applications such as gas fluidization and pneumatic conveying (Chapter 5), and multiple particles in liquids, in applications including granulation and extrusion (Chapter 6). We also explain the fundamental mechanics of particle systems, both at the bulk level, such as the development of stresses in storage and dynamics during powder flow (Chapter 7), and at the particle level, including particle–particle interaction (Chapter 8). Finally, we introduce two computational methods, namely the discrete element method (Chapter 9) and the finite element method (Chapter 10), both of which have been applied extensively in modeling the behavior of particle systems at low consolidation stresses, such as in powder flow and shearing, and at high consolidation stresses as encountered in die compaction and roll compaction. The last four chapters focus on mechanistic modeling of particle systems and are aimed primarily at Chemical Engineering students in their later years and at both Chemical Engineers and other disciplines, in industry or in academia, who need to carry out mechanical analysis and computational work in this field.

Particle Technology is a broad and inclusive subject. Any work of this kind must necessarily be very selective. This book focuses on fundamentals, particle mechanics, and computational aspects in Particle Technology; we refer readers to the volumes in the Elsevier series *Handbook of Powder Technology* for more detailed treatment of particular aspects.

We thank all those who have contributed to our understanding of the subject, particularly Mike Adams, Roland Clift, Peter Knight and Colin Thornton, and to our long-suffering families. Thanks are also due to the universities of Surrey and Birmingham, where much of this material has been used in teaching undergraduate and graduate courses.

Jonathan Seville and Charley Wu

CHAPTER

Introduction

1

Particle technology is the study of discrete elements—usually solid particles—the way in which they behave in isolation and the way in which they interact to produce a collective effect (Seville, 2001). The ultimate goal of particle technology is to obtain predictive relationships between individual particle properties and their collective behavior (i.e., behavior in bulk), which can then be used to design formulations, processes, and particulate products. Although much progress has been made toward achieving that goal, there is much still to do!

1.1 WHAT ARE PARTICLES?

Particles are endlessly fascinating, both to scientists and engineers and to children who build sand castles on the beach. Particles surround us—they form the foundations beneath our buildings, the soil in which we grow our food, very often that food itself. They are suspended in the air that we breathe and can cause us great damage when we do so. They are extracted from the ground to obtain metals and other mineral products. They make up many of the products we consume and the wastes derived from them and are liberated in the incineration of that waste. It is estimated that over two-thirds of all chemical products are sold as, or have passed through, a particulate form. Much energy is expended in their processing. For example, crushing and grinding of minerals is estimated to use 3–4% of the world's clectricity.

The first point to make about particles is the extraordinary range of scales that they occupy (Fig. 1.1)—everything from large molecules to small bricks is generally lumped into this category. The ratio between the size of the earth and the size of a football is roughly the same as that between a football and a typical nanoparticle. As in most branches of human endeavor, it is useful to have some human length scales from which to take reference: the diameter of a human hair is of order 0.1 mm (or 100 μm), while the diameter of a human blood cell is 6–8 μm. An important physical reference from the point of view of experimental observation is the wavelength of visible light—centered at about 0.5 μm.

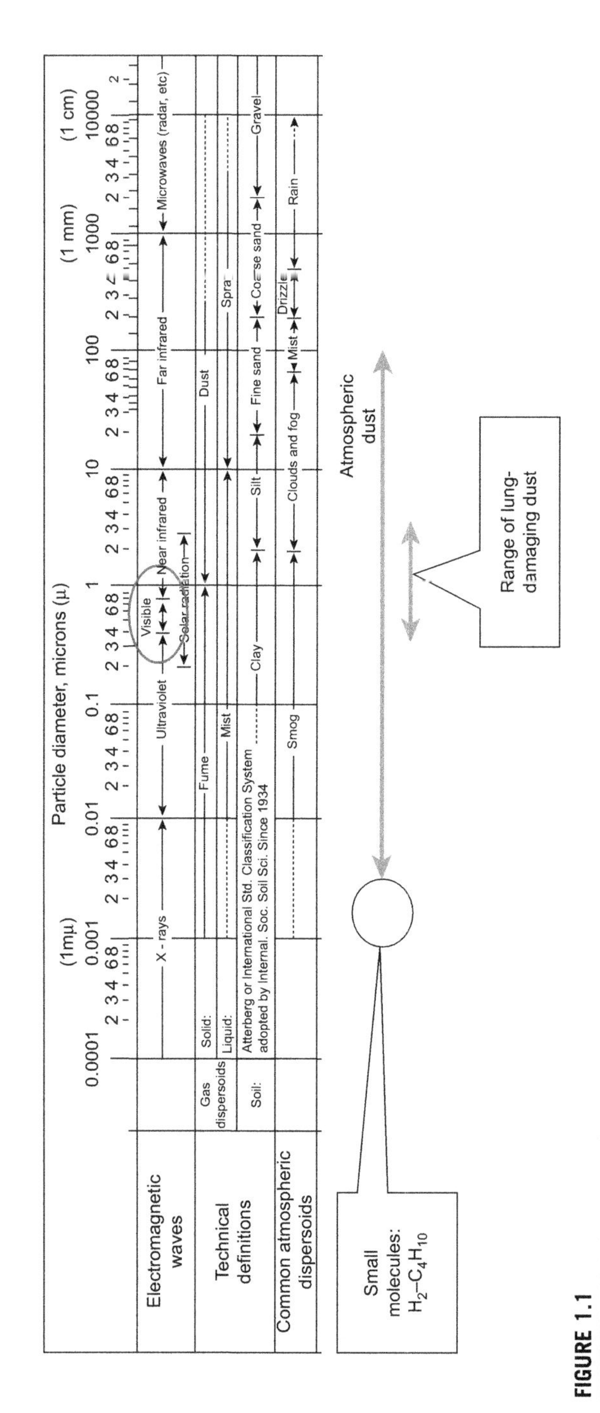

FIGURE 1.1

Particles cover a wide range of scales. (Adapted from Seville et al., 1997.)

Size has a profound effect on particle properties. Consider, for example, the family of atmospheric dispersions of water in air: rain (diameter 1 mm−1 cm); drizzle (100 μm−1 mm); and fog (1−100 μm). One obvious difference between them is the effect of gravity: the motion of raindrops is driven by gravity; they fall quickly to the ground, whereas fog droplets remain in dispersion until a rise in temperature or wind movement removes them. Another notable difference concerns the interaction with light: rain is relatively easy to see through; fog is not. Suspended particles of a size close to that of the wavelength of light are strong light-scatterers and reduce visibility. This suggests, of course, a method for measuring the concentration of particles, which is considered in Chapter 3.

Particles are not normally found singly, but in very large numbers, so that the scale of one particle is usually many orders of magnitude smaller than the scale of the container or process in which they sit. A "particulate solid" or "bulk solid" is an assembly of particles that may be surrounded by a continuous fluid phase—a gas (such as air) or a liquid. They may or may not be in contact with each other.

It would make it much easier to predict the behavior of particles if we could consider them as large molecules, for which many analytical predictive approaches have been developed, but this is bound to be an oversimplification. Particle-fluid systems present inherent experimental and theoretical difficulties (Grace, 1986) for which there are no analogies in molecular systems:

- particle shape
- particle size distribution
- surface effects, such as contamination with oxide layers
- regimes of motion, causing profound changes in drag with changes in Reynolds number[1]

in addition to secondary effects such as:

- Brownian[2] motion
- electrostatic charge distributions

Particle systems also display a "memory" of previous processing, resulting, for example, in their sticking together (agglomeration) or breaking apart (attrition), accompanied by what may be a profound change in properties.

The divided state is a constant theme in philosophical discussion—how can we exert our individuality within the collective behavior of society? More to

[1]The Reynolds number (Re) is a nondimensional group which describes the flow regime in fluid mechanics. For a particle it is defined as $\rho Ud/\mu$ where ρ and μ are the fluid density and viscosity respectively, U is the fluid velocity (relative to the particle) and d is the particle diameter. For a given particle and fluid, therefore, "low Reynolds number" implies low fluid velocity and vice versa. Low Reynolds number flows are dominated by viscosity, which is the case for micrometer-sized dust particles in air, for example. The Reynolds number can be considered as a ratio of momentum forces to viscous forces; see Chapter 4; after Irish/English Engineer Osborne Reynolds (1842−1912).

[2]Brownian motion is the random motion of small particles caused by collisions with atoms or molecules in the surrounding fluid; after Scottish botanist Robert Brown (1773−1858).

the point, given a large collection of individual people (or particles) how can we predict their collective behavior? The collective behavior of people is the province of economics and politics; particles are a little (but not much) easier to deal with.

Taking this theme further, we are surrounded by a different sort of particle—the information dot: the mark of the ink jet or the laser, the paint particle, the phosphor dot, the light-emitting diode, the tiny units of display which collectively register an effect, provided of course that we are far enough away to make sense of the resulting pattern. There is a long line of artists who have painted in this way, including, most famously, Georges Seurat,[3] the impressionist originator of "pointillism," who based his technique on a scientific study of color analysis and visual perception.

The scientist and natural philosopher Pierre-Simon Laplace[4] believed that the universe is a predictable machine (like a great clock) and that if only we could calculate the behavior of all the parts, we could predict the future. Isaac Newton[5] had made this seem possible, by showing that the motion of a set of particularly large particles obeys simple laws. At the same time, Newton knew much about matters at the scale of the wavelength of light and postulated a world of subvisible particles behaving according to analogous sets of laws. We now know, of course, that molecular systems do not behave in this way and their collective behavior can only be predicted in a statistical sense. Famously, Werner Heisenberg[6] pointed out that at the molecular scale not only can you not predict the future, but also you cannot even know the present exactly.

What has this to do with particles, which are, for the most part, large enough that at least we know where they are, even if we must take a scanning electron microscope to look at them? The answer is that computational particle technology has now reached a critical point: it appears to be able to predict anything, if only the sponsors will buy us large enough computers. Maybe there is no further need for our approximate and semi-empirical methods. Can this be true?

The purpose of this book is to introduce the newcomer to the fundamentals of particle technology and to explain the basic principles behind the emerging computational approaches to the subject.

1.2 WHAT IS KNOWN?

A short answer to this question might be: "Quite a lot about isolated spheres and spheres in contact; much less about the behavior of nonspherical particles and particles in assemblies."

[3]French painter (1859–1891).
[4]French mathematician, physicist and astronomer (1749–1827).
[5]English mathematician, physicist and astronomer (1643–1727).
[6]German physicist (1901–1976).

The problem of the drag on an isolated sphere at low Reynolds numbers was solved exactly in the late nineteenth century by George Stokes[7]—perhaps the most important contribution to the subject. The twentieth century saw a vast amount of experimental work on spherical and nonspherical particles over the full range of Reynolds numbers, brought together by Clift et al. (1978). Although the behavior of nonspherical particles is complex, simple shape-descriptors are often good enough. For example, spheroids are surprisingly good models for more complicated shapes of airborne particles, because the processes in which such small particles participate nearly always take place at low Reynolds numbers. Such processes can usually also be modeled using computational fluid dynamics (CFD) codes, because the presence of the particles has little effect on the motion of the fluid.

The difficulty arises when the solids content in the fluid becomes large enough that the fluid motion is affected by the particles. In this case, the state of the art is semi-empirical, relying on the well-known equations of Ergun and Richardson-Zaki and others (see Chapter 5). This area is by no means resolved and allegedly better approaches are continually suggested. The situation is more complex still when the particles make extensive contact, as in fluidized beds. In this case, stress can be transmitted through the particles, not just through the fluid. At the opposite end of the spectrum from the small isolated particle in suspension, then, is the case of a dense assembly of particles in contact, flowing under gravity out of a container. In this case, unless the particles are small the fluid phase is of little importance; particle motion is dominated by frictional forces between the particles and the walls of the container and internal friction between them. Particles move according to where the vacancies arise—a principle known as the "kinematic theory" (Nedderman, 1992).

It is interesting to contrast the nineteenth century work of Stokes on particle drag in fluid flow with the contact mechanics work of Heinrich Hertz[8] of about the same period. The former is well known in chemical engineering and is widely applied. The latter is much better known in mechanical engineering in connection with bearings and is little applied in particle technology. The whole field of tribology—the study of friction, lubrication, and wear at contacts between surfaces—is of great importance in particle technology and has moved on a long way from Hertz's early work on elastic contacts.

The reason that the contact mechanics developed by Hertz (1896) has not been applied in particle technology as much as in other fields is probably that Hertz's mathematical solution was obtained for the contact of two frictionless elastic spheres: that is to say for spheres which are finely polished so that the friction between them is negligible. However, real particles are rarely spherical or smooth. It was not until the 1960s (i.e., half a century later) that Mindlin and Deresiewicz (1953) considered the friction between two spheres and developed mathematical solutions for the contact

[7]Irish mathematician, physicist, politician, and theologian (1819–1903).
[8]German physicist (1857–1894).

between elastic spheres with various oblique forces. The work of Hertz, Mindlin, and Deresiewicz was further extended by Maw et al. (1976, 1981) to analyze the impact between two elastic spheres at any impact angle, for which the rebound behavior can be predicted; numerical solutions are given in Chapter 8. The application of contact mechanics in particle technology only gained its impetus in the 1990s, with the development of the discrete element method (DEM, see Chapter 9). The implementation of the theories of Hertz, Mindlin, and Deresiewicz in DEM was pioneered by Thornton (Thornton and Barnes, 1986), who demonstrated that these theories for the contact between two spheres are useful and physically sound in predicting the collective behavior of particle systems in DEM; since then they have been applied more and more frequently in particle technology. However, even these sophisticated approaches are still limited to systems of spheres; for nonspherical particles, the analysis is very complicated and mechanistic analysis is still an area for further study. Nevertheless, analysis using spheres as an approximation has built a strong theoretical base in particle technology and demonstrated the power of DEM.

Another important complication in particle mechanics arises from the many surface forces which act between particles, due to intermolecular attractions (including Van der Waals forces), electrostatics, mechanical interlocking, and the presence of surface contaminants such as free liquids. The adhesive forces which arise in the absence of any other medium, such as a liquid layer, are collectively known as "autoadhesive." The question here is not "How can autoadhesive forces be sufficiently strong as to influence the behavior of the system?" but "Why do autoadhesive forces not have a more noticeable effect on particle behavior?" Theoretical calculations of autoadhesive forces predict a force which is comparable with the particle weight for particles of order 1 mm in size (Seville et al., 2000), but the familiar consequences of interparticle force effects—increased void fractions and agglomeration into clusters—are not usually observed in practice for particles above about 100 μm in size. The reason is that adhesion between real particles is invariably limited by surface roughness (asperity contact) and contamination (Kendall, 2001).

In fluidization, for example, it is generally agreed that it is the ratio of interparticle force to particle weight which governs the transition between the characteristic fluidization behavior demonstrated by Geldart's groups A ("aeratable") and B ("bubbling") (Geldart, 1986) (Fig. 1.2). By stabilizing a structure of higher voidage than would otherwise be the case, they make it possible for fluidized beds to show a region of "uniform" expansion before bubbling, which is the most characteristic (and useful) feature of group A powders.

If there is liquid present, at high humidity for example, "liquid bridges" may form between the particles, as shown in Fig. 1.3. This introduces a range of new problems of great complexity, even if the liquid is Newtonian[9] in character and the particles are spherical. The presence of a liquid bridge between two particles introduces forces between them. If they are not in relative motion, the resulting force

[9]See Chapter 8

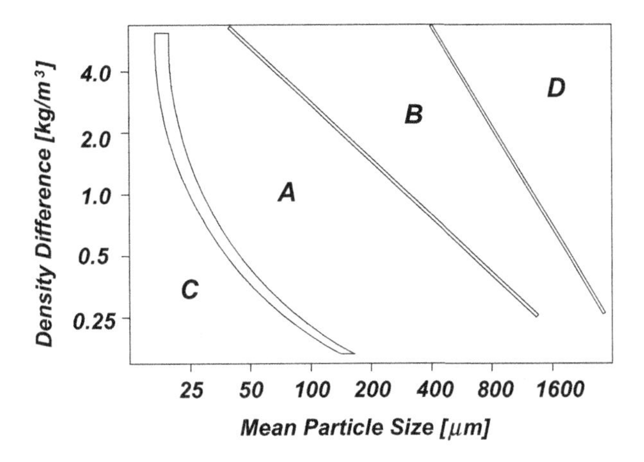

FIGURE 1.2

Geldart fluidization diagram (Geldart, 1986). *C*, cohesive; difficult to fluidize; *A*, aeratable; *B*, bubbling; *D*, spoutable.

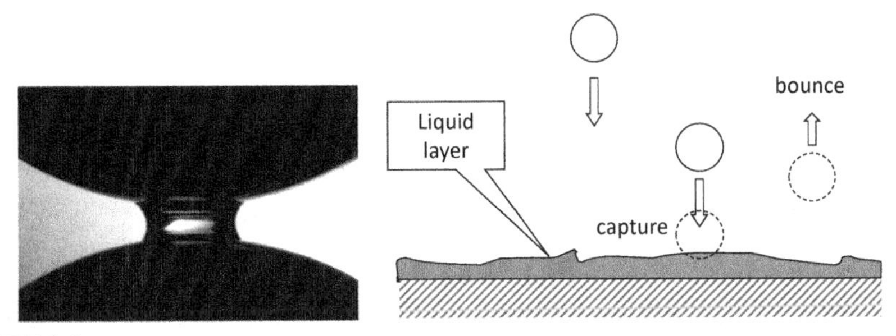

FIGURE 1.3

A liquid bridge between spherical particles (left) and bounce of a spherical particle on a liquid layer (right).

is derived from the liquid surface tension, but if there is relative movement there is a further force contribution, which dominates at higher relative velocities, due to the squeezing of liquid from the gap between the particles as they approach. As might be expected, this "lubrication" force, as it is known, is proportional to the fluid viscosity. For nondeforming smooth spheres, it is predicted to go to infinity at contact; in practice, the particles must deform or, more probably, the real contact will be limited by surface imperfections.

It is possible to derive conditions for the successful capture of particles on surfaces due to such energy dissipation mechanisms. For van der Waals' forces alone, capture depends largely on whether the surfaces deform, since van der Waals' interactions cannot dissipate energy. In outline, rebound from a surface will occur if the kinetic energy of the approaching particle exceeds the energy dissipated in the collision. In the presence of a fluid phase, energy dissipation can also occur by viscous

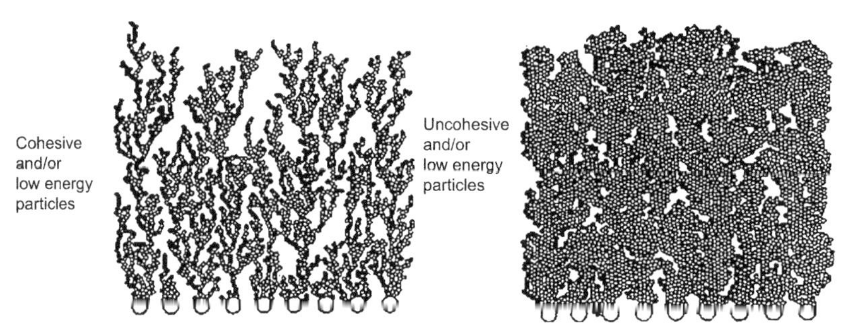

FIGURE 1.4

Filter cake structure (Houi and Lenormand, 1986).

lubrication. This problem was considered in some depth by Davis and coworkers, most notably in a simplified form by Barnocky and Davis (1988, see also Chapter 8), who demonstrated that their theory was in agreement with simple "bounce" experiments.

Many particulate products, such as detergent powders or tablets, are structured "agglomerates" (or "aggregates") of different types of particles, bound with other phases. There is a wide industrial interest in modeling the formation of such structures. Using the approaches outlined above it would appear possible to do so, but there are still many complexities to be overcome.

First, consider the case of two particles making contact at some relative velocity. Do they "stick" or do they "bounce"? If they stick, an agglomerate is formed. The simplest agglomeration case is probably that of aerosol "coagulation," in which the particles may be droplets. In this case, every contact results in coalescence and the result is a sphere of appropriate volume. This case was considered by Marian Smoluchowski[10] (see, for example, Hinds, 1982), who predicted the resulting change in the size distribution. That early work pioneered the study of population balances in particle technology, which is now a very active area of research and is also widely applied in crystallization.

One of the simplifying features of aerosol coagulation is that bounce or rebound cannot occur. A simple example in which this is not the case is filtration of solid aerosol particles from a gas. Figure 1.4 shows the results of a simple two-dimensional computation of such a filtration process. When the particles have low kinetic energy or the adhesion is high, they will stick at the first contact, resulting in the dendritic shapes seen in this figure. When the kinetic energy is high and/or the adhesion is low, particles will rebound, perhaps many times, before finding a stable position. The resulting structure is therefore denser. The general features of this behavior are observed in practice in gas filtration and are of great practical importance. A more complex example is that of diffusion-limited aggregation.

[10]Polish/Austrian physicist (1872–1917).

The situation in dense agglomeration processes such as fluidized beds and high-shear mixers is more complicated still, since the particle motion is not understood to a sufficiently high degree, and the agglomerates formed are subject to deformation, fluid movement within them, and breakage (see, for example, Iveson et al., 2001). This is an important area in the engineering of particulate products for the specialty chemical, pharmaceutical, and detergent industries. The aim is usually to form a structure with optimum properties for the delivery of a particular chemical or biological effect in a desired environment and at a desired rate. Use of a solidifying binder may help; faster solidification makes more open dendritic structures.

Formation of aggregates by application of pressure ("compacting") is an important forming operation—in pharmaceutical tableting, for example, and in the manufacture of components from powdered metals. As explained in Chapter 10, predicting the result of a compaction operation depends on friction, both internal friction between particles and each other and external friction between particles and boundaries such as walls. Briscoe and Adams (1987) describe many problems of this kind.

Both experiment and computation have shown the very interesting microstructure which develops in compacts under load. Even if an apparently uniform load is applied, the resulting stress is by no means uniform at a microscopic level, but forms a characteristic "stress fabric," in which some particles are heavily loaded and some completely redundant, as shown in Fig. 1.5. Troadec et al. (1991) carried out elegant photoelastic experiments using stressed arrays of rods, and were able to remove the unloaded rods without changing the macroscopic elastic behavior of the structure, thus demonstrating their redundancy.

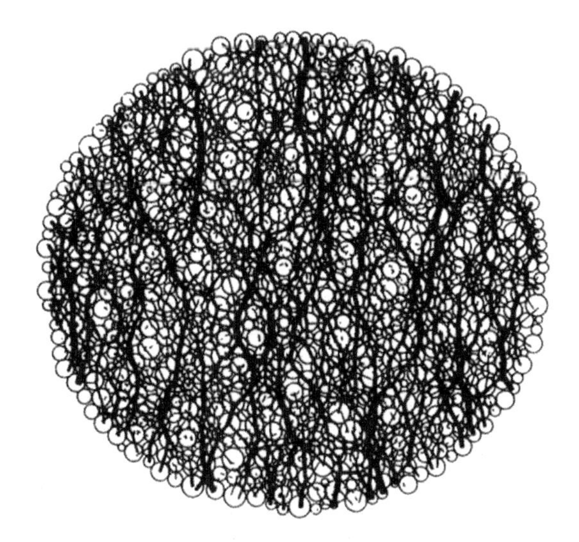

FIGURE 1.5

Stress fabric (line width indicates local contact force) (Thornton and Barnes, 1986).

As the structure is loaded, the more heavily stressed particles may break, resulting in a redistribution of the load. This is what happens in a "comminution" (grinding) process, and explains why in such a process it is necessary to carry out repeated loadings, reordering the particles each time, so as to ensure that there are always some new heavily loaded particles. Adams et al. (1994) modeled this problem using parallel columns to represent stress pathways. These fascinating structures have a natural order of their own; their study is closely related to the mathematical subject of "percolation theory."

Having formed a structure, it is often important to understand its strength. The classical model of strength was that the force needed to break an object in tension is the sum of all the inter-atomic forces acting across the failure surface. A. A. Griffith[11] showed that this is not true. Real materials are much weaker than would be predicted by this approach because they contain flaws or imperfections, from which cracks are propagated. What happens is that as the crack extends, the strain energy that is stored around the crack is dissipated, and this release of strain provides the energy required to form the new surfaces. It should not be a surprise that granular materials such as agglomerates also fail (in general) by crack propagation. In some simple cases the fracture strength can be directly related back to the energies of the individual bonds that are broken in advancing the crack. It is interesting that very small particles do not break in a brittle way, but fail plastically, because they have such a small volume that they cannot store enough strain energy for crack propagation (Kendall, 2001).

1.3 COMPUTATIONAL MODELING

The ultimate aim of many researchers in particle technology has been to relate the individual particle properties to the macroscopic behavior of the assembly. In other words, what many people would like to do is to use the properties of particles—size, shape, density, elasticity, fracture toughness, etc.—as input parameters from which the subsequent behavior of assemblies of those particles can be obtained. In the last three decades, use of the "discrete (or distinct) element method" (DEM) has gathered pace; it appears to offer particle technologists the opportunity to do just this. There are several varieties of this approach, including molecular dynamics, Newtonian dynamics, cellular automata, and stochastic techniques. Using Newtonian physics, the motion of each individual particle in a process can be determined (up to about 10^7 with current technology, including high performance computing). Individual particle properties (such as elasticity and plasticity) can be specified directly and the consequences of impacts calculated, fluid flow can be incorporated (with some approximations) and the assembly response calculated without (in principle) *a priori* assumptions. Although the number of particles is limited, periodic boundary conditions can be used to extend the physical size.

[11] English engineer (1893–1963).

Nevertheless, the maximum number is still small compared with any real process. Although particles are generally simplified as spheres, particle shape and morphology can be considered with the recent advances in DEM. The value of the technique is probably in carrying out "numerical experiments," which reveal aspects of the behavior that can be studied further using real experiments. The leading groups in the field are well represented in the volumes edited by Thornton (2000, 2008), Kishino (2001) and Wu (2012).

As a very simple indication of what can be done with DEM, some examples of computations performed with the soft sphere model (see Chapter 9 for more details) are presented below, on a device known as a "V-mixer" (Kuo et al., 2002). This is a simple mixer, widely used in the pharmaceutical industry, which consists of two cylinders, joined at 90°, as shown in Fig. 1.6. It rotates around the axis shown, thus repeatedly dividing (Λ) and recombining (V) the contents. At each rotation a small amount of material is transferred across the plane of symmetry between the two arms. The reason for studying this device is not because it is of great inherent interest in itself but because it is representative of that generic class of mixers in which the solids are repeatedly divided and recombined. Typical results of modeling are shown in Fig. 1.6.

One of the chief challenges in modeling is experimental validation, which was achieved in this case by following the trajectories of single tracer particles using a radiation-based technique called positron emission particle tracking (PEPT). Figure 1.7 shows the close agreement that can be achieved between the average

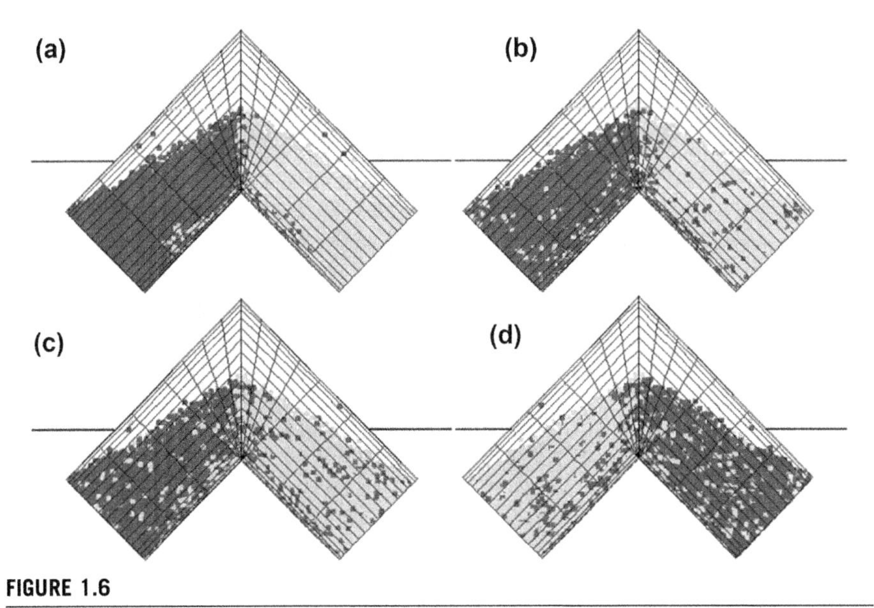

FIGURE 1.6

DEM simulations for 1 (a), 4 (b), 8 (c) revolutions in the front views and eight revolutions in the back view (d) (Kuo et al., 2002).

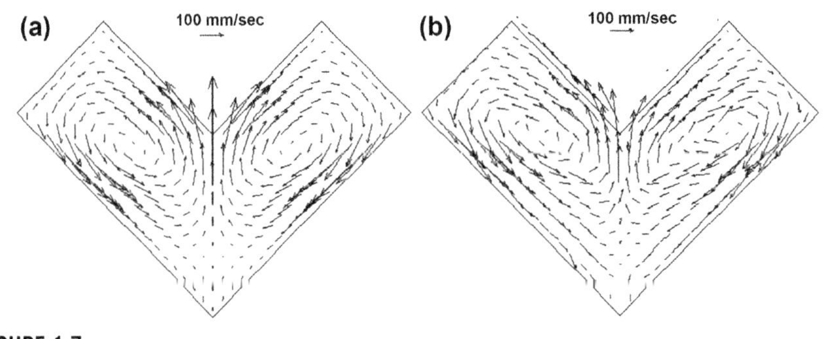

FIGURE 1.7

Velocity distributions at 20% fill, 60 rpm. (a) DEM simulations; (b) PEPT experiment (Kuo et al., 2002).

velocity distributions from both model and experiment. A more quantitative comparison between the two is provided by the exchange rate of solids across the dividing plane, which is plotted as a function of fill level in Fig. 1.8 (the exchange rate is here defined as the probability that a single particle will swop sides per revolution). Again, the agreement is good. It is well known that the exchange rate reduces as the fill level is increased, but this work also revealed the unexpected result that particle exchange during the division step is more important than during the combination step.

The agreement between model and experiment is all the more remarkable because of the very simple interaction model adopted here. Much more sophisticated (and physically realistic) models are available (see, for example, Thornton et al., 2011, 2013). No doubt it is significant that the particles used in the experiment were smooth spheres, which are sufficiently large that neither adhesion nor interstitial gas flow plays a role.

Use of the DEM family of computational techniques used to be confined to experts in the field, but advances in both hardware and software are now bringing

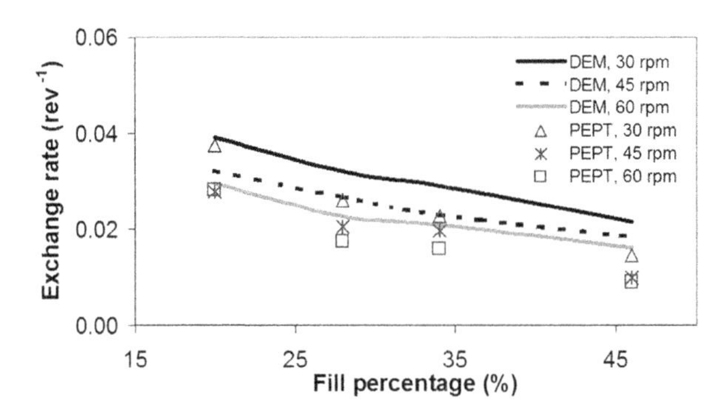

FIGURE 1.8

The exchange rate as a function of the fill percentage at different rotational speeds (Kuo et al., 2002).

it within the reach of most researchers, and there are several commercial codes. It is interesting to note that DEM modeling combines knowledge from both contact mechanics and (if fluid flow is incorporated) fluid mechanics—the two essential skills of particle technology.

1.4 WHAT ARE THE APPLICATIONS?

Particle technology has a long and fascinating history—going back at least as far as the ancient Egyptians—and an even more interesting emerging future.

Much of the early work in particle technology arose from the needs of the energy and minerals industries, particularly from the problems of handling and processing mined materials such as iron ore and coal, and from the oil industry in which large catalytic reactors were developed from the 1940s onward. Particle technology was also of importance in the development of nuclear power. Despite the growth in use of gas as a fuel for power generation—particularly in North America and Europe—in a world context, solid fuels are not declining in importance. Despite the growth in use of renewable energy sources, the International Energy Agency predicts that coal will have a 22% share of the world energy supply in 2030 and that an additional 10% of this supply will be met by biomass. There is therefore a strong incentive to develop ways of processing these fuels in an efficient way. This means minimizing carbon dioxide (and other undesirable) emissions per unit of power generated but also devising ways of removing that carbon dioxide from the effluent stream for sequestration underground. Particle technology is central to gasification and combustion: particularly in particle handling, development of fluidization-based reactors and filtration systems for removing particles and gases by "dry scrubbing." It is also central to methods of separating carbon dioxide through the development of "chemical looping" combustion. More generally, most air and water environmental protection methods involve particles.

Looking further into the future, the development of the "hydrogen economy," which many consider an attractive future direction, requires development of efficient and robust fuel cells. Again, particle technology plays an important part, in both solid oxide and polymer electrolyte type cells, and in hydrogen storage technologies.

One area of application of particle technology which is growing rapidly is the engineering of particulate products, increasingly known as Formulation Engineering. This area of work concerns the design and manufacture of microstructured products which are often sold directly to the consumer and often designed to break down in use. Examples include foods, "instant" drinks such as dried soups, pharmaceutical dosage forms such as tablets, personal and household cleaning products, paints and inks—an increasingly long list. Figure 1.9 shows the issues to be considered: product formulation and its subsequent processing is aimed at forming a specified microstructure. In use (eating, taking the dose, painting the wall, etc.) the product microstructure breaks down in a useful way, releasing flavor or active ingredients or forming a gloss layer or whatever.

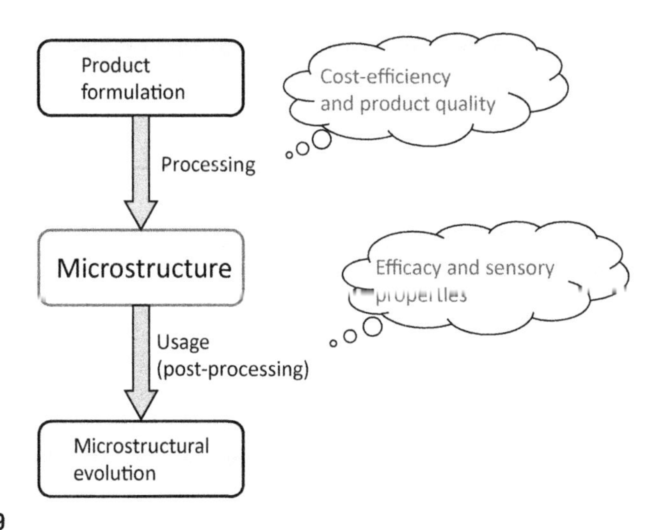

FIGURE 1.9

Manufacturing and using product microstructure (Adams, 2004).

In manufacture, cost-efficiency and product quality are of supreme importance. In use, perception of product quality will depend on the effective breakdown or evolution of microstructure.

The microstructure of the products mentioned above can be very complex, as indicated in Fig. 1.10. Apart from the complications of having many components and often several phases, there will often be liquid bridges between particles, both

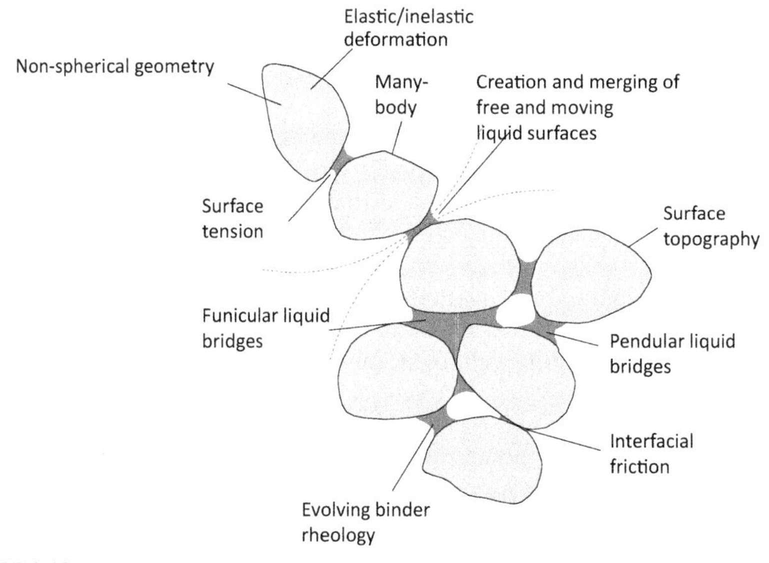

FIGURE 1.10

Some issues in understanding microstructure (Adams, 2004).

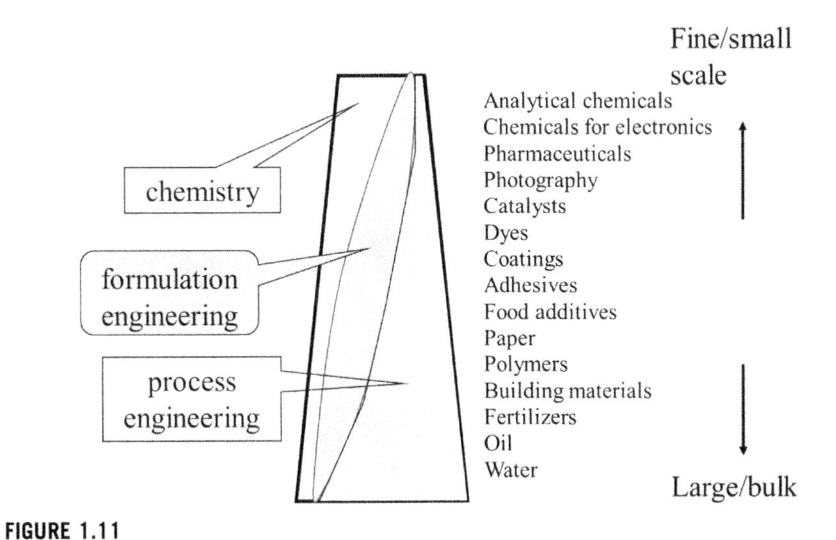

FIGURE 1.11

Formulation Engineering and the spectrum of products (Wesselingh, 2001).

filled and empty pore space, and possible dissolution of the solid so that the properties of the liquid will vary with time.

Formulation engineering sits somewhere between process engineering and product chemistry, as shown in Fig. 1.11. It has a large part to play in a spectrum of high-value-added products. In these newly emerging application areas, the fundamental challenges in particle technology might be summarized as:

- devising new ways of making (often smaller) particles;
- developing new ways of introducing greater functionality into particulate products;
- understanding scaling issues in modeling, enabling us to "bridge the gaps" in scale between…
  - the molecule and the single particle
  - the single particle and many particles
  - many particles and industrial processes
- new measurement techniques for following particle behavior at all these scales.

REFERENCES

Adams, M.J., 2004. Inaugural Lecture. University of Birmingham.

Adams, M.J., Mullier, M.A., Seville, J.P.K., 1994. Agglomerate strength measurement using a uniaxial confined compression test. Powder Technology 78, 5–13.

Barnocky, G., Davis, R.H., 1988. Elastohydrodynamic collision and rebound of spheres: experimental verification. Physics of Fluids 31 (6), 1324–1329.

Briscoe, B.J., Adams, M.J., 1987. Tribology in Particulate Technology. Adams Hilger, Bristol, Philadelphia.

Clift, R., Grace, J.R., Weber, M.E., 1978. Bubbles, Drops and Particles. Academic Press, New York.

Geldart, D. (Ed.), 1986. Gas Fluidization Technology. Wiley, Chichester.

Grace, J.R., 1986. Contacting modes and behaviour classification of gas—solid and other two-phase suspensions. The Canadian Journal of Chemical Engineering 64, 353–363.

Hertz, H., 1896. On the contact of elastic solids and on hardness. In: Hertz, H. (Ed.), Miscellaneous Papers. Jones & Schott, Macmillan and Co., London.

Hinds, W.C., 1982. Aerosol Technology. Wiley, New York.

Houi, D., Lenormand, R., 1986. Particle accumulation at the surface of a filter. Filtration and Separation 238–241.

Iveson, S.M., Litster, J.D., Hapgood, K., Ennis, B.J., 2001. Nucleation, growth and breakage phenomena in agitated wet granulation processes: a review. Powder Technology 117, 3–39.

Kendall, K., 2001. Molecular Adhesion and Its Applications. Kluwer/Plenum, New York.

Kishino, Y. (Ed.), 2001. Powders and Grains 2001. Balkema, Lisse.

Kuo, H.P., Knight, P.C., Parker, D.J., Tsuji, Y., Adams, M.J., Seville, J.P.K., 2002. The influence of DEM simulation parameters on the particle behaviour in a V-mixer. Chemical Engineering Science 57, 3621–3638.

Maw, N., Barber, J.R., Fawcett, J.N., 1976. The oblique impact of elastic spheres. Wear 38, 101–114.

Maw, N., Barber, J.R., Fawcett, J.N., 1981. The role of elastic tangential compliance in oblique impact. Transactions of the ASME, Series F: Journal of Lubrication Technology 103, 74–80.

Mindlin, R.D., Deresiewicz, H., 1953. Elastic spheres in contact under varying oblique force. Transactions of the ASME, Series E: Journal of Applied Mechanics 20, 327–344.

Nedderman, R., 1992. Statics and Kinematics of Granular Materials. Cambridge University Press.

Seville, J.P.K., 2001. Plenary Lecture, The 6th World Congress of Chemical Engineering. Melbourne, Australia.

Seville, J.P.K., Willett, C.D., Knight, P.C., 2000. Interparticle forces in fluidization: a review. Powder Technology 113, 261–268.

Seville, J.P.K., Tüzün, U., Clift, R., 1997. Processing of Particulate Solids. Blackie Academic and Professional, London, pp. 330–348.

Thornton, C., Barnes, D.J., 1986. Computer simulated deformation of compact granular assemblies. Acta Mechanica 64, 45–61.

Thornton, C. (Ed.), 2000. Powder Technology. Special issue, vol. 109, pp. 1–3.

Thornton, C. (Ed.), 2008. Special issue on discrete element modelling of fluidised beds. Powder Technology, vol. 184, pp. 132–265.

Thornton, C., Cummins, S.J., Cleary, P.W., 2011. An investigation of the comparative behaviour of alternative contact force models during elastic collisions. Powder Technology 210, 189–197.

Thornton, C., Cummins, S.J., Cleary, P.W., 2013. An investigation of the comparative behaviour of alternative contact force models during inelastic collisions. Powder Technology 233, 30–46.

Troadec, J.D., Bideau, D., Dodds, J.A., 1991. Compression of two-dimensional packings of cylinders made of rubber and plexiglas. Powder Technology 65, 147–151.

Wesselingh, J.A., 2001. Structuring of products and education of product engineers. Powder Technology 119, 2–8.

Wu, C.-Y., 2012. Discrete Element Modelling. Special Issue for Powder Technology. Elsevier, London.

CHAPTER

Bulk Solid Characterization

2

Bulk solids, also termed particulate solids or particulate materials, consist of assemblies of solid particles. The assembly needs to be large enough (i.e., it must consist of a sufficiently large number of particles) that its properties are statistically representative and do not depend on the number of particles present. Bulk solids are ubiquitous in nature, in our daily life, and in industrial applications. Based upon the sizes of the constituent particles, bulk solids can generally be classified into *granular materials*, in which the particles generally have a large size (say >1 mm), and *powders*, in which the particles are relatively small (say < 1 mm), as stated in the British Standard (1993). Rice, coffee, sugar, sand, and coal are typical granular materials, while flour, salt, and cement are typical examples of powders.

Bulk solids are widely manufactured and used in various industries (Richard et al., 2005) and over 2/3 of the products of the chemical industries are either in the physical form of bulk solids or manufactured using some bulk solids. Bulk solids are unique in the sense that they can, according to the circumstances, show characteristics reminiscent of solids, liquids, or gases. For example, compacted bulk solids (such as tablets and ceramic parts) behave like a solid. Bulk solids can also flow from a storage hopper or container like a liquid, while highly agitated bulk solids in a confined space, e.g., under intensive vibration, behave similarly to a gas. Nevertheless, bulk solids exhibit distinctive properties that cannot simply be characterized using the methods developed for solids, liquids, and gases. In this chapter, some fundamental properties of bulk solids are defined, including density, surface area, flowability, compressibility and compactibility, and the techniques and methods which can be used to characterize these properties are also introduced.

2.1 DENSITY

Density is a fundamental property of any material and determines its physical behavior in various processes. Density is generally defined as the ratio of the mass to the volume of a material, but for bulk solids, several density definitions are introduced to reflect how the volume of the bulk solids is defined. Some of the most widely used densities for bulk solids are:

1. Solid density ρ_s
2. Bulk density ρ_b
3. Tapped density ρ_t

Particle Technology and Engineering.

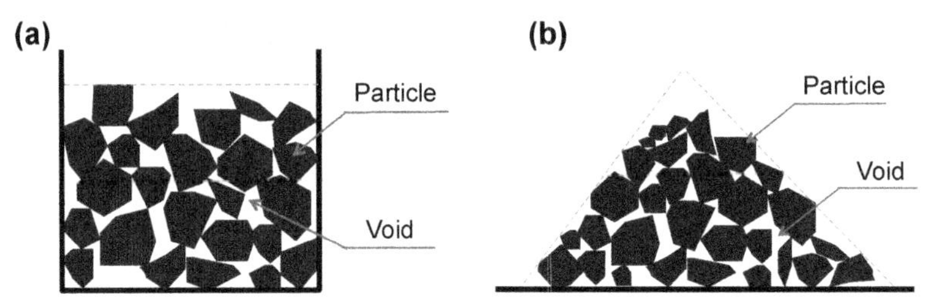

FIGURE 2.1

Illustration of a bulk solid in (a) a container and (b) a pile.

2.1.1 SOLID DENSITY

Solid density, ρ_S, also called true density, absolute density, or particle density, is defined as the ratio of mass to volume of the solid particles (i.e., excluding the volume of open and closed voids):

$$\rho_S = \frac{m}{V_S} \tag{2.1}$$

where m and V_S are the mass and volume of the solid particles (colored in Fig. 2.1), respectively.

The solid density is a material constant so it does not depend on the packing state of the bulk solid. For example, the bulk solid made of the same material shown in Fig. 2.1(a) and (b) should have the same solid density, even though the packing states are different. The solid density is not readily determined using conventional methods for measuring the densities of solids, liquids, and gases, as it is not a trivial task to measure the volume of the solid particles V_S directly because of the complex particle shape and morphology, especially if the particles are very small.

For bulk solids, especially granular materials, the solid density can be determined using the liquid displacement method. In this method, the liquid needs to be carefully chosen so that the bulk solids are immersed but do not dissolve in the liquid. The chosen liquid is poured into a measuring cylinder and its volume, V', is measured. The bulk solid of a given mass, m, is then added into the measuring cylinder and the total volume of the liquid and the bulk solid, V'', is then measured. The difference between V' and V'' (i.e., the displaced liquid volume) gives the volume of the solids, V_S, i.e.,

$$V_S = V'' - V' \tag{2.2}$$

The solid density can then be determined using Eq. (2.1).

The solid density of powders can be measured by gas pycnometry (Webb and Orr, 1997), which uses a similar principle to the liquid displacement method (Box 2.1). However, instead of using a liquid as the displacement medium, a gas pycnometer uses an inert gas (normally helium). Furthermore it indirectly measures the volume from the ideal gas law, i.e.,

$$PV = nRT \tag{2.3}$$

where P and V are the pressure and volume of the gas, respectively, n is the amount of substance of the gas (i.e., number of moles), R is the ideal gas constant with a value of 8.314 J/K mol, and T is the temperature of the gas in Kelvin.

BOX 2.1 THE PRINCIPLE OF GAS PYCNOMETRY

The operating principle of gas pycnometry is illustrated in Fig. 2.2. A gas pycnometer typically consists of a measuring chamber and an expansion chamber of known volumes V_m and V_e, respectively. These two chambers are connected with a valve. The pressure in the measuring chamber is measured using a pressure transducer. Once a powder of known mass, m, but unknown solids volume, V_s, is placed in the measuring chamber, the air in the chamber is removed and the chambers are sealed with the interconnecting valve being closed. A gas (normally helium) is then pumped into the measuring chamber. After the system is sealed, the pressure within the measuring chamber, P_{m0}, is measured, as illustrated in Fig. 2.2(a). Using Eq. (2.3), we have

$$P_{m0}(V_m - V_s) = nRT \tag{2.4}$$

The valve connecting the two chambers is then opened and the gas will expand into the expansion chamber until the pressure in both chambers equilibrates (Fig. 2.2(b)). The pressure is then

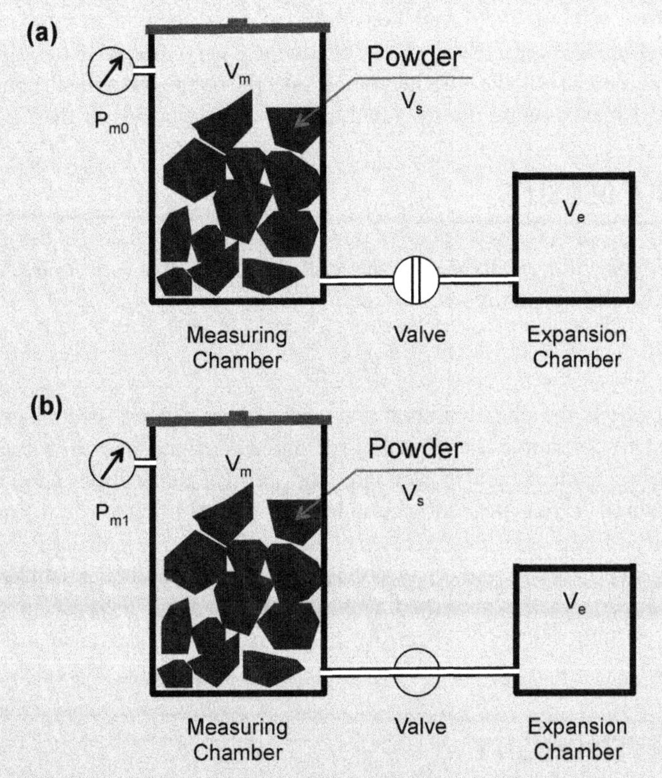

FIGURE 2.2

Illustration of the operating principle of a gas pycnometer (a) when the interconnecting valve is closed (b) when the interconnecting valve is opened.

measured as P_{m1}. During the measurement, the temperature of the whole system is maintained constant and the amount of gas is also constant as there is no net loss or gain of gas. We therefore have

$$P_{m1}(V_m + V_e - V_s) = nRT \tag{2.5}$$

From Eqs (2.4) and (2.5), we obtain

$$P_{m0}(V_m - V_s) = P_{m1}(V_m + V_e - V_s) \tag{2.6}$$

Solving Eq. (2.6) for V_s gives

$$V_s = V_m - \frac{p_{m1}}{p_{m0} - p_{m1}} V_e \tag{2.7}$$

Substituting Eq. (2.7) into Eq. (2.1), the solid density ρ_s can then be determined.

Gas pycnometry is a very useful technique for measuring the solid density of powders, especially dry fine powders. As shown above, gas pycnometry is based upon the ideal gas law and assumes a constant number of gas molecules during the measurements, which is not always valid. For example, for moist powders, water vapor may be released in the chambers during measurements, leading to an overestimated value of P_{m1}; consequently the actual solid volume V_s is underestimated (see Eq. (2.6)) and the solid density is overestimated (Sun, 2005).

Most bulk solids consist of a mixture of two or more substances (i.e., components). Before mixing, their true densities can be directly measured using the displacement methods discussed above. Frequently it is necessary to calculate the effective true density of the mixture based on that of the individual components. This can be achieved using the so-called mixing rule described in Box 2.2.

2.1.2 BULK DENSITY

Bulk density, ρ_b, also termed apparent density or packing density, is defined as the mass of the bulk solid, m, divided by the total volume, V, that it occupies, including the volume of all interparticle voids (see Fig. 2.1):

$$\rho_b = \frac{m}{V} \tag{2.8}$$

Bulk density is the most common and useful characteristic of bulk solids, as it can be used to determine wall loading in hopper design, to size storage spaces and volumetric feeders, and to estimate flowability (see Section 2.3). It can be readily measured using a measuring cylinder or cup in a similar way to measurement of the density of liquids. Bulk density depends upon how the powder is handled or packed; the interparticle voids may collapse when the material is shaken, which leads to reduction in the measured volume and hence an increase in measured bulk density. Bulk density is therefore not a material property but depends on how the material is handled in, for example, filling or packing processes.

2.1.3 TAPPED DENSITY

Tapped density is defined as the mass of the bulk solids divided by the volume that is occupied after tapping. The occupied volume also includes that of the voids which still exist after the tapping. The tapped density is generally obtained by mechanically tapping a measuring cylinder or vessel. The sample powder is first filled into the cylinder or vessel, and the initial powder volume is measured. The cylinder is then tapped and the volume reading is taken, from which the tapped density can be determined. Obviously the tapped density depends upon the tapping process (i.e., the number of taps and intensity of the tapping). It is generally recommended that the bulk volume used in calculating the tapped density should be the lowest volume that can be achieved through tapping, i.e., little further volume change is induced with further tapping. Because of the dependence on the tapping process, tapped density is not an

BOX 2.2 THE MIXING RULE (WU et al., 2006)

Products such as pharmaceutical tablets are generally formed from a mixture of several ingredients, each with their own properties, including density. For such a multicomponent mixture it is useful in practice to work with an *effective density* for the mixture. Let us first consider a powder mixture consisting of two different components: A and B, as illustrated in Fig. 2.3. Component A has a solid density of ρ_{sA} and the volume fraction of A in the mixture is

$$\xi_A = \frac{V_{sA}}{V_{sM}} \tag{2.9}$$

where V_{sA} and V_{sM} are the volumes of solid particles of component A and of the mixture, respectively.

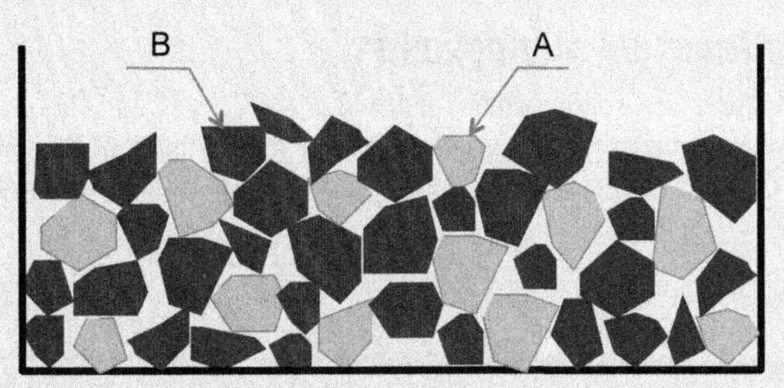

FIGURE 2.3

A powder mixture consisting of components A and B.

Component B has a solid density of ρ_{sB} and its volume fraction is

$$\xi_B = \frac{V_{sB}}{V_{sM}} \tag{2.10}$$

where V_{sB} is the volume of solid particles of component B. As the mass of the mixture is the sum of the masses of A and B, we have

$$m_M = m_A + m_B \tag{2.11}$$

Equation (2.11) can be rewritten as

$$m_M = \rho_A V_{sA} + \rho_B V_{sB} \tag{2.12}$$

The effective solid density of the mixture ρ_M can then be determined using Eq. (2.1), i.e.,

$$\rho_M = \frac{m_M}{V_{sM}} \tag{2.13}$$

Substituting Eq. (2.12) into Eq. (2.13) and using Eqs (2.10) and (2.11), we obtain

$$\rho_M = \frac{m_M}{V_{sM}} = \frac{\rho_A V_{sA} + \rho_B V_{sB}}{V_{sM}} = \rho_A \xi_A + \rho_B \xi_B \tag{2.14}$$

It is clear that the effective solid density of binary mixtures can be determined using Eq. (2.14), if the solid density and volume fraction of each component are known. Note, however, that this approach is only applicable if the mixture stays mixed! If the two components start to separate (known as *segregation*), it should be used with caution.

Equation (2.14) can also be extended to mixtures of multiple components as (Wu et al., 2006)

$$\rho_M = \sum_{i=1}^{n} \rho_{si} \xi_i \tag{2.15}$$

where n is the number of components in the mixture, ρ_{si} and ξ_i are the solid density and volume fraction of component i, respectively.

In practice, it is much easier to determine the mass fraction of each component in the mixture than the volume fraction, because the mass fraction of the component i, ζ_i, is the ratio of the mass of component i to the total mass of the mixture, for which the mass can be measured directly using a balance. The mixing rule can also be expressed in term of the mass fraction, i.e.,

$$\frac{1}{\rho_M} = \sum_{i=1}^{n} \frac{\zeta_i}{\rho_{si}} \tag{2.16}$$

inherent material property. The tapped density is generally higher than the bulk density for a given bulk solid; the difference between them depends on both the initial state and the ease with which particles can move relative to each other and rearrange. It is therefore a parameter that can be used to estimate the packing and flowability of powders (see Section 2.3).

2.1.4 VOID FRACTION AND POROSITY

In bulk solids there are solid particles and voids (See Fig. 2.1). The proportion of the total volume not occupied by particles (i.e., the total volume of the voids divided by the total volume) is termed the *voidage*, also referred to as the *void fraction*, ε, i.e.,

$$\varepsilon = \frac{V - V_s}{V} = 1 - \frac{V_s}{V} \tag{2.17}$$

In Eq. (2.17), the ratio of the solid volume V_s to the total volume V is the solid fraction D, hence

$$\varepsilon = 1 - D \tag{2.18}$$

The term voidage expresses the volume of space *between* particles in relation to the total volume occupied. It is important to distinguish this from the space (if any) *within* individual particles, sometimes termed the *porosity*. Void fraction is an important structural parameter that dominates the behavior of bulk solids and powder compacts. For instance, for powder compacts, the tensile strength is determined primarily by the void fraction (Wu et al., 2006). The voidage or porosity is related to the bulk density and the solid density by

$$\rho_b = (1 - \varepsilon)\rho_s + \varepsilon\rho_g \tag{2.19}$$

where ρ_g is the density of the fluid which surrounds the particles. If the surrounding fluid is a gas at near atmospheric pressure, $\rho_s >> \rho_g$ so the second term above can be neglected.

2.2 SURFACE AREA

The surface area of a bulk solid is an important property that dominates the product quality and performance in many industrial applications. For example, in chemical reactions involving bulk solids either as a reactant or a catalyst, the surface area plays an important role in controlling the reaction rate. A faster reaction can be achieved with finer powders rather than coarse granular materials or a single lump of the same mass, because fine powders have a greater surface area (Tichnor and Saluja, 1990). In pharmaceutical applications, for example, the surface area is one of the key material attributes that affect the formulation and manufacturing processes, such as purification, blending, tableting, and coating, as well as the dissolution rate. Knowing the surface area is therefore very important in optimizing the formulation and product performance.

The surface area of a bulk solid can be measured by a variety of physical and chemical techniques, but is typically determined by gas adsorption. It is usually expressed as a *specific* surface area, i.e., as surface area per unit mass of solid.

If the size distribution of the particles is accurately known it is possible in principle to calculate the surface area directly. In general, however, variations in shape and surface irregularities make this method unreliable in all but the simplest of cases.

As mentioned above, adsorption is a commonly used technique for measuring surface areas of bulk solids. In gas adsorption, a bulk solid is exposed to a gas (typically nitrogen, krypton, or argon) under controlled conditions, and the adsorption of the gas on the surface of the solid particles is evaluated, from which the surface area is determined. Before exposing the bulk solids to the adsorptive gas, the solid sample needs to be pretreated to remove surface contaminants, including adsorbed gases; this is the so-called degassing or outgassing stage. Degassing is a very important step in measuring the surface area using gas adsorption because the adsorption is very sensitive to the surface condition of the bulk solid. If an area of the particle surface is occupied by molecules of contaminants that are not removed before the specific surface area measurement, the measured area will be reduced, leading to significant measurement error.

The pretreated bulk solid is then cooled to cryogenic temperature under vacuum and dosed with an adsorptive gas (normally nitrogen) at incrementally increasing steps in pressure (Box 2.3). At each increment, the pressure is allowed to equilibrate and the quantity of gas adsorbed is determined. From the gas volume adsorbed at each pressure, the volume quantity of gas required to form a monolayer over the surface of the solids can be calculated, from which the surface area can then be determined as explained in Box 2.3 (Webb and Orr, 1997).

2.3 FLOWABILITY

In many industrial applications, such as packing a bulk material into a container or discharging it from a silo or a hopper, the design engineer needs to know how well the bulk material can flow. The term *flowability* is introduced to quantify the flow performance of bulk solids. Flowability is not an intrinsic characteristic of bulk solids; it is a process-dependent property that is related not only to the physical properties of the material itself (size, shape, density, surface area, etc.), but also to the specific process. Consequently, various indices are proposed to define flowability, including the angle of repose, Hausner ratio, Carr index, flow function, and critical fill speed. Some of the commonly used flowability indices are introduced in this section.

2.3.1 ANGLE OF REPOSE

The angle of repose is defined as the slope angle of the free surface of a bulk solid settled under gravity (McGlinchey, 2005). There are three possible variants of the angle of repose, depending on the way that the free surface is created: (1) poured

BOX 2.3 BET SURFACE AREA

For gas adsorption, the data are commonly analyzed according to the BET adsorption isotherm equation (Brunauer et al., 1938):

$$\frac{1}{V_a\left(\frac{P_0}{P}-1\right)}=\frac{C-1}{V_m C}\cdot\frac{P}{P_0}+\frac{1}{V_m C} \tag{2.20}$$

where V_a is the volume of gas adsorbed at a temperature of 273.2 K and standard atmospheric pressure (1.013×105 Pa), V_m is the volume of gas adsorbed to produce a monolayer on the surface of the bulk solid at the same condition, P_0 is the saturated pressure of adsorbed gas, P is the partial vapor pressure of adsorbate gas in equilibrium with the surface at the boiling point of liquid nitrogen (i.e., 77.4 K), and C is a dimensionless constant related to the enthalpy of adsorption.

In gas adsorption experiments, V_a is measured at each relative pressure P/P_0, then the BET value $\frac{1}{V_a\left(\frac{P_0}{P}-1\right)}$ is plotted against P/P_0, which should yield a straight line (Fig. 2.4); the slope $\bar{k}$ and the intercept $\bar{c}$ can be determined from linear regression analysis. From Eq. (2.20), we have

$$\bar{k}=\frac{C-1}{V_m C} \tag{2.21a}$$

$$\bar{c}=\frac{1}{V_m C} \tag{2.21b}$$

From Eq. (2.21), we obtain

$$V_m=\frac{1}{\left(\bar{k}+\bar{c}\right)} \tag{2.22a}$$

$$C=\frac{\bar{k}}{\bar{c}}+1 \tag{2.22b}$$

The specific surface area, S is then calculated using the following equation

$$S=\frac{V_m \cdot N \cdot a}{22400\, m} \tag{2.23}$$

where $N(=6.022 \times 10^{23}\ \mathrm{mol}^{-1})$ is the Avogadro constant, a is the effective cross-sectional area of an adsorbate molecule (for argon $a=0.146\ \mathrm{nm}^2$; for nitrogen $a=0.162\ \mathrm{nm}^2$; for krypton $a=0.195\ \mathrm{nm}^2$) and m is the mass of the test sample.

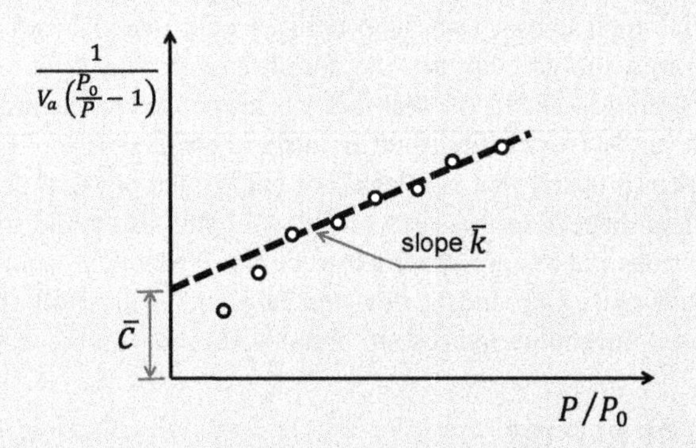

FIGURE 2.4

Illustration of the BET method for analyzing gas adsorption data.

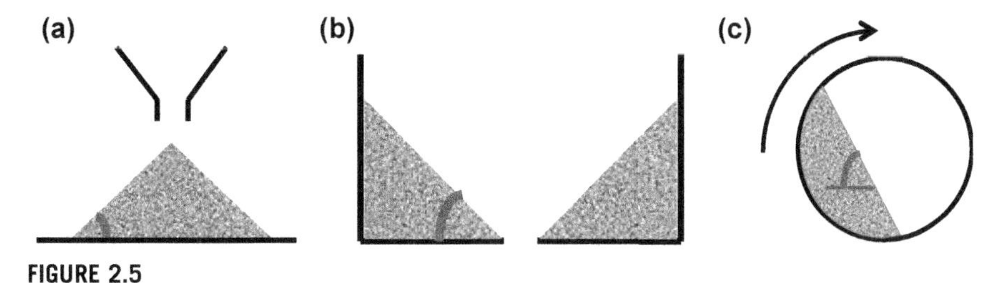

FIGURE 2.5

(a) Poured angle of repose, (b) drained angle of repose, (c) dynamic angle of repose.

angle of repose (Fig. 2.5(a)), (2) drained angle of repose (Fig. 2.5(b)), and (3) dynamic angle of repose (Fig. 2.5(c)). The poured angle of repose is referred to as the slope angle of a heap of the bulk solid poured from a specific height, which can be a useful parameter in storage design, such as sizing of piles of iron ore or agricultural products. The drained angle of repose can be determined by measuring the slope of the free surface once the bulk solid is discharged from a silo. This can give an indication of what the best hopper angle should be in order to promote mass flow (see Chapter 7). The dynamic angle of repose is generally referred to as the maximum angle of a stable free surface of the bulk material developed in a rotating drum or kiln, i.e., the slope angle immediately prior to the initiation of avalanche or cascading, which is useful in designing rotating drums and kilns and predicting flow behavior of bulk materials in these types of equipment (Prescott and Barnum, 2000).

Generally, a smaller angle of repose indicates better flowability (easier flow). The angle of repose depends upon the physical properties of the bulk material, such as particle shape, size, density, and moisture content, and also on the way in which the free surface is developed (McGlinchey, 2005). The angle of repose is an easy and direct method to assess flow properties, but it is only applicable for bulk materials with low and intermediate cohesion, for which reproducible values can be obtained. Caution needs to be taken in assessing the flowability using the angle of repose for more cohesive powders and those with a strong tendency towards segregation. Segregation is a common phenomenon during handling and processing of bulk solids, in which a well-mixed blend of particles with different properties (size, shape, density, and composition) becomes spatially non-uniform, in such a way that the concentration of the different components varies according to their position in the mixture. Segregation can be induced by differences in particle size, density, shape and composition.

2.3.2 HAUSNER RATIO AND CARR INDEX

Evaluating the change in packing density of bulk materials subjected to tapping can also be used to assess the flowability. The Hausner ratio and Carr index are two indices that are determined from the measurements of bulk and tapped densities

(Sections 2.1.2 and 2.1.3). The Hausner ratio H is defined as the ratio of the tapped density ρ_t to the bulk density ρ_b, i.e.,

$$H = \frac{\rho_t}{\rho_b} \tag{2.24}$$

The Carr index Ψ is the ratio of the difference between the tapped and bulk densities to the tap density, i.e.,

$$\Psi = \frac{\rho_t - \rho_b}{\rho_t} \tag{2.25}$$

A higher Hausner ratio or Carr index generally implies that the bulk material is more cohesive and less free flowing. This is because more cohesive powders can show higher initially stable voidage, which is reduced more by tapping than would be the case for a free-flowing powder. Similarly to the measurement of angle of repose, these are simple methods based upon relatively quick and easy measurements and are reproducible and effective in assessing how well the bulk material flows.

2.3.3 JENIKE FLOW INDEX

A flow index was first introduced by Jenike (1964) to express how easily the flow of a bulk solid can be initiated. In other words, it provides a quantitative measure for incipient flow of a bulk solid. The concept of the Jenike flow index can be illustrated using the confined and unconfined uniaxial compression tests shown in Fig. 2.6. Let us assume that a bulk solid is loaded into a smooth mold and consolidated under a stress σ_1. Under consolidation, particles in the mold will re-arrange themselves, and some voids inside the bulk solid will collapse or shrink. At the same time, the material will try to expand laterally. Due to the constraint of the mold walls, a lateral stress σ_2 will be induced. σ_2 can be regarded as the stress transmitted to the mold walls as a result of the consolidation stress σ_1. Assuming the mold surface is smooth enough, the friction between the bulk solid and the mold surface will be very small and can be ignored. Because σ_1 is generally greater than σ_2, for the stress state acting on the bulk solid in Fig. 2.6(a), σ_1 will be the

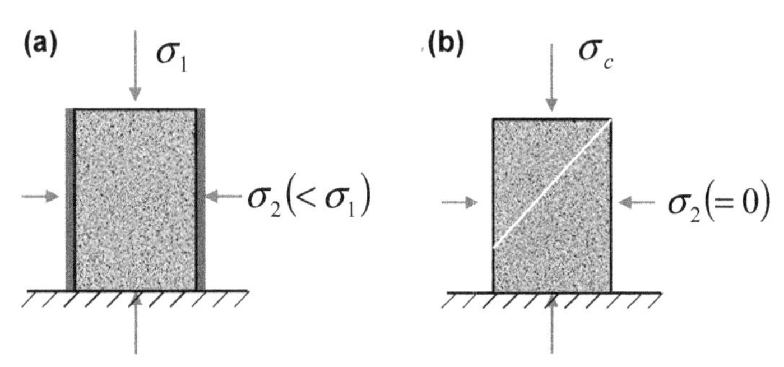

FIGURE 2.6

Illustration of (a) confined uniaxial compression and (b) unconfined uniaxial compression.

maximum principal stress while σ_2 is the minimum principal stress. If we further assume that a coherent test specimen of the bulk solid can be produced under the consolidation stress σ_1 and it can support itself once the mold (i.e., the side wall) is removed, the test specimen thus produced can then be subjected to an unconfined compression test as illustrated in Fig. 2.6(b) (where $\sigma_2 = 0$). The test specimen will fail (i.e., yield or break) once the compression stress reaches σ_c, i.e., the unconfined stress at which the bulk solid "fails" and starts to "flow." σ_c is termed the unconfined yield stress causing incipient flow of the bulk solid at a maximum consolidation stress of σ_1. Hence, for a given σ_1, a lower value of σ_c indicates that it is easier to initiate flow of the material.

The Jenike flow index (ff_c) is defined as the ratio of the maximum consolidation stress σ_1 to the corresponding unconfined yield stress σ_c, i.e.,

$$ff_c = \frac{\sigma_1}{\sigma_c} \tag{2.26}$$

The flowability of bulk materials can then be classified according to the ff_c value as given in Table 2.1 (Schwedes, 2003), a larger value of ff_c representing better flowability.

Table 2.1 Classification of Flowability Using Jenike Flow Index (ff_c) (Schwedes, 2003)

ff_c Value	Flowability
$ff_c < 1$	Hardened
$1 \leq ff_c < 2$	Very cohesive
$2 \leq ff_c < 4$	Cohesive
$4 < ff_c < 10$	Easy flowing
$ff_c < 10$	Free flowing

Although the confined and unconfined uniaxial compression tests illustrated in Fig. 2.6 are widely used to determine the failure/flow conditions for bulk solids (hopper flow, for example), they are not applicable at low loadings, as it is impossible to produce a coherent test specimen for the unconfined uniaxial compression test under such low loadings. In practice, the Jenike flow index is generally obtained by a different method: the shear cell test.

Shear cell tests can be performed using a Jenike shear tester (see Fig. 2.9(a), Jenike, 1964), a ring shear tester (Fig. 2.9(b), Schulze, 1996), or other similar apparatus. Typically shear cell testing involves two steps (Schwedes, 2003): (1) preshear and (2) shear, as illustrated in Fig. 2.9(c). In preshear, a specified normal stress σ_{ss} is applied to the bulk solid, and the sample is continuously sheared. During this process, the shear stress will initially increase from zero and eventually reaches a plateau with a constant shear stress τ_{ss}, at which so-called steady state flow is obtained. The stress state at steady state flow during preshear gives one data point (σ_{ss}, τ_{ss}) on the yield locus in the σ-τ space (Box 2.4). After preshear, the shear stress τ is reduced to zero and the normal stress is decreased from σ to $\sigma'(\sigma' < \sigma_{ss})$. In the second step, shear, the shear stress is gradually increased until shear failure

BOX 2.4 MOHR'S CIRCLE AND YIELD LOCUS

Mohr's circle is a graphical approach for stress analysis, which is widely used in solid mechanics and soil mechanics to represent the stress state at a specific point in the material and to determine the stress components acting on any plane passing through that point, regardless of orientation. Mohr's circles are presented in a coordinate system with the abscissa showing the normal stress σ while the ordinate shows the shear stress τ; hence, each point on a Mohr's circle represents the stress state (normal stress σ, shear stress τ) on a plane at a specific orientation.

A Mohr's circle can be constructed for a known stress state at a given point in the material as follows. Let us assume that a 2D element inside a bulk solid is subjected to compressive normal stresses σ_{xx} and σ_{yy} ($\sigma_{xx} > \sigma_{yy}$) and shear stresses τ_{xy} and τ_{yx} ($\tau_{xy} = -\tau_{yx}$), as shown in Fig. 2.7(a). The corresponding Mohr's circle can be constructed by the following steps:

1. Draw a Cartesian coordinate system (the σ-τ space) with normal stress as the horizontal axis, and shear stress the vertical axis (Fig. 2.7(b)).

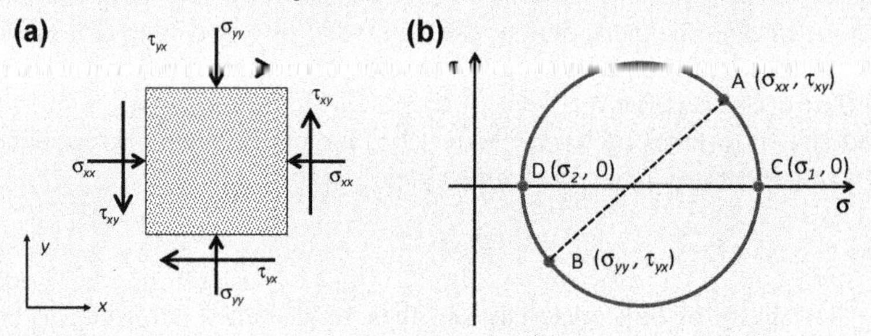

FIGURE 2.7

(a) The stress state on a two-dimensional element and (b) the corresponding Mohr's circle.

2. In the σ-τ space, the stress component in the x direction is represented by point A with coordinates (σ_{xx}, τ_{xy}) while that in the y-direction is represented by point B with coordinates (σ_{yy}, τ_{yx}). Note the sign conventions for normal stresses: compression is positive and tensile stress is negative; for shear stresses, anticlockwise is positive and clockwise is negative.
3. Draw a straight line to connect points A and B, which will be the diameter of the Mohr's circle, and the intersection with the horizontal axis will be the center of the circle.
4. Draw a circle that passes through points A and B and intersects with the horizontal (σ) axis at points C and D. The normal stresses at points C and D are the maximum and minimum principal stresses σ_1 and σ_2, respectively.

Using a similar method and assuming that the stresses are distributed uniformly inside the bulk solid under confined and unconfined uniaxial compression as shown in Fig. 2.6, the stress state for confined uniaxial compression (Fig. 2.6(a)) can be represented using a Mohr's circle centered at coordinate $\left(\frac{\sigma_1+\sigma_2}{2}, 0\right)$ that intersects with the σ axis at (σ_1, 0) and (σ_2, 0) with σ_1, σ_2 being the maximum and minimum principal stresses, respectively. The stress state for the unconfined compression can be represented with the Mohr circle centered at coordinate $\left(\frac{\sigma_c}{2}, 0\right)$ and passing through the origin (0,0) and (σ_c, 0), as illustrated in Fig. 2.8.

The yield locus is then obtained by plotting a straight line tangent to both Mohr's circles (shown as the dashed line in Fig. 2.8). The yield locus defines the maximum shear stress τ that a bulk solid can sustain when consolidated at a certain normal stress σ (Schwedes, 2003).

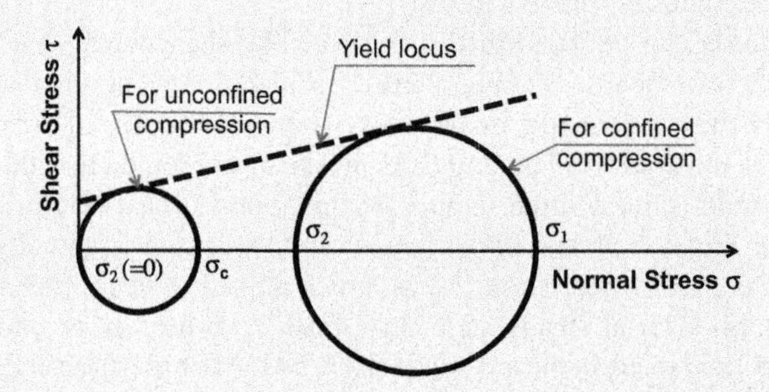

FIGURE 2.8

The Mohr's circles and yield locus for confined and unconfined uniaxial compression shown in Fig. 2.6.

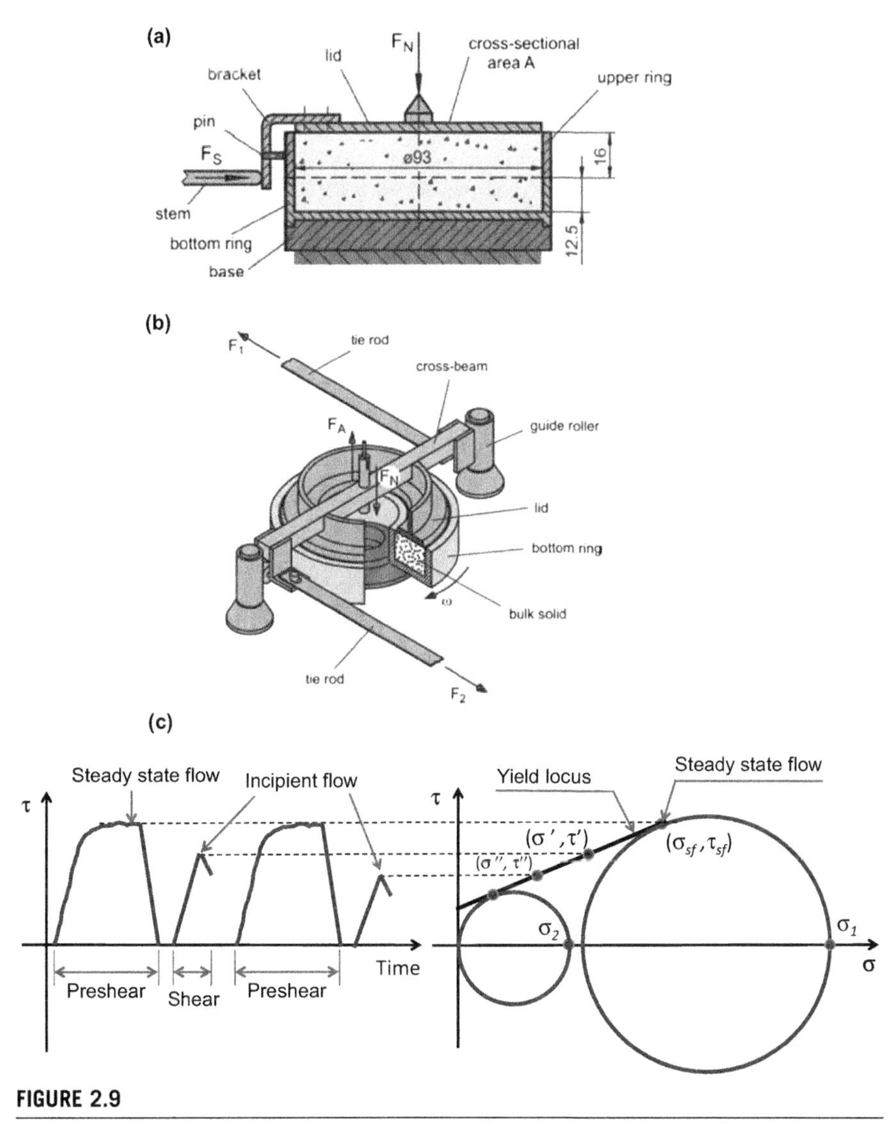

FIGURE 2.9

(a) The Jenike shear tester, (b) the ring shear tester (Schulze, 1996), and (c) the procedure to obtain a yield locus using the Jenike shear tester (Schwedes, 2003).

(i.e., incipient flow) is induced, which is generally indicated by a drop in the shear stress value; the maximum shear stress τ' at the point of incipient flow is then obtained. This provides another point (σ', τ') on the yield locus in the σ-τ space. More data points on the yield locus can be obtained by repeating the above procedure. Note that preshear needs to take place at the identical normal stress σ_{SS} but the shear process should be performed at a different normal stress values, e.g., σ'', σ''', on either a fresh sample (using the Jenike shear tester, see Fig. 2.9(b)) or the same sample (using the ring shear tester, see Fig. 2.9(b)). These yield points are then approximated using a straight line or curve that makes up the so-called yield locus. Once the yield locus is determined, two Mohr's circles can then be constructed, for each of which the yield locus is a tangent and which pass through the origin (0, 0) and the steady state (σ_{ss}, τ_{ss}), respectively. The intersections of these two Mohr's circles with the abscissa axis give the values of σ_c and σ_1, respectively. The Jenike flow index can then be calculated using Eq. (2.26).

The Jenike flow index gives a good indication of the flowability of bulk solids, especially incipient flow. In addition, the angle of internal friction (defined as the slope angle of the yield locus) and cohesion (the intersection of the yield locus with the vertical axis) can be determined, which provides useful information about the bulk material at incipient failure and can be used to predict flow/no-flow situations. The Jenike and Schulze shear testers are versatile and can be used to examine free flowing and cohesive powders over a wide range of stress levels. However, the link between the Jenike flow index and transient or dynamic flow behavior is not clear. Furthermore, these shear testers cannot currently be used to predict powder flow in loosely packed, aerated, or fluidized states.

2.3.4 FLOW METERS AND POWDER RHEOMETERS

Flow meters have also been developed to assess the flowability directly, especially for dynamic flows. These include the Hall and Flodex flow meters. Using these devices, the time taken to discharge a certain amount of bulk solid from a funnel or a vessel with a well-defined orifice is measured. The Hall flow meter is widely used in the powder metallurgy industry and measures the time required to discharge 50 g of powder through an orifice of diameter 2.5 mm situated at the apex of a conical funnel. Based upon these measurements, the mass flow rate is then determined. Hall flow meters can normally only be used to assess the flowability of free-flowing powders. Nevertheless, some variants have also been developed for cohesive powders. For instance, fluidization is introduced in some Hall flow meters to assist the flow of fine powders in the funnel.

The Flodex flow meter uses the same principle as in the Hall device—flow through an orifice—the main difference being that the Flodex flow meter consists of a cylinder which can be fitted at its base with a series of replaceable disks of different opening sizes. The aperture of the Flodex flow meter can be controlled using a mobile shutter. When the shutter is opened, the flow/no-flow condition at the minimum orifice diameter can be determined as a flowability differentiator.

Some more sophisticated instruments have also been developed for measuring powder flowability, exemplified by the FT4 Powder Rheometer (Freeman Technology, Worcestershire, UK) and the Powder Flow Analyser (Stable Micro Systems Ltd, Surrey, UK). These powder rheometers have a rotating blade that moves down through a column of powder. The torque applied to rotate the blade, or the force on the sample cell base, can be measured continuously, and the work done during the test cycle can be determined to assess powder flowability (Prescott and Barnum, 2000; Freeman, 2001). Using these instruments, the flow behavior of a wide range of powders under low consolidation or in the aerated state can be assessed.

Although the methods briefly described above have been designed to assess and classify flowability, the intrinsic complexity of bulk solids/vessel systems means that these methods may give different results in terms of flowability for a given material. This is due to the fact that each test imposes a different process state condition on the powder, i.e., the powder will be subjected to different stress levels and flow conditions in the different tests and the flow behavior of the powder depends strongly on the process state condition (Wu et al., 2012). This condition defines the totality of the forces acting on the powder at any given instance within a powder test or process environment. Ideally, the most appropriate flowability test should be the one which mimics the real condition that the powder will experience in the process of interest (Wu et al., 2003).

2.4 COMPRESSIBILITY

When a bulk solid is compressed, it will deform, which may be accompanied by rearrangement of constituent particles, resulting in collapse of voids if the compression pressure is low, and deformation of individual particles if the pressure is high. The extent of deformation under a certain compression pressure depends on the material properties. For example, fluffy soft particle systems exhibiting significant interparticle forces will deform more than hard and large particles if they are compressed at the same pressure. To characterize the ability of bulk solids to deform or consolidate during compression, the concept of compressibility is introduced.

Knowledge of compressibility is very useful in understanding densification and compaction processes. In particular, it describes how the bulk density (solid fraction or porosity) of a bulk solid changes with the applied pressure. Various mathematical models and empirical fits to data have been developed and a comprehensive review of these can be found in Alderborn and Nystrom (1996). Among these, the Heckel, Kawakita, and Adams equations are widely used to describe compression processes and will be introduced in this section.

2.4.1 HECKEL EQUATION

Heckel (1961) assumed that the densification process under pressure can be approximated as a first-order rate process, i.e., the rate of change in the voidage ε with respect to pressure P is directly proportional to the voidage. Mathematically, this gives

$$-\frac{d\varepsilon}{dP} = k\varepsilon \tag{2.27}$$

Integrating Eq. (2.27) and using Eqs (2.17) & (2.18), we have

$$ln\left(\frac{1}{1-D}\right) = kP + A \tag{2.28}$$

where D is the relative density of the bulk solid at the applied pressure P and is equal to $1 - \varepsilon$, and k is a material-related parameter and is inversely related to the mean yield pressure of the bulk solid. A larger value of k indicates a smaller mean yield

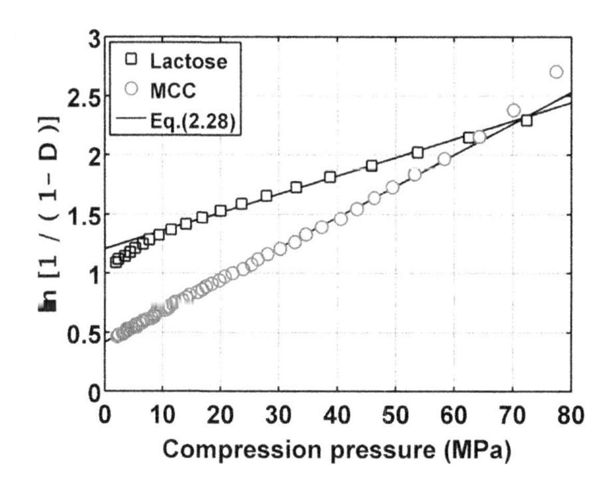

FIGURE 2.10

Heckel plots for lactose and microcrystalline cellulose (Avicel PH102).

pressure and hence a greater degree of plasticity. A is a parameter related to the densification due to die filling (e.g., initial powder packing) and rearrangement of particles.

According to the Heckel analysis, if $\ln\left(\frac{1}{1-D}\right)$ is plotted against the compression pressure P, a linear relationship can be obtained; this is the so-called Heckel plot. A typical Heckel plot is given in Fig. 2.10, in which the compression data for lactose monohydrate and microcrystalline cellulose (MCC) of grade Avicel PH102 are shown. It is clear that at intermediate and high pressures (10–60 MPa) straight lines are obtained, which can be fitted using the Heckel equation (Eq.(2.28)). The parameters k and A can then be determined using linear regression. It can be seen in Fig. 2.10 that the slope for MCC is greater than that for lactose, indicating that MCC has a higher degree of plasticity than lactose. This implies that the Heckel equation can be used to approximate the densification behavior at intermediate to high pressures, where plastic deformation prevails. At low pressures, the densification behavior is dominated by particle rearrangement. The parameter k is not dependent on the initial packing as it is normally determined after the particle rearrangement phase is completed (i.e., for the plastic deformation phase at intermediate to high pressures).

2.4.2 KAWAKITA EQUATION

Kawakita and Ludde (1971) also analyzed the volume reduction of bulk solids with applied pressure during compression. By assuming that the product of the compression pressure and the volume of the bulk solid under compression is constant. The Kawakita equation is given as:

$$C = \frac{V_0 - V}{V_0} = \frac{abP}{1 + bP} \tag{2.29}$$

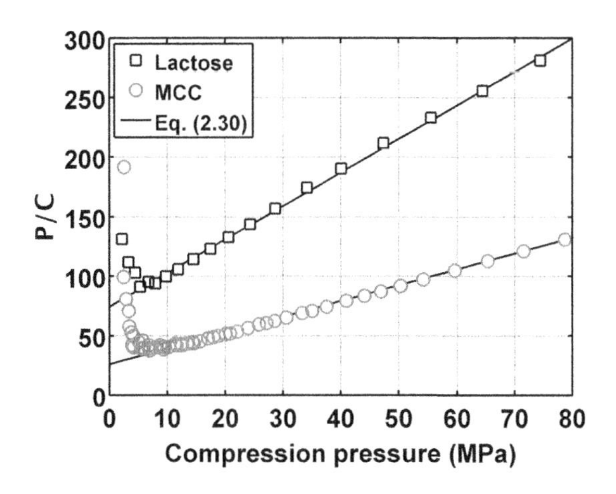

FIGURE 2.11

Kawakita plots for lactose and microcrystalline cellulose (Avicel PH102).

where C is the degree of volume reduction, V_0 is the initial volume of the bulk solid, V is the volume of the bulk solid at pressure P, and a and b are two material constants with a indicating the initial powder porosity before compression (i.e., the total proportion of reducible volume at maximum pressure), and b being a constant related to the yield stress of particles.

Equation (2.29) can be expressed as

$$\frac{P}{C} = \frac{P}{a} + \frac{1}{ab} \tag{2.30}$$

For given compression data, one can plot P/C as a function of the compression pressure, which is often referred to as the Kawakita plot. The Kawakita plots for the same compression data as shown in Fig. 2.10 are presented in Fig. 2.11. It can be seen that a linear relationship is again obtained. The constants a and b can be determined from linear regression of the straight line portion of the compression curve. In particular, the slope from the linear part of the plot gives the reciprocal of constant a, while the intercept of the fitted straight line at the P/C axis gives the value of $1/ab$. It should be noted that, in contrast to the Heckel analysis, the Kawakita analysis is very sensitive to the initial packing, as the degree of volume reduction C is determined by the initial packing.

2.4.3 ADAMS EQUATION

Adams et al. (1994) performed a first-order lumped-parameter analysis of the compression process in which the load is considered to be distributed among a set of percolating force chains in the bed of the bulk solid, reflecting the picture revealed by Discrete Element Analysis (see Fig. 1.5 and Chapter 7). The bed is then modeled as a series of parallel load-bearing columns. A mathematical model was developed according to the Mohr-Coulomb failure criterion, assuming that the shear failure stress τ is the sum of the cohesive strength τ_0 and the frictional stress $\alpha\sigma_r$:

$$\tau = \tau_0 + \alpha\sigma_r \tag{2.31}$$

where α is a pressure coefficient and σ_r is the lateral stress.

A relationship between the compression pressure P and the bed height h was proposed as follows

$$dP = -c_1\tau\frac{dh}{h} \tag{2.32}$$

where c_1 is a constant. Equation (2.32) is a first-order differential equation relating the compression pressure P to the strain. Assuming a constant axial-to-lateral stress transmission ratio and introducing the natural strain ϵ that is defined as

$$\epsilon = ln\left(\frac{h_0}{h}\right) \tag{2.33}$$

Equation (2.32) can be integrated to give

$$lnP = ln\left(\frac{\tau_0}{c_0}\right) + c_0\epsilon + ln\left(1 - e^{-c_0\epsilon}\right) \tag{2.34}$$

where τ_0 is the apparent single particle strength and c_0 is a constant related to the friction between particles. At high natural strains, the last term in Eq. (2.34) can be neglected, which leaves a linear function between lnP and ϵ.

$$lnP = ln\left(\frac{\tau_0}{c_0}\right) + c_0\epsilon \quad \text{(for large strains)} \tag{2.35}$$

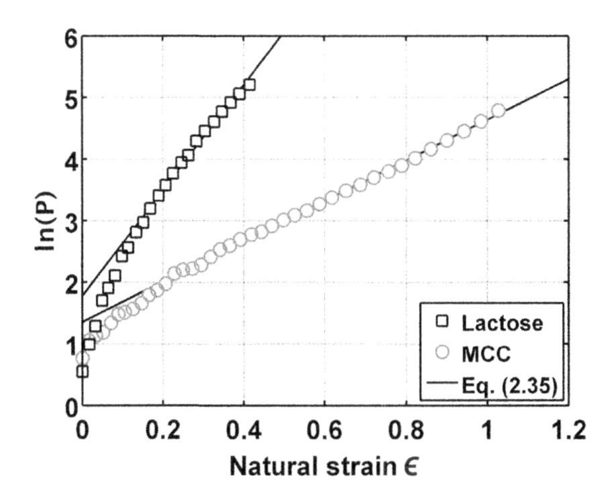

FIGURE 2.12

Typical compression pressure versus strain relationship during uniaxial compaction of lactose and MCC powders.

The intercept and the slope of the straight line fitted to the linear part of the relationship are used to determine τ_0 and c_0. Alternatively, multivariate fitting can also be used to obtain the values of τ_0 and c_0. The result is illustrated in Fig. 2.12, in which the same data are used as in Figs 2.10 and 2.11, for lactose and MCC.

2.5 COMPACTIBILITY

Powder compaction is a commonly-used manufacturing route for products such as pharmaceutical and detergent tablets, metal and ceramic components. If a bulk solid shows good compressibility, this does not guarantee that a coherent compact will result. Compactibility is introduced to describe the ability of a bulk solid to form coherent compacts. It is used to quantify how the mechanical strength of the produced compacts changes with process variables, such as maximum compression pressure. The mechanical strength of powder compacts, in particular the tensile strength, can be characterized using conventional mechanical testing approaches, including the three-point bending flexural test or the direct tensile test. However, as most powder compacts are too fragile to be evaluated using these methods, the diametrical compression test (also known as the Brazilian test) is commonly used to determine the tensile strength (Fell and Newton, 1970).

In the diametrical compression test, bulk solids are compressed into cylindrical specimens: cylinders or disks. The specimens are then placed between two platens (Fig. 2.13) and compressed across a diameter until they break or crush. As illustrated in Fig. 2.13, when a disk is loaded in this way, a tensile stress (σ_x) is induced along the loading plane, which is responsible for the fracture of the specimen.

The tensile strength σ_t is determined from the maximum crushing force F together with the dimensions of the specimens as follows (Fell and Newton, 1970):

$$\sigma_t = \frac{2F}{\pi dt} \tag{2.36}$$

where d and t are the diameter and thickness of the specimen, respectively. This is a simple procedure that can be used to measure the tensile strength of powder compacts made from a wide variety of materials.

The tensile strength of powder compacts will obviously depend on the microstructure. Many efforts have been made to explore this dependency. Ryshkewitch (1953) investigated how the tensile strength is related to the porosity for sintered alumina and zirconia, showing that the logarithm of the tensile strength is inversely proportional to the porosity. Further to Ryshkewitch's analysis Duckworth (1953) derived the following equation for correlation of the tensile strength with the porosity:

$$ln\left(\frac{\sigma_t}{\sigma_0}\right) = c_1(D - 1) \tag{2.37}$$

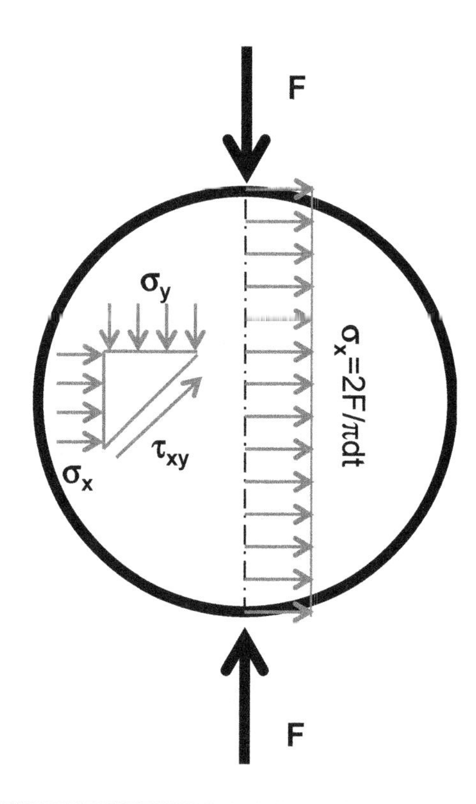

FIGURE 2.13

Stress distributions in a cylinder under diametrical compression (Alderborn and Nystrom, 1996).

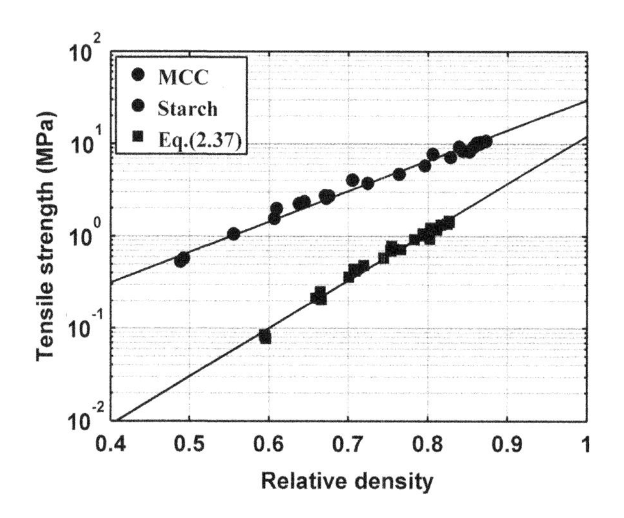

FIGURE 2.14

Tensile strength as a function of the relative density for lactose and MCC.

where σ_0 is the tensile strength of the same material at full density (i.e., D = 1), and c_1 is a constant related to the bonding capacity. A higher value of c_1 implies stronger bonding between primary particles (Steendam and Lerk, 1998). The Ryshkewitch–Duckworth equation (Eq. (2.37)) provides a good correlation for the variation of tensile strength with porosity for a wide range of porous materials (Steendam and Lerk, 1998; Wu et al., 2006), as exemplified in Fig. 2.14.

REFERENCES

Alderborn, G., Nystrom, C., 1996. Pharmaceutical Powder Compaction Technology. Marcel Dekker Inc., New York.

Adams, M.J., Mullier, M.A., Seville, J.P.K., 1994. Agglomerate strength measurement using a uniaxial confined compression test. Powder Technology 78, 5–13.

British Standard BS2955, 1993. Glossary of Terms Related to Particle Technology. BSI, ISBN 0 580 21953 4.

Brunauer, S., Emmett, P.H., Teller, E., 1938. Adsorption of gases in multimolecular layers. Journal of the American Chemical Society 60, 309.

Duckworth, W., 1953. Discussion of Ryshkewitch paper. Journal of the American Ceramic Society 36, 68.

Freeman, R., 2001. An insight into the flowability and characterisation of powders. American Laboratory 33, 13-14–13-16.

Fell, J.T., Newton, J.M., 1970. Determination of tablet strength by the diametrical compression test. Journal of Pharmaceutical Sciences 59, 688–691.

Heckel, R.W., 1961. An analysis of powder compaction phenomena. Transactions of the Metallurgical Society of AIME 221, 1001–1008.

Jenike, A.W., 1964. Storage and Flow of Solids, Bulletin 123, Engineering and Experiment Station, University of Utah, USA.

Kawakita, K., Ludde, K.H., 1971. Some considerations on powder compression equations. Powder Technology 4, 61–68.

McGlinchey, D., 2005. Characterisation of Bulk Solids. Blackwell Publishing Ltd, Oxford, UK.

Prescott, J.K., Barnum, R.A., 2000. On powder flowability. Pharmaceutical Technology 24 (10), 60–84.

Richard, P., Nicodemi, M., Delannay, R., Ribière, P., Bideau, D., 2005. Slow relaxation and compaction of granular systems. Nature Materials 4 (2), 121–128.

Ryshkewitch, E., 1953. Compression strength of porous sintered alumina and zirconia. Journal of the American Ceramic Society 36, 65–68.

Sun, C.C., 2005. True density of microcrystalline cellulose. Journal of Pharmaceutical Sciences 94, 2132–2134.

Schwedes, J., 2003. Review on testers for measuring flow properties of bulk solids. Granular Matter 5, 1–43.

Schulze, D., 1996. Flowability and time consolidation measurements using a ring shear tester. Powder Handling and Processing 8, 221–226.

Steendam, R., Lerk, C.F., 1998. Poly(DL-lactic acid) as a direct compression excipient in controlled release tablets: part I. Compaction behaviour and release characteristics of poly(DL-lactic acid) matrix tablets. International Journal of Pharmaceutics 175, 33–46.

Ticknor, K.V., Saluja, P.P.S., 1990. Determination of surface areas of mineral powders by adsorption calorimetry. Clays and Clay Minerals 38 (4), 437−441.

Wu, C.Y., Best, S.M., Bentham, A.C., Hancock, B.C., Bonfield, W., 2006. Predicting the tensile strength of multi-component pharmaceutical tablets. Pharmaceutical Research 23 (8), 1898−1905.

Webb, P.A., Orr, C., 1997. Analytical Methods in Fine Particle Technology. Micromeritics Instrument Corporation.

Wu, C.Y., Armstrong, B., Vlachos, N., 2012. Characterisation of powder flowability for die filling. Particulate Science and Technology 30 (4), 378−389.

Wu, C.Y., Dihoru, L., Cocks, A.C.F., 2003. The flow of powder into simple and stepped dies. Powder Technology 134 (1−2), 24−39.

CHAPTER

Particle Characterization

3

Many properties of bulk solids described in Chapter 2 are determined by the characteristics of their constituent particles, which also influence the behavior of bulk solids during handling and processing. The important characteristics of particles are shape, size, and mechanical properties, such as Young's modulus and Poisson's ratio. There is a plethora of literature demonstrating how particle shape, size, and mechanical properties affect the bulk properties (bulk and tap densities, flowability, compressibility, and compactibility) and their relationship to various processes including flow, packing, compaction, sintering, fluidization, pneumatic conveying, reaction, dissolution, and dispersion. For instance, large and round or blocky particles, such as salt and glass beads, typically flow better than small and irregularly shaped ones like corn flours and chocolate powders, while small particles generally dissolve faster in liquids than large particles but may be more difficult to disperse.

Characterization of particle properties is necessary for product quality control and process monitoring purposes, as the quality of many products is determined by their particle properties. For example, in the pharmaceutical industry, particle size, shape, and mechanical properties affect the dispersion efficiency of dry powder inhalers and the weight and dosage variation of tablets. Particle properties of food and beverage products influence color and flavor. The size and shape of metallic and ceramic particles affect the physical properties of the finished products, such as surface finish and mechanical strength. In process monitoring, control, and optimization, typical applications include crystallization, milling (size reduction), and granulation (size enlargement), for which the aim in each case is to produce particles or granules with desired attributes. Monitoring the change in particle properties (especially size and shape) during the process is therefore critical and requires robust and reproducible characterization.

In addition to quality control and process monitoring, particle characterization enables a better understanding of the correlations between particle characteristics, process performance, and product quality and also enables improvement and optimization of the manufacturing efficiency, improvement in process performance, and increased productivity. In most cases the particle characteristics of interest are difficult to measure, especially for particles smaller than 100 μm in size.

3.1 PARTICLE SHAPE

Particles generally have complex geometric features, summarized under the term *shape* but including overall *form*, the presence of *reentrant* features, and surface *irregularity*. Shape is therefore difficult to define. Although the literature on particle shape is extensive and a number of shape factors and descriptors are proposed, there is no universal agreement on how to define particle shape and therefore no agreement on how it can be measured properly. Only a small selection of shape factors and descriptors will be discussed in this chapter, but readers are referred to Hawkins (1993) and Endoh (2006) for more comprehensive discussion on the subject.

Particle shapes may be qualitatively characterized by description of their visual appearance, such as *spherical, round, angular, dendritic, platy, equidimensional, rodlike*, and *acicular* or *needle-shaped*. Although these descriptions are simple, easily comprehensible, and convenient to use (Endoh, 2006), they are vague terms that cannot be used to distinguish between particles of similar shapes; quantitative descriptions with clear physical interpretations are needed.

Although many observation methods, such as microscopy, result in a two-dimensional (2D) image, strictly speaking, shape is a three-dimensional (3D) attribute of particles and should be characterized quantitatively based on the analysis of the data in 3D, such as 3D images. Although it is now possible to obtain 3D images of particles of different sizes using advanced imaging techniques, such as X-ray computed tomography (XRCT), it is still problematic and challenging to make 3D measurements and to ensure that the data so obtained are representative. Most particle shape characterization methods are therefore based upon 2D data in the form of projected images obtained using various imaging techniques or sectioned images obtained using metallography or XRCT (Hentschel and Page, 2003). The 2D methods are simple but useful in practice and can to some extent provide representative description of particle shapes as a large number of particles can be projected in random orientations, which provides statistically meaningful information. Therefore, only shape characterization using 2D data will be discussed in this book.

3.1.1 BASIC MEASUREMENT

In order to define a shape factor to represent particle shape, some basic measurements of the particle dimensions are needed, which includes length between different boundary features, perimeter, and area. With modern imaging techniques, 2D projected images of particles are normally digitized so that these measurements are generally obtained in the form of pixels, as illustrated in Fig. 3.1, which shows the scanning electron microscopy (SEM) image of a lactose particle and the corresponding projected image used for shape characterization.

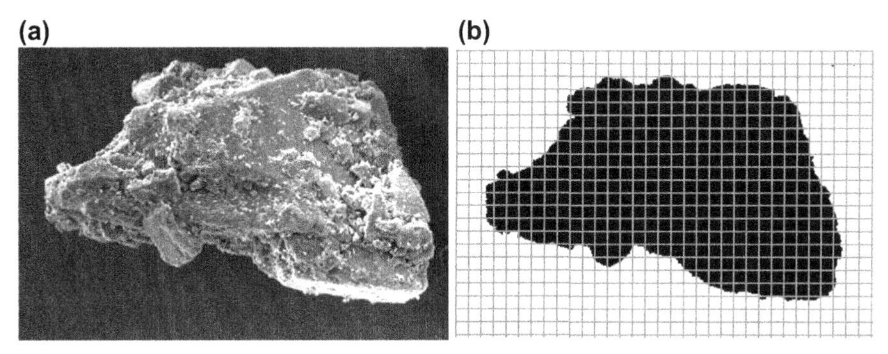

FIGURE 3.1

Scanning electron microscopy (SEM) image of a lactose particle (a) and the corresponding projected image (b).

Feret dimensions are the most commonly used length measurements. Feret length ℓ (also known as Feret diameter) is the distance between two tangents to the contour of the particle in a specified direction (see Fig. 3.2). In other words, it corresponds to the measurement by a caliper or slide gauge; hence, it is also called the caliper diameter. Depending on the direction in which the Feret measurement is taken, various Feret lengths can be obtained for the same particle. The commonly used ones are listed in Table 3.1.

The *perimeter*, *P*, is the total length of the outline surrounding the projection of the particle, and it is most easily determined by counting the number of boundary pixels (see Fig. 3.3). The area, *A*, is calculated from the total number of pixels in the particle image (i.e., shaded pixels in Fig. 3.1(b)). It is clear that the accuracy of the perimeter and area measurements is governed by the resolution of the particle images.

FIGURE 3.2

Feret length of the lactose particle shown in Fig. 3.1 in a random direction.

Table 3.1 Commonly Used Feret Measurements (Hentschel and Page, 2003)

Feret Length	Measurement Method
Maximum Feret length, ℓ_{max}	Maximum Feret length measured over all possible orientations
Minimum Feret length, ℓ_{min}	Minimum Feret length measured over all possible orientations
Orthogonal to maximum Feret length, ℓ'_{max}	Feret length measured at an angle of 90° to that of the maximum Feret diameter
Orthogonal to minimal Feret length, ℓ'_{min}	Feret length measured at an angle of 90° to that of the minimum Feret diameter
Mean Feret length, $\bar{\ell}$	Mean value of the Feret diameters over all orientations

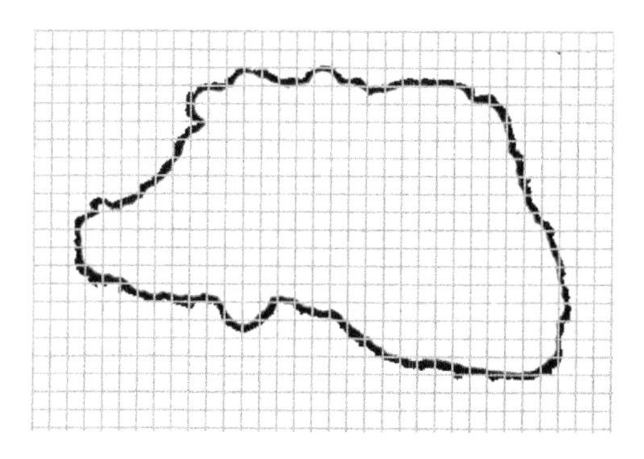

FIGURE 3.3

Perimeter of the lactose particle shown in Fig. 3.1.

3.1.2 SHAPE FACTORS

Using the basic measurements discussed in Section 3.1.1, a number of shape factors can be defined. The simplest is the *aspect ratio*, S_A, which is defined as the ratio of the minimum to the maximum Feret length, i.e.,

$$S_A = \frac{\ell_{min}}{\ell_{max}} \tag{3.1}$$

The aspect ratio is a parameter indicating degree of elongation. A lower value of aspect ratio represents more elongated particles. Another shape factor sensitive to elongation is the *roundness*, S_R:

$$S_R = \frac{4\pi A}{P^2} \tag{3.2}$$

The roundness typically has a value less than 1. For a perfect sphere (circular projection), $S_R = 1$.

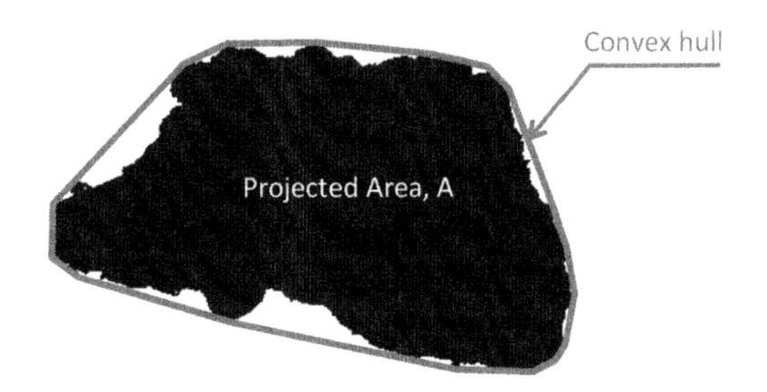

FIGURE 3.4

Illustration of the circumscribed convex hull for the lactose particle shown in Fig. 3.1.

The *sphericity*, S_s, is defined as the ratio of the perimeter of the equivalent circle, P_c, to the perimeter of the projection of the particle. The equivalent circle refers to the circle with the same area as the projection of the particle. Hence,

$$S_s = \frac{P_c}{P} = \frac{2\sqrt{\pi A}}{P} = \sqrt{S_R} \tag{3.3}$$

The value of sphericity varies between 0 and 1, for a spherical particle. The more irregular the particle shape is, the smaller the value of sphericity since, for the same projected area, the perimeter increases with increasing departure from a circle.

The *convexity* S_C (also known as solidity) describes the degree of compactness of a particle (Endoh, 2006); it is defined as the ratio of actual area to the area bounded by a convex hull around the projection of the particle (as illustrated in Fig. 3.4), i.e.,

$$S_c = \frac{A}{A_h} \tag{3.4}$$

where A_h is the area of the circumscribed convex hull. The convexity has a value between 0 and 1, equaling 1 if there are no concave or *reentrant* features.

3.2 PARTICLE SIZE

Bulk solids consist of many billions of particles of different size and shape. Determining a distribution of particle size is therefore problematic, since only spheres and cubes can be uniquely defined with a single number such as diameter or the side length. Much theoretical work in particle engineering takes the particles to be spherical but real particles are very rarely spherical. Thus, the use of the term *diameter* to define the particle size is somehow ambiguous; this difficulty is overcome by use of the concept of *equivalent diameter*, which is the diameter of a sphere having the same value of a particular physical attribute, such as volume, surface area, or projected area. Many equivalent diameters can be defined, some of which are discussed in this section. Methods for measuring these equivalent diameters and analyzing the resulting data are also introduced. Commercial instruments for particle size characterization may measure different equivalent diameters, as discussed later in this chapter, so that two instruments may not necessarily give exactly the same results for measurements on the same particulate sample. Nevertheless, it is often possible to convert one equivalent diameter to another if some knowledge of the particle shape is available. Conversely, independent measurements of two different equivalent diameters can sometimes be used to infer information about the particle shape.

3.2.1 PARTICLE SIZE DEFINITION

The equivalent diameters in common use are equivalent project area diameter d_A, equivalent surface area diameter d_S, equivalent volume diameter d_V, Stokes diameter d_{St}, and aerodynamic diameter d_a (see Fig. 3.5).

1. Equivalent projected area diameter, d_A, is defined as the diameter of a circle with the same area as the projected area of the particle, A, i.e.,

$$d_A = \left(\frac{4A}{\pi}\right)^{1/2} \tag{3.5}$$

   Note that for a nonspherical particle, the projected area depends on orientation, as does the equivalent projected area diameter.
2. Equivalent surface area diameter, d_S, is the diameter of a sphere with the same surface area, S, as the particle:

$$d_S = \left(\frac{S}{\pi}\right)^{1/2} \tag{3.6}$$

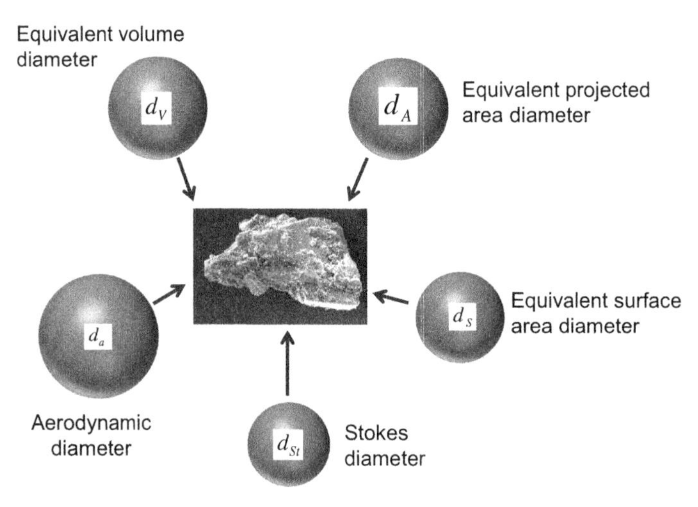

FIGURE 3.5

Various equivalent diameters for defining the particle size.

3. Equivalent volume diameter, d_V, is defined as the diameter of a sphere with the same volume, V, as the particle, i.e.,

$$d_V = \left(\frac{6V}{\pi}\right)^{1/3} \tag{3.7}$$

4. Stokes diameter, d_{St}, is the diameter of a sphere with the same settling velocity (see Chapter 4) as the particle in the Stokes regime (see section 4.3). Unlike the other definitions introduced above, this equivalent diameter is defined in terms of the aerodynamic behavior of the particle instead of its geometrical properties. It is expressed as

$$d_{St} = \left[\frac{18\mu V_t}{C_s(\rho_p - \rho)g}\right]^{1/2} \tag{3.8}$$

where V_t is the particle terminal velocity, C_s is the Cunningham slip correction factor (which can be neglected unless the particles are sub-micron; Seville et al., 1997), μ is the viscosity of the fluid, g is the gravitational acceleration and ρ_p and ρ are the densities of the particle and the fluid, respectively.

5. Aerodynamic diameter, d_a, is the diameter of a sphere with a density of 1.0×10^3 kg/m^3 and the same settling velocity as the particle in the Stokes regime. It is similar to the definition of the Stokes diameter given in Eq. (3.8), but instead of using the particle density ρ_p, it uses a fixed density $\rho_0 = 1.0 \times 10^3$ kg/m^3, i.e.,

$$d_a = \left[\frac{18\mu V_t}{C_a(\rho_0 - \rho)g}\right]^{1/2} \tag{3.9}$$

where C_a is the appropriate Cunningham slip correction factor for d_a.

As mentioned earlier, it is possible to construct relationships between the different equivalent diameters, which depend on shape. For example, the Stokes diameter can be related to the equivalent volume diameter (Seville et al., 1997):

$$d_{St} = \left[\frac{3\pi d_V^3}{c}\right]^{1/2} \tag{3.10}$$

where c is the hydrodynamic resistance of the particle ($c=3\pi d$ for a sphere). For an irregularly shaped particle, d_{St} is always greater than d_V.

Using Eqs (3.8) and (3.9), it can be shown that the aerodynamic diameter is related to the Stokes diameter as follows

$$d_a = d_s\left[\frac{C_s(\rho_p - \rho)}{C_a(\rho_0 - \rho)}\right]^{1/2} \tag{3.11}$$

For particles in a liquid, $C_s = C_a = 1$, so that Eq. (3.11) becomes

$$d_a = d_s \left(\frac{\rho_p - \rho}{\rho_0 - \rho} \right)^{1/2} \tag{3.12}$$

For particles in a gas, $C_s \approx C_a$, $\rho_p \gg \rho$, and $\rho_0 \gg \rho$, so that Eq. (3.12) reduces to

$$d_a = d_s \left(\frac{\rho_p}{\rho_0} \right)^{1/2} \tag{3.13}$$

The choice of equivalent diameter depends on the actual application, i.e., what the data are intended for. For example, if the separation efficiency of a cyclone is of interest, it is appropriate to use the Stokes diameter, as it best describes the behavior of particles suspended in a fluid. If particle size data are requested for designing dry powder coating processes, the use of the equivalent project area diameter or the equivalent surface area diameter is more appropriate.

3.2.2 MEASUREMENT METHODS

Many methods and instruments for measuring particle size and concentration have been developed, in response to the increasing demand from researchers and practitioners. Nevertheless, it remains a grand challenge to develop a robust online instrument that has a fast response and can accurately measure the particle size distribution most closely related to the phenomenon under investigation. The purpose of this section is to provide a brief introduction to some common measurement methods for characterizing particle sizes substantially above 1 μm, and to explain which equivalent diameter they measure in practice. For sub-micron particles, since the physical laws governing their behavior are generally different, the methods used for measuring the size of these particles form a distinct subject, which will not be considered here.

3.2.2.1 Sieving

As a method of particle separation, a stack of sieves arranged in such a way that mesh sizes decrease with height in the stack can be used to sort particles by size. The powder is placed on the top sieve, and the apparatus is shaken. The powder mass collected in each sieve is then weighed, so that a distribution of the mass of particles with diameters between each sieve size is obtained, and a size distribution by mass can be determined. An advantage of this method is that the sorted fractions can be retained and used for other purposes or for further analysis. In sieving, particles are sorted by the two smallest dimensions (because particles can align to pass through the mesh apertures), so the result may become complex if both size and shape of the particles vary. In practice, an automated sieve shaker is normally used to perform sieving in order to reduce operator bias. Nevertheless, measurement errors can be induced as a result of "blinding" (i.e., sieve blocking), particle breakage (especially for fragile particles), and mesh stretching due to overloading.

3.2.2.2 Microscopy

Particles can be deposited onto a viewing surface and directly observed and measured using either *optical microscopy* or *scanning electron microscopy* (SEM). For particles greater than about 10 μm in size, optical microscopy may be sufficient, while SEM is more appropriate for smaller papers, with which high-magnification and high-resolution images can be taken and the size can be measured more accurately. The diameter obtained is the equivalent projected area diameter, d_A, in whatever orientation the particles are deposited.

Using this method, as only a limited number of particles can be measured, sampling (see Section 3.3) and preparation are critical to ensure that the selection of particles is representative and each particle can be viewed individually (i.e., to avoid overlap between particle projections). Fortunately, advances in image analysis software enable rapid and automatic data analysis to be performed.

3.2.2.3 Dynamic Image Analysis

The image analysis involved in microscopy discussed above is primarily based on static images, i.e., the images are taken when the sample is stationary. Similar principles can also be applied to analyze dynamic images that are captured when dispersed particles flow past a suitable detector. This is so-called *dynamic image analysis*.

In dynamic image analysis, particles are dispersed in a gas or liquid and flow past a light source, such as a light-emitting diode (LED). The projection of the particles is then captured with a digital detector (or camera), such as a charge-coupled device and complementary metal-oxide-semiconductor. From the captured images, each particle is identified and analyzed to obtain its size and shape. Similarly to microscopy, dynamic image analysis measures the equivalent projected area diameter.

In most dynamic image analysis systems, a single light source and a single camera are used. Depending on the specification (brightness, focus, resolution) of the light source and camera, only a limited range of particle sizes can be measured. This method has recently been improved by Retsch Technology (Haan, Germany) and the resulting two-camera system is illustrated in Fig. 3.6. This consists of two pulsed LED light sources and two cameras: one camera (i.e., the basic camera in Fig. 3.6) is used to detect large particles and capture more particles in a large-view field so that good statistics can be obtained; the other camera, of high resolution (i.e., the zoom camera), is optimized in order to capture small particles. The two LED light sources are also optimized with appropriate brightness, pulse length, and field of illumination, so that the optical paths of both LED-camera pairs intersect in the measurement area. Using this technique, a wide size range (1 μm ~ 8 mm) can be measured with reliable detection of small

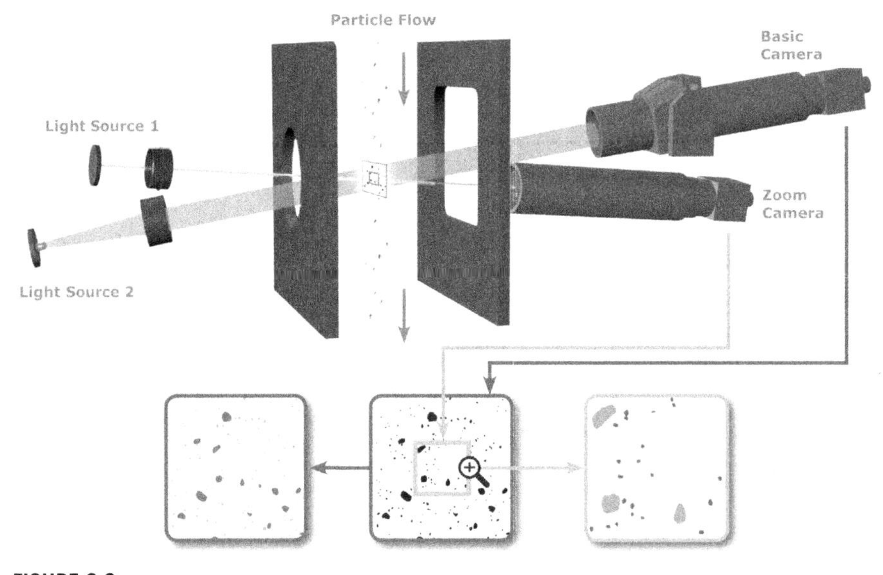

FIGURE 3.6

Principle of dynamic imaging with a two-camera system.

Courtesy of Retsch Technology GmbH, reproduced with permission.

amounts of small and large particles simultaneously. In addition, the particle shape is also recorded and analyzed for each particle.

3.2.2.4 The Coulter Principle

The *Coulter principle*, also known as the "electrical sensing zone" or "electrozone" method, is named after its inventor. In this method, particles are dispersed in a weak electrolyte solution that is forced to flow through a small (10–400 μm diameter) aperture. On each side of the aperture there is an immersed electrode, between which an electric current can flow (Fig. 3.7), so that an "electric sensing zone" will be created once a voltage is applied across the aperture. When the suspension is sufficiently dilute, particles will pass through the aperture one by one. As each particle goes through the aperture, it displaces its own volume of the conducting liquid and hence changes the electrical resistance across the aperture. With a constant current, the change of impedance generates a voltage pulse whose amplitude is directly proportional to the volume of the conducting liquid being displaced, i.e., the particle volume. As a known volume of particle suspension is drawn through the orifice, the resulting voltage pulses are electronically scaled and registered, from which the number of particles in the suspension can be counted, and a particle size distribution can be determined by scaling the pulse magnitudes in volume units. This method therefore measures the equivalent volume diameter, and the measurement is independent of particle orientation and particle shape. As it counts every single particle, it provides a direct measurement of the particle size distribution.

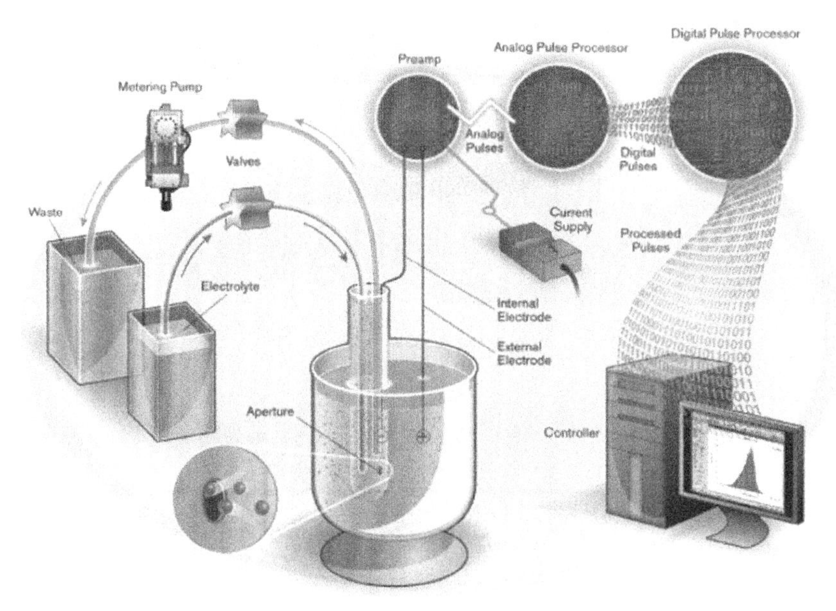

FIGURE 3.7

Illustration of the Coulter principle.

Courtesy of Beckman Coulter Inc. Reproduced with permission.

The Coulter method requires particles to be well dispersed in the electrolyte, so that an appropriate aqueous or organic electrolyte needs to be used. To establish the correlation between the magnitude of the induced voltage pulse and the particle volume, the instrument must be calibrated using particles of known size, which can be achieved using standard dispersions of polymer latex since the calibration does not need to use the same material as the particles of interest.

3.2.2.5 Aerodynamic Methods

As discussed in detail in Chapter 4 (see also Seville et al., 1997), when dispersed in a fluid, particles of different sizes (and densities) will respond in different ways to any movement of the fluid. This forms the basis of the aerodynamic methods for sizing particles. The inertial impactor is one of the devices that measure particle size using this principle (Fig. 3.8(a)). In the *inertial impactor*, the gas containing the particles is forced through the impactor nozzle. If an impaction plate is placed close to the nozzle, the jet streamlines bend sharply, as shown. Larger particles, having higher inertia, deviate from the streamlines, impact on the plate, and are collected; small particles are carried away with the gas stream. A typical collection efficiency curve for a single stage of an inertial impactor is illustrated in Fig. 3.8(b), showing the variation of collection efficiency with particle diameter; this is known as a *grade efficiency* curve.

In practice, a set of inertial impactors is generally stacked to form a *cascade impactor* (Fig. 3.8(c)), in which the nozzle size is made progressively smaller

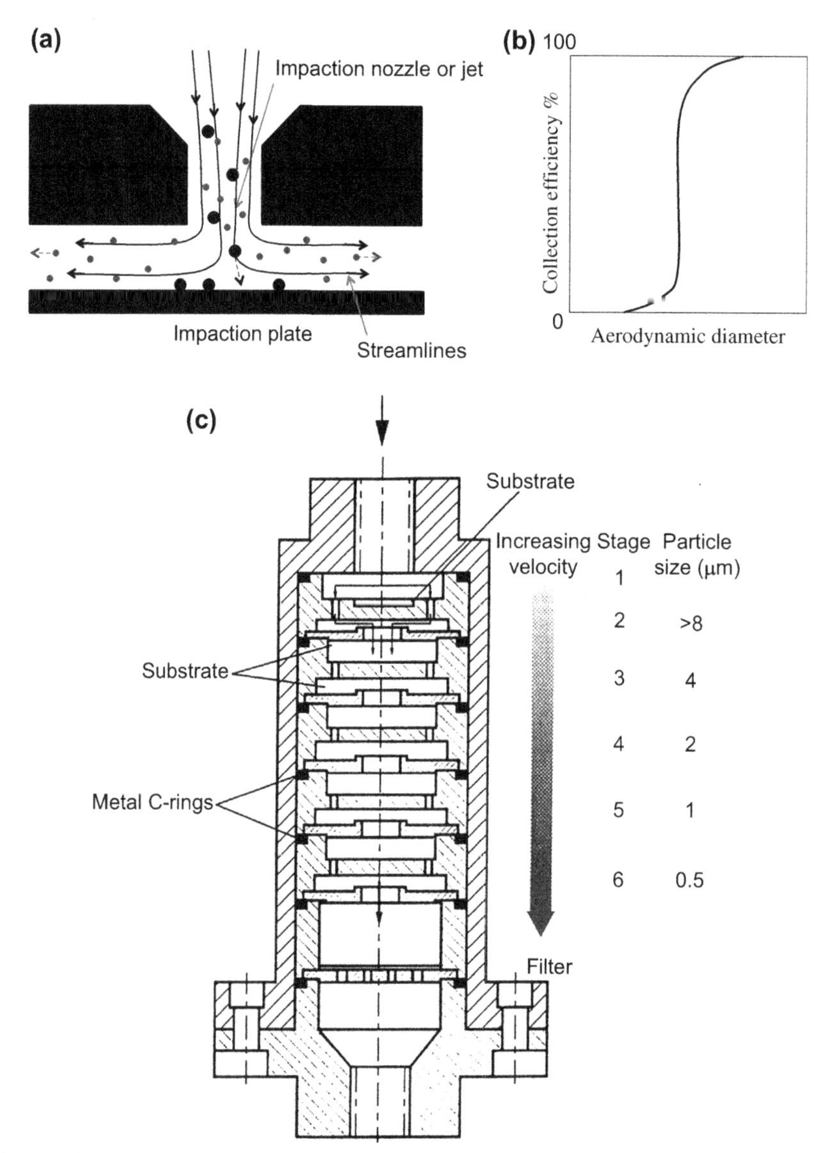

FIGURE 3.8

Schematic diagram of (a) an Inertial Impactor and (b) a typical grade efficiency for a single stage of the inertial impactor, and (c) a cascade impactor.

from stage to stage, resulting in an increase in the jet speed, so that successively smaller particles can be collected and separated on each stage. Particles collected on each impaction plate are weighed so that a size distribution by mass can be obtained. These impactors measure the aerodynamic diameter and are widely used in pharmaceutical and environmental applications, for measuring the particle size distribution of dry powder inhalation formulations and air pollutants, for example.

The aerodynamic diameter (or the Stokes diameter) of a particle can be determined by any device that can measure its response to suitably rapid changes in gas velocity. One device based upon this principle is called the aerodynamic particle sizer (APS), manufactured by TSI (Shoreview, MN, USA). In the APS, a suspension of particles flows through a fine nozzle, where the gas is accelerated. Particles suspended in the gas are also accelerated and their rates of acceleration are determined by their aerodynamic size (Fig. 3.9). Large particles accelerate more slowly due to their greater inertia, while fine particles accelerate at a rate closer to that of the gas. When particles leave the nozzle they pass through two closely spaced and partially overlapping laser beams, so that their exit velocities are measured using "time of flight" (Fig. 3.10). Based upon this measurement of individual particle exit velocity, each particle's aerodynamic diameter can be determined and a size distribution can be built up. Similarly to the Coulter method, the APS counts individual particles and provides an absolute measurement of particle size. Although there is a good theoretical basis for the measurement, calibration is still necessary to correlate the measured exit velocity with the particle size, which can be achieved using particles of known size. Furthermore, the suspension

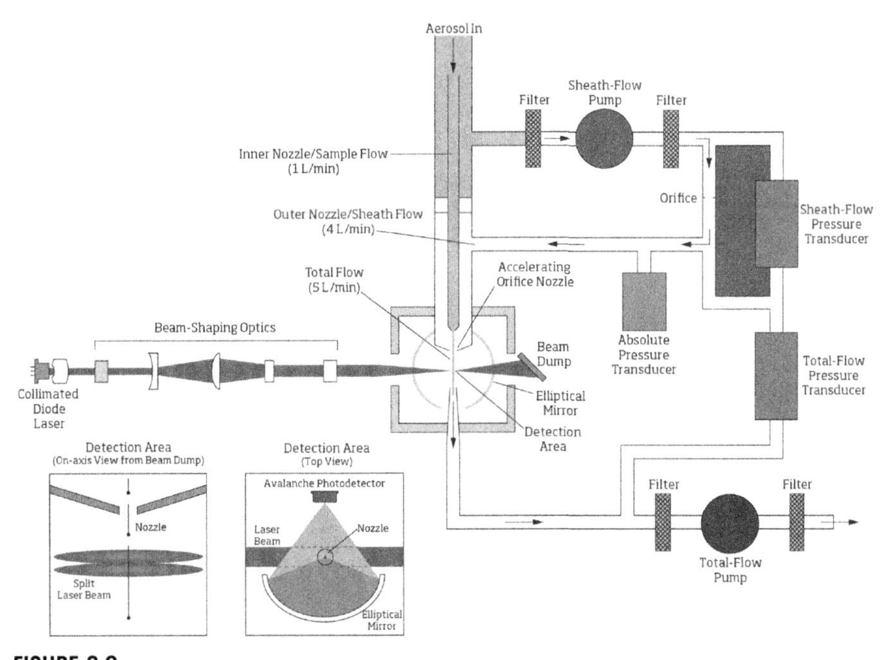

FIGURE 3.9

Schematic design of an aerodynamic particle sizer.

Courtesy of TSI Inc., MN, USA, www.TSI.com.

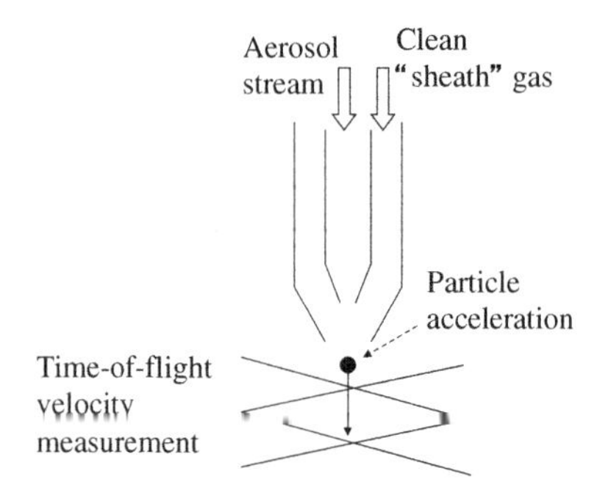

FIGURE 3.10

Illustration of the measuring principle of the aerodynamic particle sizer.

entering the instrument must be sufficiently diluted (e.g., by as much as 100:1) to ensure that only one particle is measured at a time.

3.2.2.6 Laser Diffraction

Instead of building up a particle size distribution from measurements on individual particles as discussed above, laser diffraction determines the size distribution for the whole sample simultaneously. When a light wave encounters a particle, light is diffracted, the extent of the diffraction depending on its size: large particles scatter light at small angles, while small particles scatter most strongly at large angles. The intensity distribution of the scattered light is usually captured using a multi-element or position-sensitive photo-detector. A laser beam (such as that from a low-power He/Ne laser) is focused into the center of the detector. When particles pass through the laser beam, the scattered light is detected by concentric elements of the detector, and the intensity distribution is then measured. The angular variation in the intensity of scattered light is then analyzed to determine the particle size distribution.

The measured intensity data can be converted mathematically to a particle size distribution using either the Fraunhofer or the more general Mie theory of light scattering, assuming a volume equivalent sphere model (Seville et al., 1997). It is well recognized that the Mie theory provides a better accuracy, especially for measuring small particles, but it requires knowledge of the optical properties (such as the refractive index) of both the sample being measured and the dispersant. The Fraunhofer approximation is a relatively simple approach that does not require knowledge of the optical properties of the sample. It is generally suitable for larger particles (say >10 μm).

Laser diffraction can be used to measure particle sizes in suspension in a liquid (i.e., wet) or a gas (i.e., dry) and reports the equivalent volume diameter. The measurement is fast and can be used *in situ* and for online measurement. Table 3.2 in Box 3.1 summarizes various size measurement methods, the indicative diameter measured, measurement range, and distribution type.

BOX 3.1 COMPARISON OF VARIOUS SIZE MEASUREMENT METHODS

Table 3.2 Methods for Particle Size Measurement

Method	Diameter Measured	Range (μm)	Distribution Type	Comments
Sieving	"Sieve diameter"	>20	Mass	Slow; cheap; errors due to blinding, attrition, and fragmentation.
Optical microscopy	d_A	0.8–150	Number	Can be interfaced to image analysis computer;
Scanning electronic microscopy	d_A	0.01–20	Number	Depth of focus problem; identification possible; chemical analysis possible.
Dynamic imaging analysis	d_A	1–3000	Number	Fast; shape determination.
Laser diffraction	d_V	0.1–10000	Volume	Fast; online measurement and process control
Aerodynamic (impactor)	d_S or d_a	0.5–50	Mass	Errors due to bounce, breakage; re-entrainment; cheap
Electrozone or Coulter counter	d_V	0.3–200	Number	Particles must be dispersed in electrolyte; shape-independent

3.2.3 SIZE DISTRIBUTION—PRESENTATION AND ANALYSIS

Most powders handled in industrial processes have a wide size distribution, and the fractions of different sized particles vary from batch to batch and from process to process. Powders showing a distribution of particle size are termed *polydisperse*. For some powders, such as polymer latex spheres formed under zero gravity in industry and plant pollen in nature, the particles are of almost the same size. These powders are termed *mondisperse*. The following section presents methods for presenting measured particle size data and approaches to describing the particle size distribution.

3.2.3.1 How to Present Particle Size Data

Table 3.3 gives an example of typical particle size data, in which column 4 represents the measured (or "raw") data. In this example, the sizes of 600 particles are measured and the numbers of particles in discrete size intervals are counted and given as the count frequency. The size distribution is here based on number or count and is therefore a *number distribution*. Some particle size measurement systems allow the user to specify the limits of the sizing intervals. In this case, it is recommended to keep the resolution (the interval width divided by the mean interval size) approximately constant.

The simplest representation of the size data given in Table 3.3 is a frequency histogram showing the variation of measured count frequency in each size range (Fig. 3.11). It should be noted that this representation can be misleading because the shape of the histogram depends on the choice of the sizing intervals. In order to avoid this problem, it is preferable to divide the count in each interval by the total number of counts to obtain the fraction in each interval (i.e., in each size class, as shown in the 5th column in Table 3.3), and then to divide that fraction by the interval width (usually in micrometers) to obtain the fraction in unit interval width (fraction/μm, see column 6 in Table 3.3). When the fraction/μm is plotted against particle size in a histogram as shown in Fig. 3.12, the resulting representation possesses some unique characteristics: (1) the area under each rectangle represents the fraction of particles in that size interval; and (2) the total area is equal to one. This can be expressed mathematically as follows:

$$q_0(i) = \frac{n_i}{N} = h_i \Delta d_i \tag{3.14}$$

$$\sum q_0(i) = \sum_i \frac{n_i}{N} = \sum_i (h_i \Delta d_i) = 1 \tag{3.15}$$

where n_i is the number of particles in each interval (i.e., particle counts), N is the total number of particles, h_i is the height of the ith interval, of width Δd_i, and $q_0(i)$ is known as the frequency distribution function based on number. $q_0(i)$ is defined as the fraction of the total number of particles with diameters between d_i and d_{i+1} (i.e., the ith interval).

The frequency distribution can be shown either in a discrete form, as in Fig. 3.12, or as a continuous distribution using a smooth curve through q_0 (the tops of the frequency rectangles) at the mean value of each interval, as shown in Fig. 3.13. For the mean value of an interval, it is usual to take the geometric mean, i.e., $(d_i \cdot d_{i+1})^{1/2}$. However, if the interval width is small, the arithmetic

Table 3.3 An Example of a Particle Size Distribution

Size Interval Number, i	Size Range (μm)		Count Frequency	Fraction (Percent in Range), $q_0(i)$	Fraction per μm, $q_0(i)/\Delta d$	Cumulative Frequency	Cumulative Percent below Upper Bound
	Lower	Upper					
1	0	20	7	1.17	0.0583	7	1.17
2	20	30	35	5.83	0.5833	42	7.00
3	30	50	150	25.00	1.2500	192	32.00
4	50	80	216	36.00	0.9000	408	68.00
5	80	120	132	22.00	0.5500	540	90.00
6	120	180	43	8.00	0.1333	588	98.00
7	180	300	12	2.00	0.0167	600	100.00
Total			600	100			

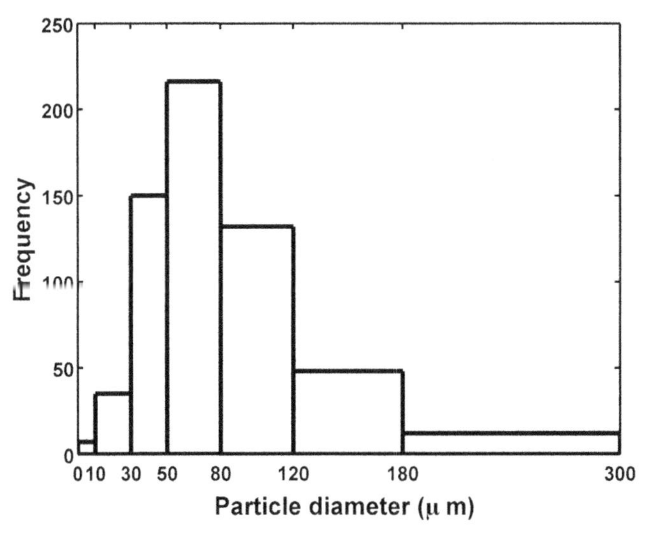

FIGURE 3.11

Number frequency distribution.

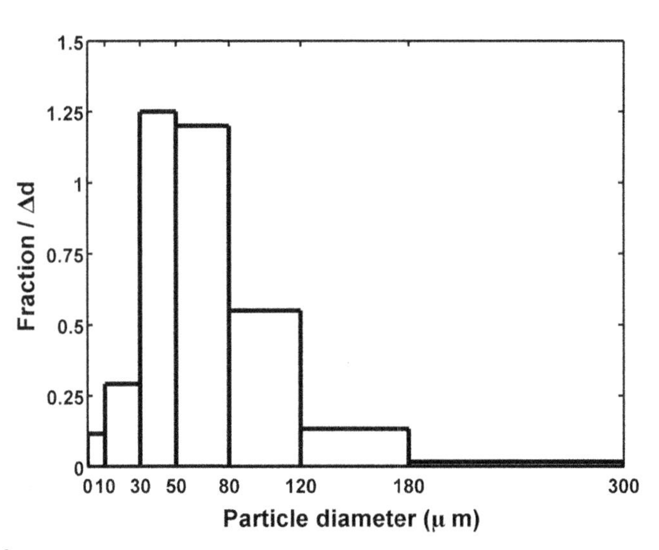

FIGURE 3.12

Discrete number distribution: fraction per μm versus particle size.

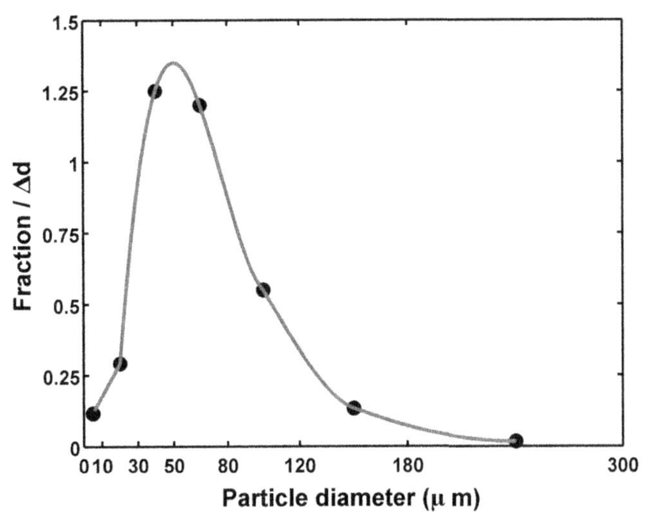

FIGURE 3.13

Continuous number distribution: fraction/μm versus particle size.

mean, $(d_i + d_{i+1})/2$, can be used as these two differ very little. The arithmetic mean is used in Fig. 3.13.

For a continuous distribution, the fraction of the total number of particles in the size interval $[a,b]$ is given as

$$q_0^{ab} = \int_a^b q_0(x)\mathrm{d}x \tag{3.16}$$

where $q_0(x)$ is the continuous frequency distribution function based on number. Equation (3.16) indicates that the fraction is the integral under the distribution curve between the limits of the interval. The total area under the distribution curve is again equal to one:

$$\int_0^\infty q_0(x)\mathrm{d}x = 1 \tag{3.17}$$

Alternatively, particle size data can also be represented as the *cumulative* distribution, as shown in Fig. 3.14 for the same data given in Table 3.3. In continuous form, the cumulative number distribution function, $Q_0(d)$, is defined as the fraction of the total number of particles with diameters smaller than d:

$$Q_0(d) = \int_0^d q_0(x)\mathrm{d}x \tag{3.18}$$

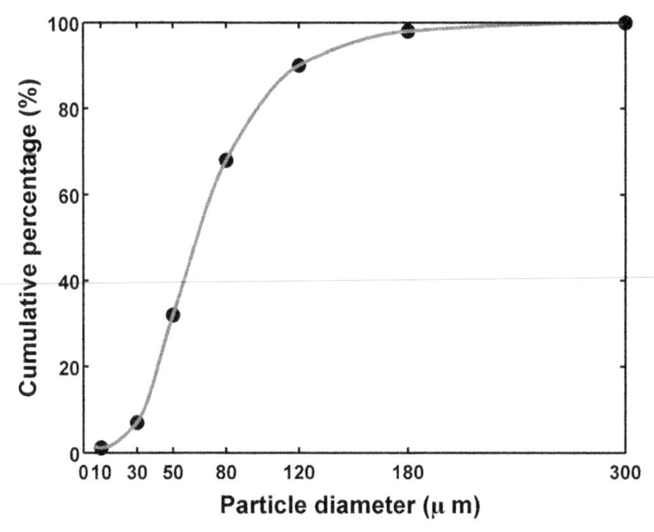

FIGURE 3.14

Cumulative number distribution.

or

$$q_0(x) = \frac{dQ_0(x)}{dx} \tag{3.19}$$

Equation (3.19) shows that the frequency function at any point $q_0(x)$ can be determined from the slope of the cumulative distribution function $Q_0(x)$.

3.2.3.2 Attributes of a Particle Size Distribution

As shown in Table 3.3 and Section 3.2.3.1, a large amount of information is needed to define a particle size distribution completely. In practice, it is useful to be able to approximate the distribution by some form of mathematical function or by a few (preferably two) parameters that can represent the distribution to some extent, especially for classification and comparison purposes. These functions and parameters are referred to as the *characteristics* of a particle size distribution. A number of such attributes are introduced for characterizing the particle size distribution, which can be classified into two groups: one for defining the location of the distribution and one for defining its width or breadth. The first group is usually some form of "average," such as the mean (strictly arithmetic mean), geometric mean, median, or mode:

The arithmetic mean, $\overline{d}_p$, is determined by dividing the sum of all the particle diameters by the total number:

$$\overline{d}_p = \frac{\sum n_i d_i}{N} \tag{3.20}$$

In continuous form, it is given as

$$\overline{d}_{\mathrm{p}} = \int_0^\infty d_i q_0(x_i)\mathrm{d}x_i \tag{3.21}$$

The geometric mean, d_{g}, is calculated as follows

$$d_{\mathrm{g}} = \left[d_1^{n1} d_2^{n2} d_3^{n3} \ldots d_i^{ni}\right]^{1/N} \tag{3.22}$$

where ni is the number of particles in the ith size class with a diameter of d_i. In practice, it is more convenient to determine d_{g} by converting Eq. (3.22) to natural logarithms, i.e.,

$$\ln d_{\mathrm{g}} = \frac{\sum(n_i \ln d_i)}{N} \tag{3.23}$$

so that

$$d_{\mathrm{g}} = \exp\left[\frac{\sum(n_i \ln d_i)}{N}\right] \tag{3.24}$$

The *median*, often denoted as d_{50}, is the diameter for which 50% of particles are larger, and the other 50% are smaller. It divides the frequency distribution into equal areas (see Fig. 3.13) and is the diameter which corresponds to $Q_0 = 0.5$ on the cumulative distribution curve (Fig. 3.14). The median is the most frequently used attribute in describing the particle size distribution, as it is less sensitive to *skewness* (lack of symmetry) of the distribution than the mean.

The *mode* defines the most frequent size, and it is the size corresponding to the highest point on the frequency curve (see Fig. 3.13).

For distributions that are skewed toward larger sizes as shown in Fig. 3.15, mode $<$ median $<$ mean. In addition to these four averages, it is also possible to define some other attributes that are related to the mass, surface area, or volume of the particles. For example, in Fig. 3.15, the *diameter of average mass* is defined as the diameter of a sphere of the same average mass as the whole sample, i.e.,

$$\overline{d}_{\mathrm{m}} = \left[\frac{\sum n_i d_i^3}{N}\right]^{1/3} \tag{3.25}$$

To describe the width or breadth of a particle size distribution, attributes such as variance, quantiles, and span, are introduced. *Quantiles* are particle size values which divide the distribution such that there is a given proportion below the quantile value. For example, the median divides the distribution such that 50% of the distribution lies below it. Apart from the median, two of the most frequently used quantiles are the lower decile d_{10}, which is defined as the size for which 10% of particles are smaller, and the upper decile d_{90}, which is the diameter for which 90% of the particles are smaller. These can be readily determined from the cumulative size distribution curve (see Fig. 3.14).

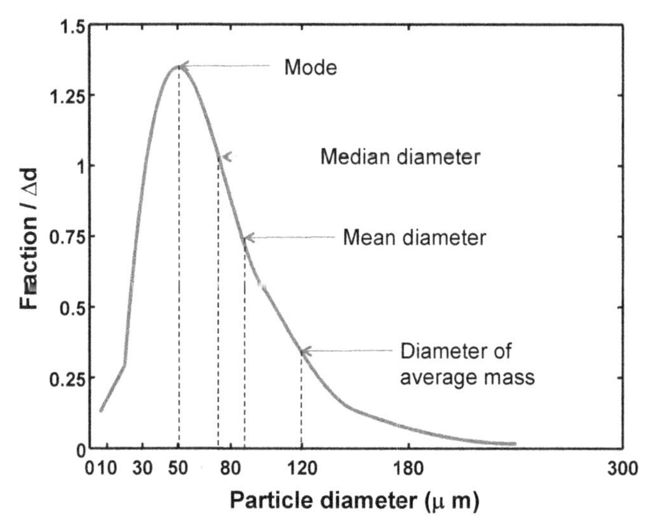

FIGURE 3.15

Illustration of some attributes of the particle size distribution.

The *span* Ψ is one of the parameters introduced as a measure of the breadth of a distribution and is defined as

$$\Psi = \frac{d_{90} - d_{10}}{d_{50}} \tag{3.26}$$

The breadth of a particle size distribution can also be represented by the *variance* σ^2, which is defined as

$$\sigma^2 = \sum \left[q_0(d) \left(d - \overline{d} \right)^2 \right] \tag{3.27}$$

where $\overline{d}$ is the mean and σ is the standard deviation.

3.2.3.3 Weighted Distributions

In many applications, some *weighted* distribution, such as the distribution by mass or by surface or volume, is more relevant than the number or count distribution discussed so far. For instance, the size distribution of coarse particles is commonly measured using sieving (see Section 3.2.2.1), from which a size distribution by mass (i.e., a mass distribution) is obtained. Just as a number distribution defines the fraction of the total *number* of particles in any size class, the mass distribution defines the fraction of the total *mass* of particles in any size class.

It is important to note that, for the same sample, the graphical representations and the values of the attributes for these two distributions are generally different. A product that is 80% by number within a desired size range may contain only 20% of the desired material by weight! Moreover, as discussed in the previous section, measuring instruments may give distributions by number or distributions which are weighted in some way. It is weighted distributions, such as mass, volume, and surface, which are generally of most practical use.

The number mean diameter given in Eq. (3.20) can be rewritten as

$$\bar{d}_{\mathrm{p}} = \frac{\sum n_i d_i}{N} = \sum \left[\frac{n_i}{N} d_i\right] \tag{3.28}$$

Similarly, the mass mean diameter, $\bar{d}_{\mathrm{mm}}$, is given by

$$\bar{d}_{\mathrm{mm}} = \sum \left[\frac{m_i}{M} d_i\right] \tag{3.29}$$

where m_i is the mass of particles in the ith size class and M is the total mass. The ratio (m_i/M) can be regarded as a weighting factor in the averaging process. If particle shape is not a function of particle size, then

$$m_i = \mathrm{k} d_i^3 \tag{3.30}$$

where k is a constant for all values of d_i, then

$$\bar{d}_{\mathrm{mm}} = \sum \left[\frac{m_i}{M} d_i\right] = \frac{\sum m_i d_i}{M} = \frac{\sum n_i d_i^4}{\sum n_i d_i^3} \tag{3.31}$$

$\bar{d}_{\mathrm{mm}}$ indicates the size of those particles constituting the bulk of the sample volume, so it is also known as the volume moment mean diameter or the mass moment mean diameter. Its value is very sensitive to the presence of large particles.

The surface mean diameter (also known as the "volume–surface mean" or the "Sauter mean") is defined in a similar way:

$$\bar{d}_{\mathrm{sm}} = \frac{\sum s_i d_i}{S} = \frac{\sum n_i d_i^3}{\sum n_i d_i^2} \tag{3.32}$$

where s_i is the surface area of particles in the ith size class and S is the total surface area. The surface mean diameter is often denoted as $\bar{d}_{32}$ and is the appropriate diameter to use in calculation of the pressure drop through a packed bed of particles at low Reynolds numbers (see Chapters 4 and 5).

In fact, Eqs (3.30) and (3.32) can be further generalized to obtain all the means of a population distribution as follows

$$\bar{d}_{ab}^{a-b} = \frac{\sum n_i d_i^a}{\sum n_i d_i^b} \tag{3.33}$$

where a and b are integer variables. Thus, when $a = 1$ and $b = 0$, Eq. (3.33) gives $\overline{d}_{10}$, that is the number length mean. Similarly

$a = 2$ and $b = 0$ give the number surface mean $\overline{d}_{20}$;
$a = 3$ and $b = 0$ produce the number volume mean $\overline{d}_{30}$;
$a = 3$ and $b = 2$ give the surface volume mean $\overline{d}_{32}$;
$a = 4$ and $b = 3$ give the volume moment mean $\overline{d}_{43}$;
and so on.

It is possible to represent the same particle size data using different weighted distributions through an appropriate conversion (Box 3.2), which generally requires certain assumptions about the form and physical properties of the particle. The number-based size distribution can also be converted to surface-, mass-, and volume-based size distributions, and vice versa. It is important to note that different weighted distributions are likely to have different attributes (averages, span, etc., see Section 3.2.3.2) even for the same sample or the same particle size data.

BOX 3.2 CONVERSION BETWEEN WEIGHTED DISTRIBUTIONS

As an example, let us convert the count size distribution shown in Table 3.3 to a mass distribution. Defining the fraction of the total number of particles in the ith size class as $q_0(d_i)$, the mass of these particles is then

$$m_i = Nq_0(d_i)\frac{\pi\overline{d}_i^3}{6}\rho_p \tag{3.34}$$

where $\overline{d}_i$ is the mean diameter of the size class.

The cumulative mass distribution function $Q_3(d_i)$ is then given by:

$$Q_3(d_i) = \frac{\sum_1^{i-1} Nq_0(d_i)\left(\frac{\pi\overline{d}_i^3}{6}\right)\rho_p}{\sum_1^I Nq_0(d_i)\left(\frac{\pi\overline{d}_i^3}{6}\right)\rho_p} = \frac{\sum_1^{i-1} q_0(d_i)\overline{d}_i^3}{\sum_1^I q_0(d_i)\overline{d}_i^3} \tag{3.35}$$

where I is the total number of intervals. In fact it is not essential for the particles to be spherical, but only necessary for their shape to be independent of size, i.e., for Eq. (3.30) to be valid. Similarly, to convert a mass distribution, represented by $q_3(d_i)$, to a cumulative number distribution, $Q_0(d_i)$:

$$Q_0(d_i) = \frac{\sum_1^{i-1}\left[\frac{q_3(d_i)}{\overline{d}_i^3}\right]}{\sum_1^I\left[\frac{q_3(d_i)}{\overline{d}_i^3}\right]} \tag{3.36}$$

The data of Table 3.3 (showing the number distribution) can now be retabulated to obtain the mass distribution, as shown in Table 3.4. The converted cumulative mass distribution is compared with the measured cumulative number distribution in Fig. 3.16, clearly showing the difference between these two distributions.

BOX 3.2 CONVERSION BETWEEN WEIGHTED DISTRIBUTIONS—cont'd

Table 3.4 Conversion From Count Distribution to Mass Distribution

Size Interval Number, i	Size Range (μm) Lower	Upper	Mean Size d_i, (μm)	Count Frequency, n_i	Fraction, $q_0(i)$	$q_0(i)d_i^3$	Cumulative Mass Percent Below Upper Bound, $Q_3(i)$
1	0	20	5.00	7	1.17	1.46E+02	0.000
2	20	30	20.00	35	5.83	5.60E+04	0.003
3	30	50	40.00	150	25.00	2.05E+06	0.115
4	50	80	65.00	216	36.00	1.87E+07	1.139
5	80	120	100.00	132	22.00	9.00E+07	6.074
6	120	180	150.00	48	8.00	3.31E+08	24.208
7	180	300	240.00	12	2.00	1.38E+09	100.000
Total				600	100	1.82E+09	

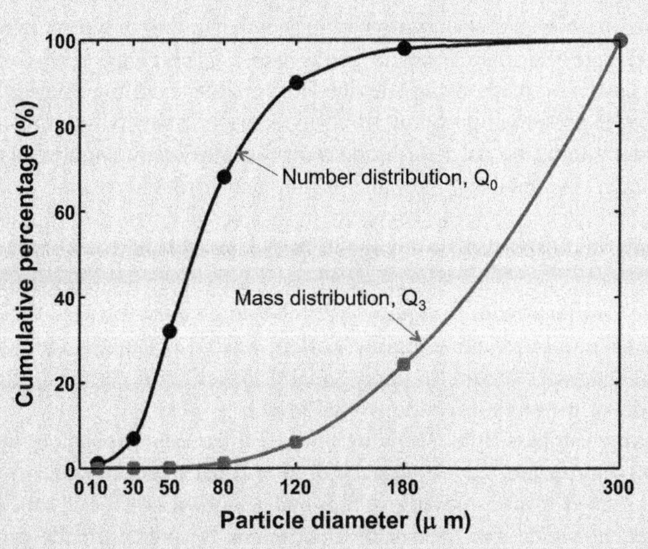

FIGURE 3.16

Comparison of converted cumulative mass distribution with the measured cumulative number distribution.

3.3 SAMPLING FOR SHAPE AND SIZE CHARACTERIZATION

Sampling is required for almost all particle characterization techniques before making a measurement, as it is very rare that one can perform a shape-and-size analysis on the entirety of the powder. It is a critical step in characterizing particle shape and size because only a small amount of sample (normally a few grams) is generally needed for the measurement, and the sample needs to be representative of the whole stock or the production stream (kilograms or even tonnes), which is itself often inhomogeneous in time or space or both. Sampling requires careful design and preparation to ensure that the sample is truely representative of the whole. Poorly designed sampling is primarily responsible for unreliable measurements.

Almost all bulk solids are highly polydisperse systems consisting of particles of different sizes, shapes and densities. Due to the differences in these physical properties, the particles tend to *segregate* (i.e., particles of different properties separate from each other) very easily during transportation and handling. For example, when particles are tipped onto surfaces or into bins, the large particles tend to roll down the sloping sides of the tip, resulting in a greater concentration of fine particles near the center. For powders stored in a container, vibration during transportation may cause smaller particles to sift through the gaps or voids between the large particles, leading to a higher concentration of fines at the bottom of the container. Segregation is therefore the main obstacle in the way of achieving good sampling. A detailed discussion on methods and devices for reliable sampling in these and other circumstances is presented in Allen (1990). Generally a device such as a *riffler* or sample divider should be used to divide the bulk into many smaller samples that should be further divided using the same technique, in order to obtain a small sample for analysis. Whenever possible, samples should be taken from a moving stream, while the bulk solid is in motion, i.e., before being packed and stored for subsequent usage.

When the powder stream is moving, the following "golden rules" should be followed to achieve appropriate sampling (Allen, 1990): A sample should be taken from the whole powder stream for many equally spaced short periods of time, rather than from part of the stream for the whole time.

When taking samples from a moving powder stream or suspension, two general measurement approaches can be employed: (1) offline measurement, in which the sample is extracted from a moving stream and analyzed outside it, and (2) online measurement, in which the measuring instrument is fitted into the process line and the measurement is carried out *in situ*. For the offline measurement, the sample is physically contained in the fraction of the total flow that is extracted. For the online measurement, the sample is not physically separated. For example, when optical methods are used, the sample is the fraction of the powder captured by the beam. For both methods, measurements should be taken at different points within the moving stream in order to minimize measurement errors arising from spatial variations in flow velocity and/or particle loading. The advantages and disadvantages of offline and online measurements are compared in Table 3.5.

Table 3.5 Advantages and Disadvantages of Offline and Online Sampling Methods (Seville et al., 1997)

Method	Advantages	Disadvantages
Offline	**1.** A wide range of equipment can be used; **2.** A portion of the sample extracted can be retained; **3.** Various analyses can be performed and repeated if necessary.	**1.** Slow measurement process as the analysis results are not immediately available; **2.** The sample may not be representative, as extractive equipment can disturb material flow; **3.** Agglomeration may occur in sampling, and conversely de-agglomeration can take place during measurement, resulting in unrepresentative samples for size analysis.
Online	**1.** Sample extraction equipment is not required; **2.** Particles can be monitored as they are being processed; **3.** Rapid data return **4.** Measurement can be used for process control	**1.** Calibration can be difficult; **2.** No sample is retained; **3.** Further analysis is not possible.

REFERENCES

Allen, T., 1990. Particle Size Measurement, 4th ed. Chapman and Hall, London.

Endoh, S., 2006. Particle shape characterisation. In: Masuda, H., Higashitani, K., Yoshida, H. (Eds.), Powder Technology Handbook, third ed. CRC Press. http://dx.doi.org/10.1201/9781439831885.ch1.3.

Hentschel, M.L., Page, N.W., 2003. Selection of descriptors for particle shape characterization. Particle & Particle Systems Characterization 20, 25–38. http://dx.doi.org/10.1002/ppsc.200390002.

Hawkins, A.E., 1993. The Shape of Powder-Particle Outlines. Wiley, New York.

Seville, J.P.K., Tüzün, U., Clift, R., 1997. Processing of Particulate Solids. Chapman and Hall/Kluwer, London.

CHAPTER

Particles in Fluids

4

The earlier chapters in this book are mostly concerned with the properties of solid particles, but in particle-processing operations the particles are surrounded by a fluid—a gas or a liquid—or perhaps by a mixture of fluids. Much of particle technology is concerned with the interaction between particles and fluids, which is a vast topic in itself (see, for example, Clift et al., 1978). This chapter concerns only single isolated particles—spheres and other shapes—in a continuous fluid.

4.1 SETTLING OF A SINGLE ISOLATED PARTICLE IN A CONTINUOUS FLUID

Consider first a single isolated sphere held stationary in a stationary continuous fluid, as shown in Fig. 4.1. What are the forces acting on it?

These are its weight (particle density × particle volume × acceleration due to gravity), and its buoyancy, which according to Archimedes' principle is equal to the weight of fluid displaced (fluid density × particle volume × acceleration due to gravity). Both act vertically.

If the particle is released from rest it will either fall (if its density is above that of the fluid), rise (if the opposite is true), or remain where it is (if the particle and fluid densities are the same). More precisely, if the densities are unequal, weight and buoyancy are unbalanced, and there is a net force on the particle which gives rise to an acceleration. The particle will then move, upward or downward, relative to the fluid, giving rise to a third force, termed the drag force, which acts in the opposite direction to the direction of relative motion. The particle accelerates to its terminal settling velocity, a constant velocity at which the three forces are in balance (Fig. 4.2).

In this condition it is possible to write:

$$\text{Drag force} = \text{Weight} - \text{Buoyancy} \quad (4.1)$$

or

$$F_D = \rho_p g V - \rho g V = (\rho_p - \rho) g V \quad (4.2)$$

where F_D is the drag force, ρ_p and ρ are the densities of the particle and the fluid, respectively, V is the particle volume, and g is the acceleration due to gravity.

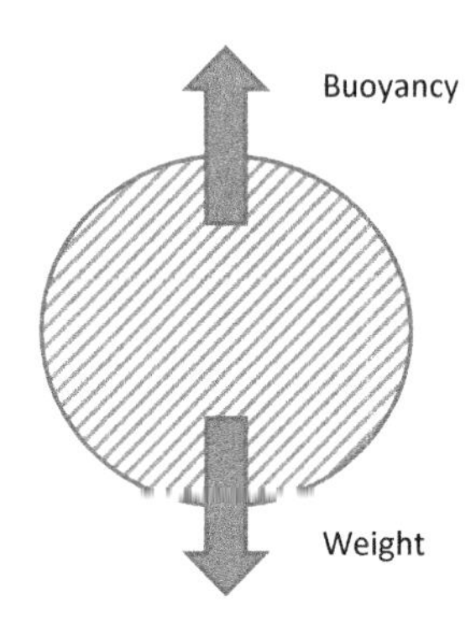

FIGURE 4.1

An isolated stationary sphere in a continuous fluid.

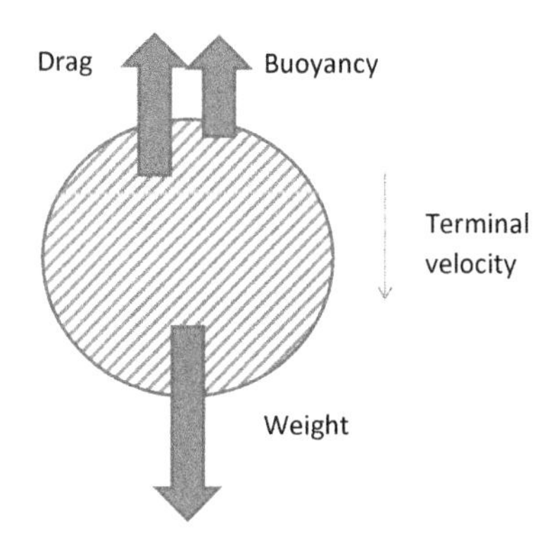

FIGURE 4.2

An isolated sphere at its terminal settling velocity (particle denser than the fluid).

The question then arises of how to calculate the terminal settling velocity, which should be possible if the drag force is known as a function of particle velocity. (More precisely, this should be *relative* velocity, since it is the particle velocity relative to that of the fluid which is important. In the case just considered, the fluid is stationary, but in general both the particle and the fluid will be in motion.) Box 4.1 considers how the drag force arises and how it might be calculated.

4.2 DRAG FORCE ON AN ISOLATED PARTICLE (SEVILLE ET AL., 1997)

A relationship between F_D and the terminal velocity u_t is required in order to solve Eq. (4.2), but there is no universally applicable theoretical result for this. It is therefore necessary to use some form of empirical correlation. Such correlations (and

BOX 4.1 A MORE COMPLEX SITUATION—DRAG, BUOYANCY, AND LIFT ON A NONSPHERICAL SHAPE

Settling of a sphere in a stationary fluid is relatively simple—it is a one-dimensional problem because the forces are all aligned. A much more complex situation is shown in Fig. 4.3, where the particle is nonspherical and the fluid flow direction is not aligned with gravity.

Any element dS of the surface of the particle will experience both a shear stress and a normal stress, acting respectively tangential and normal to the surface. The integral of the shear stress over the surface is the *skin friction drag* of the fluid on the particle.

The normal stress includes two components: (1) the hydrostatic pressure, whose integral over the surface gives the buoyancy force acting on the particle, which as noted before is equal to the weight of the fluid displaced and acts vertically upward, and (2) the "reduced" pressure caused by the flow, whose integral is known as *form drag* (The "reduced" pressure represents the difference between the actual pressure in the flowing fluid at dS and the hydrostatic pressure.).

The total drag on the particle, F_D, is the sum of the skin friction drag and the form drag and acts parallel to the fluid approach velocity U. Any component of the total hydrodynamic force normal to U is known as *lift*.

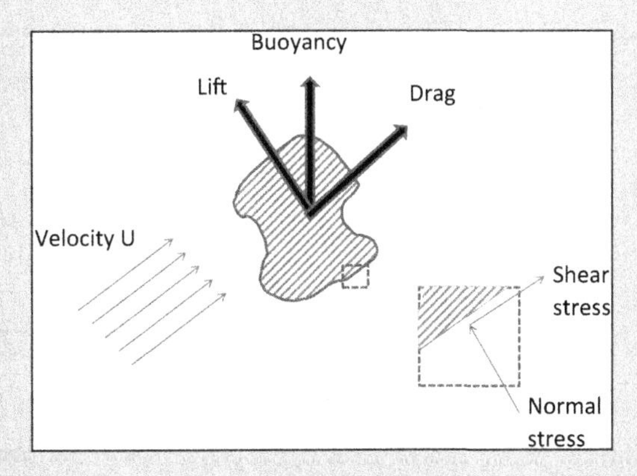

FIGURE 4.3

Forces on a stationary particle in a flowing fluid.

there are many available; see Clift et al., 1978) are commonly expressed in terms of two dimensionless variables, the particle Reynolds number,

$$Re_p = \rho U d/\mu \tag{4.3}$$

and the drag coefficient

$$C_D = F_D/\left[A\left(\rho U^2/2\right)\right] \tag{4.4}$$

where A is the cross-sectional area of the particle, and therefore proportional to d^2. For a spherical particle, A is $\pi d^2/4$ so that

$$C_D = 8F_D/\pi\rho U^2 d^2 \tag{4.5}$$

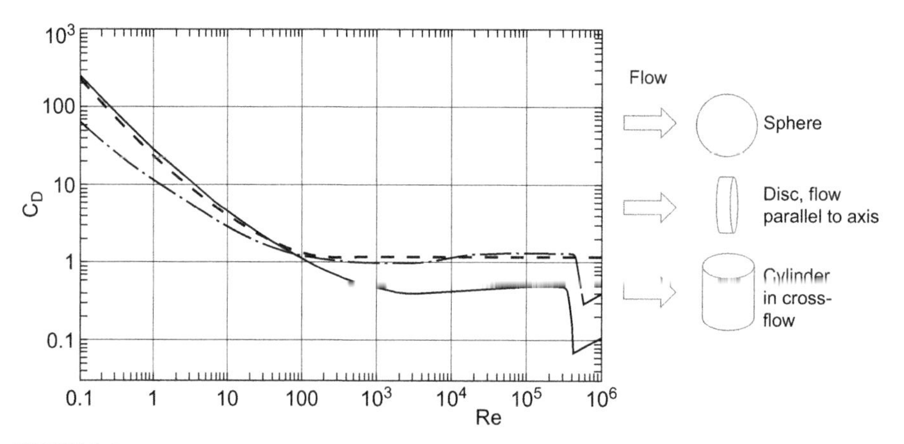

FIGURE 4.4

Drag coefficient C_D as a function of Reynolds number (continuous line—sphere; dashed line—disc with flow parallel to axis; and chain-dashed line—cylinder in cross flow).

After Seville et al. (1997).

Experimental data are correlated as $C_D = f(\mathrm{Re_p})$. A simple but reliable expression is due to Turton and Levenspiel (1986):

$$C_D = \frac{24}{\mathrm{Re_p}}\left[1 + 0.173\mathrm{Re_p^{0.657}}\right] + \frac{0.413}{1 + 16300\mathrm{Re_p} - 1.09} \tag{4.6}$$

which is shown in Fig. 4.4 with corresponding results for a disc with flow parallel to the axis and a cylinder in cross flow (In these two cases, the disc or cylinder diameter is d, and in the case of the cylinder, the drag is calculated per unit length.). Equation (4.6) applies for $\mathrm{Re_p} < 3 \times 10^5$ only, but this covers the range of interest here. Note that C_D goes through a minimum around $\mathrm{Re_p} = 4000$.

As shown in Fig. 4.4, at low $\mathrm{Re_p}$, C_D is inversely proportional to $\mathrm{Re_p}$, while at high $\mathrm{Re_p}$, C_D is approximately constant, and these general observations are true not only of spheres but also of discs and cylinders in these given orientations, with differences in the numerical values.

At low $\mathrm{Re_p}$ ($\mathrm{Re_p} < 1$), an exact theoretical result known as Stokes' law applies with

$$C_D = 24/\mathrm{Re_p} \tag{4.7}$$

or

$$F_D = 3\pi\mu dU \tag{4.8}$$

so that, in this range, the drag force is directly proportional to U.

The fluid streamlines are shown schematically in Fig. 4.5. At low $\mathrm{Re_p}$, the particle is described as being in *creeping flow*, i.e., its motion is dominated by viscous effects in the fluid. Note that according to Eq. (4.8), at low Re_p, the drag force depends on the fluid *viscosity* but not on its *density*. Nevertheless, of the total drag at low Re_p, two-thirds arises from the pressure distribution over the sphere and one-third from shear stresses; i.e., even in this creeping flow range, form drag on a sphere is twice the skin friction drag.

The changes in drag coefficient shown in Fig. 4.4 correspond to changes in the fluid flow around the sphere:

- At $Re_p = 20$, flow separates from the surface to form a closed recirculatory wake which becomes larger as Re_p increases (Fig. 4.5(b)).
- At higher Re_p, the downstream tip of the wake oscillates until, for $Re_p > 100$, the fluid is shed from the wake. Throughout this range, the point at which flow separates from the surface moves forward with increasing Re_p.
- For $Re_p > 6000$, the separation point reaches a steady position a little forward of the equator. In this range, skin friction drag is negligible and form drag depends most strongly on the position of flow separation. Therefore, the drag coefficient varies very little with Re_p; over the range $750 < Re_p < 3.4 \times 10^5$, C_D only varies by $\pm 13\%$ about a value of 0.445 (or 4/9). This is usually known as the *Newton's law* range, because Newton's concepts of fluid dynamic drag implied a constant drag coefficient.

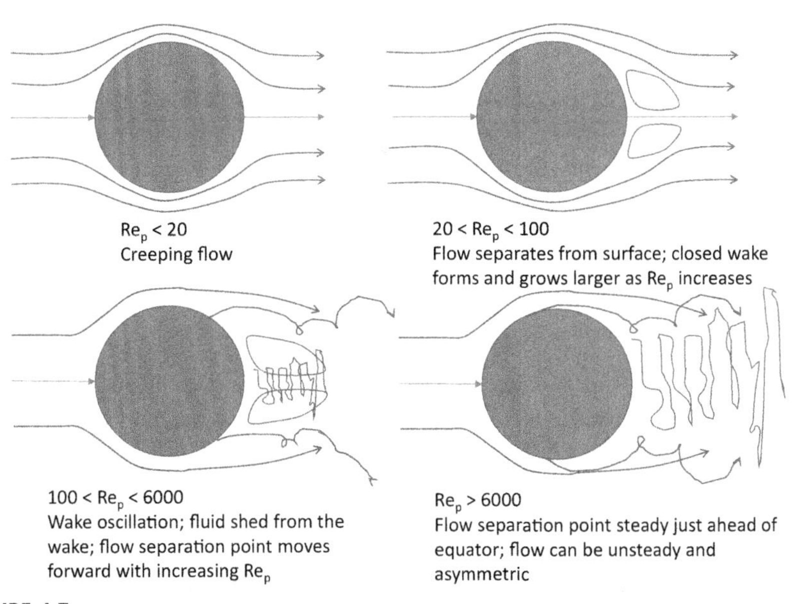

FIGURE 4.5

Flow past a sphere at different Reynolds numbers.

- For Re_p above 10^5, turbulence intensifies and turbulence bursts occur over the greater part of the surface of the sphere, until a "critical transition" occurs, beyond which the turbulence reattaches and C_D reduces strongly. This range is not usually encountered in particle technology but is important in the behavior of projectiles and balls in sports such as golf.

4.3 CALCULATION OF THE TERMINAL VELOCITY

Using the drag force relationships above, it is straightforward to calculate the terminal settling velocity, U_t, in the Stokes' law and Newton's law ranges as follows:

For $Re_p < 1$, from Eqs (4.2) and (4.8),

$$F_D = (\rho_p - \rho)gV = 3\pi\mu dU_t \tag{4.9}$$

Substituting for V ($=\pi d^3/6$) and rearranging gives

$$U_t = d^2(\rho_p - \rho)g/18\mu \tag{4.10}$$

For $10^4 < Re_p < 2 \times 10^5$, Newton's law applies with $C_D \approx 0.445$ and from Eq. (4.5)

$$U_t = 1.73\left[d(\rho_p/\rho - 1)g\right]^{1/2} \tag{4.11}$$

In the intermediate regime between the Stokes' and Newton's law ranges, calculation of the terminal velocity is not straightforward because the equations are implicit and iteration is therefore needed. It is therefore preferable to express the drag relationship in a different form:

$$U_t^* = f[d^*] \tag{4.12}$$

where

$$\begin{aligned} U_t^* &= [4Re_t/3C_D]^{1/3} \\ &= U_t\left[\frac{\rho^2}{\mu(\rho_p - \rho)g}\right]^{1/3} \end{aligned} \tag{4.13}$$

and

$$\begin{aligned} d^* &= \left[3C_D Re_p^2/4\right]^{1/3} \\ &= d\left[\frac{\rho(\rho_p - \rho)g}{\mu^2}\right]^{1/3} \end{aligned} \tag{4.14}$$

The resulting relationship is shown in Fig. 4.6, and correlations for the full range are tabulated in Table 4.1.

Figure 4.7 shows the terminal velocities of three sizes of particle as functions of temperature at pressures of 1 and 10 bar. The dependence on temperature and pressure illustrates the difference between "small" and "large" particles outlined above. For small particles/low Re_p, the terminal velocity depends on μ but, for particles in gases with $\rho << \rho_p$, not on ρ; therefore, U_t decreases with increasing temperature but is essentially unaffected by pressure. For relatively large particles/high Re_p, ρ matters but not μ; i.e., the gas flow and the particle motion are dominated by inertial effects, not by viscosity, and the dominant drag term is form drag. The terminal velocity therefore increases as temperature increases (so ρ decreases) but decreases as pressure increases (so ρ increases).

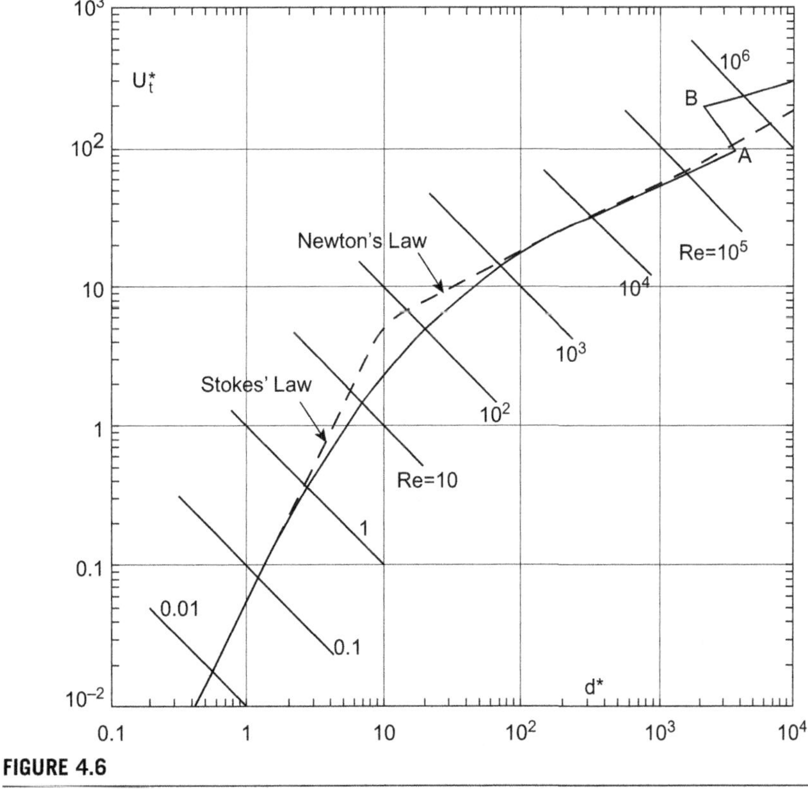

FIGURE 4.6

The relationship between the dimensionless terminal velocity and dimensionless particle size.

After Seville et al. (1997).

Table 4.1 Correlations for Drag and Terminal Velocity (Turton and Levenspiel, 1986; Seville et al., 1997)

1. Spheres

a. Drag

Range of Re	Correlation
$\leq 3.38 \times 10^5$	$C_D = \frac{24}{Re}(1 + 0.173Re^{0.657}) + \frac{0.413}{1 + 1.63 \times 10^4 Re^{-1.09}}$
3.38×10^5 to 4×10^5	$C_D = 29.78 - 5.3 \log_{10} Re$
4×10^5 to 10^6	$C_D = 0.1 \log_{10} Re - 0.49$
$\geq 10^6$	$C_D = 0.19 - 8 \times 10^4 Re^{-1}$

b. Free fall or rise

Correlations for dimensionless terminal velocity U_{ts}^*, as a function of dimensionless diameter, d^*, after Grace (1986). These are defined in Eqs (4.13) and (4.14).

Range of			Correlation
d^*	U_{ts}	Re_p	
≤ 3.8	≤ 0.624	≤ 2.37	$U_{ts}^* = (d^*)^2/18 - 3.1234 \times 10^{-4}(d^*)^5 + 1.6415 \times 10^{-6}(d^*)^8 - 7.278 \times 10^{-10}(d^*)^{11}$
3.8–7.58	0.624–1.63	2.37–12.4	$x = -1.5446 + 2.9162w - 1.0432w^2$
7.58–227	1.63–28	12.4–6370	$x = -1.64758 + 2.9478w - 1.09703w^2 + 0.17129w^3$
226–3500	28–93	6.37×10^3 – 3.26×10^5	$x = 5.1837 - 4.51034w + 1.687w^2 - 0.189135w^3$

Note: $w = \log_{10} d^*$; $x = \log_{10} U_{ts}^*$.

2. Cylinders $E = L/d$ = aspect ratio

a. Drag $(E \gg 1): C_D' = 9.689Re^{-0.78}$

$0.1 < Re < 5$ $\quad C_D = C_D'(1 + 0.147Re^{0.82})$

$5 < Re < 40$ $\quad C_D = C_D'(1 + 0.227Re^{0.55})$

$40 < Re < 400$ $\quad C_D = C_D'(1 + 0.0838Re^{0.82})$

b. Free fall or rise $(E \geq 1, d^* < 20, U_t^* < 25, Re < 100)$

$$\log_{10} U_t^* = A_0 + A_1 w + A_2 w^2 + A_3 w^3, \text{ where}$$

$$A_0 = -0.66140 - 0.46143/E - 0.03246/E^2$$

$$A_1 = 1.38545 + 1.37542/E - 0.45526/E^2$$

$$A_2 = 0.21234 - 1.27355/E - 0.57592/E^2$$

$$A_3 = -0.03436 - 0.37525/E - 0.18969/E^2$$

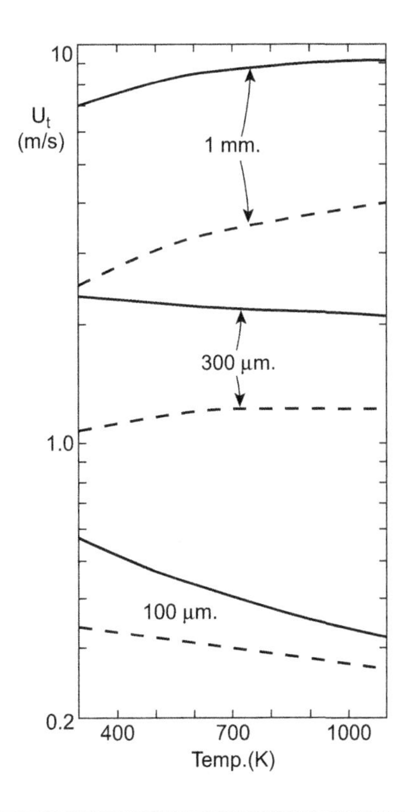

FIGURE 4.7

Terminal velocity of particles of density 2650 kg/m^3 in air (continuous line—1 bar; dashed line—10 bar).

After Seville et al. (1997).

4.4 OTHER SHAPES

Figure 4.4 shows the drag–velocity relationships for discs and cylinders as well as spheres, illustrating some interesting effects of shape.

An infinitesimally thin disc in a fluid moving parallel to its axis experiences only form drag because there is no surface parallel to the direction of flow and hence no skin friction. At very low Re_p, this reduces the drag by comparison with a sphere. Wake formation occurs at a lower Re value for a disc, around $Re_p = 1$, because the sharp edge encourages the flow to separate. Disc wakes start to oscillate at about $Re_p = 100$ but, because the position of flow separation is fixed, the drag coefficient is constant, at $C_D = 1.17$, for $Re_p > 133$. At high Re_p, flow patterns continue to be qualitatively similar to those around a sphere except that, because the separation position is fixed by the particle shape, there is nothing equivalent to the critical transition.

A long cylinder in cross flow shows similar features, but the numerical values of the transitions, such as flow separation, are different. Over the range $70 < Re_p < 2500$, wake shedding occurs alternately from the two sides of the cylinder to form a regular series of vortices alternating in their sense of rotation; this is often termed a "von Kármán vortex street."[1]

In order to estimate the drag on a nonspherical particle, it is usual to compare its shape with the standard shapes for which drag coefficient–Reynolds number relationships are known. (The sphere is special in having a unique dimension. The question of how a length dimension is defined for a nonspherical particle is addressed in Chapter 3.) In addition to the sphere, disc, and cylinder, the other common "ideal" shape is the spheroid, which comes in two types: prolate (a sphere distorted by pulling at the poles; a shape like a rugby ball) and oblate (a sphere distorted by pushing at the poles; a shape like a bread bun). Drag relationships for these are given by Clift et al. (1978). In the creeping flow range, the drag on a particle is always less than or equal to that on a body which encloses it and greater than or equal to that on any body contained within it (Hill and Power, 1956). At low Re_p, the drag force is most sensitive to the particle surface, as in the disc–sphere comparison. At intermediate and high Re_p, drag is most likely to be influenced by edges which define the position of flow separation. Such effects can, of course, be reduced by "streamlining," i.e., profiling the particle shape to prevent boundary layer separation, as in the design of vehicles for high Re number operation.

So far, little mention has been made of orientation. In practice, particles in fluids are usually free to find their own orientation. At Reynolds numbers intermediate between the Stokes' and Newton's law ranges—the range of common practical interest—a nonspherical particle in free motion tends to adopt a preferred orientation with its maximum area presented to the fluid, i.e., in a horizontal plane, for a falling particle. However, particles in gases can maintain other orientations for long times and large distances of travel. The subject of particle orientation is highly complex and dense with studies on different shapes under different conditions. Anyone who has observed leaves falling from trees or coins falling in water will be familiar with some extremes of this behavior. A comprehensive study is presented by Clift et al. (1978).

4.5 ACCELERATION AND UNSTEADY MOTION[2]

As shown earlier (Eq. (4.10)) the terminal velocity of an isolated particle in free fall under gravity in the Stokes' law range is given by

$$U_t = \tau g \tag{4.15}$$

where

$$\tau = (\rho_p - \rho)d^2/18\mu \tag{4.16}$$

[1]After Theodore von Kármán, Hungarian–American mathematician, engineer and physicist (1881–1963).
[2]The authors acknowledge the example of Hinds (1982) in preparing this section.

τ is known as the "relaxation time," so-called because it characterizes the time taken for a particle to adjust to any changes in flow velocity or direction. This approach can be generalized to give the terminal velocity of a particle subjected to any constant external force, F:

$$U_{\mathrm{tF}} = \tau \frac{F}{m} \tag{4.17}$$

During acceleration in a gas, the drag force may still be approximated by Stokes' law, using the instantaneous velocity, so that for acceleration from rest in a gravity field

$$m\frac{\mathrm{d}U(t)}{\mathrm{d}t} = mg - 3\pi\mu U(t)d \tag{4.18}$$

or

$$\tau\frac{\mathrm{d}U(t)}{\mathrm{d}t} = \tau g - U(t) \tag{4.19}$$

which can be solved to give

$$U(t) = U_{\mathrm{t}}\left(1 - \mathrm{e}^{-t/\tau}\right) \tag{4.20}$$

By similar reasoning, if U_0 is the initial particle velocity at $t = 0$ and U_{f} is the final (steady-state) velocity:

$$U(t) = U_{\mathrm{f}} - (U_{\mathrm{f}} - U_0)\mathrm{e}^{-t/\tau} \tag{4.21}$$

Integrating this equation, the distance traveled, x, can be obtained:

$$x(t) = U_{\mathrm{f}}t - (U_{\mathrm{f}} - U_0)\tau\left(1 - \mathrm{e}^{-t/\tau}\right) \tag{4.22}$$

In the case of a particle projected into a stagnant gas at velocity U_0, so that $U_{\mathrm{f}} = 0$, this reduces to

$$x(t) = U_0\tau\left(1 - \mathrm{e}^{-t/\tau}\right) \tag{4.23}$$

so that the "stopping distance" or total distance traveled (for $t >> \tau$) is simply

$$x_{\mathrm{s}} = U_0\tau \tag{4.24}$$

From this, it is now possible to estimate how far a particle can be "thrown" into a stagnant gas. (For a 1 μm particle of density 10^3 kg/m^3 in air, x_{s} is about 3.6 μm per (m/s) initial velocity, i.e., a small multiple of its diameter.)

It should be noted that all of the above applies only for particles in gases (or more generally, for particles which are much denser than their surrounding fluid). If this is not the case, as for most particles in liquids, the calculation is more complex (Box 4.2).

BOX 4.2 DRAG IN UNSTEADY MOTION

Earlier sections of this chapter considered steady motion of a particle in a fluid at a constant velocity or under simple acceleration, and it has been shown how dimensional analysis leads to two dimensionless groups, e.g., C_D and Re_p, or d^* and U_t^*, which are sufficient to describe the situation completely. In general, unsteady motion is much complex, because the inertia of the particle is important as well as its immersed weight; i.e., ρ_p is significant as well as $(\rho_p - \rho)$. A detailed discussion is given by Clift et al. (1978) but some general points will be summarized here.

In the creeping flow range, the drag on a sphere in unsteady motion through a stagnant fluid can be written

$$-F_D = 3\pi\mu dU + \frac{1}{2}\rho\left[\frac{\pi d^3}{6}\right]\frac{dU}{dt} + \frac{3}{2}d^2\sqrt{\pi\rho\mu}\int_{-\infty}^{t}\left[\frac{dU}{dt}\right]_{t=s}\frac{ds}{\sqrt{t-s}} \tag{4.25}$$

where $U(t)$ is the instantaneous particle velocity.

The total instantaneous drag is made up of three terms, of which the first is the steady (Stokes) drag at the instantaneous velocity. The second term on the right of Eq. (4.25) is known as the "added mass" or "virtual mass" term. It arises from the entrained fluid which accelerates with the particle; for a sphere, this "added mass" of fluid corresponds to a volume equal to half that of the sphere. The final term, known as the "Bassett term" or "history term," involves an integral over all past accelerations of the sphere. The form of this term arises from the generation of vorticity at the surface of the particle, which diffuses into the surrounding fluid. The history term can be thought of as describing the state of development of the flow around the particle: if it is small, then the relaxation time of the flow is sufficiently small that the instantaneous flow field differs little from steady flow at the instantaneous value of U.

Equation (4.25) can be developed to give an equation of motion of the sphere, for example, to describe the initial motion of a particle released from rest; although its form only applies in creeping flow, it can be extended semiempirically to describe accelerated motion at higher Reynolds numbers; see Clift et al. (1978). Its solution is obviously complicated by the history integral, but some useful general conclusions include:

1. For particles in gases, with $\rho_p >> \rho$, the added mass and history terms are often small, so that the drag can be approximated by the steady-state drag at the instantaneous velocity. However, this is by no means universally true, even for particles in gases. It is therefore advisable to estimate $U(t)$ by making the quasi-steady approximation and then evaluate the two other terms to verify the validity of the approximation.
2. For particles in liquids, the quasi-steady assumption is almost never justified. Therefore, the kind of calculations made for particles in gases, for example, in inertial devices for separating particles from gases (see Chapter 5), cannot be applied to particles in liquids.
3. The history term is usually larger than the added mass term. Therefore, it is not possible to ignore the history term but to include the added mass term.

4.6 CURVILINEAR MOTION

In the examples considered so far, the forces acting on the particle and the direction of particle motion have all been aligned and the motion has been in a straight line. What happens if particles are subjected to forces acting simultaneously in more than one direction or when there is a change in the direction of the fluid streamlines?

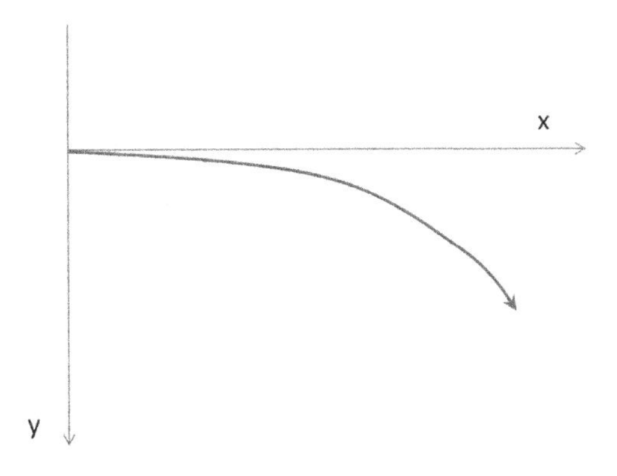

FIGURE 4.8

Path of a particle projected horizontally into a gravity field.

A typical example of the latter is when the fluid has to move around an obstacle, such as a fiber in a filter (see Chapter 5). If the fluid is carrying particles with it, will these follow the curving gas streamlines or carry straight on and impact on the obstacle?

A common example of the case of two forces acting simultaneously in different directions is what happens when a particle is thrown horizontally in a gravitational field, as shown in Fig. 4.8.

This problem is much easier to solve if the motion is at low $\mathrm{Re_p}$, i.e., within the Stokes' regime, because in that case it is possible to analyze the motion *independently* in the two orthogonal directions. For a particle projected horizontally at a velocity U_0 (from Eq. (4.23)):

$$x(t) = U_0\tau\left(1 - e^{-t/\tau}\right) \tag{4.26}$$

$$y(t) = U_t t - U_t\tau\left(1 - e^{-t/\tau}\right) \tag{4.27}$$

If the particle motion is outside the Stokes regime, the motion along one axis affects the motion along other axes and a more complicated analysis is required.

For flow around an obstacle, it is first necessary to define the flow field, using appropriate assumptions, if necessary. The equation of motion of the particle is then solved numerically at successive time intervals (Box 4.3).

4.7 ASSEMBLIES OF PARTICLES

Only single particles have been considered so far, but what happens if there are many particles. For example, do many particles close together settle at the same settling velocity as a single particle?

BOX 4.3 THE STOKES NUMBER, St

Curvilinear and other motion of particles in fluids with curved streamlines or of rapidly changing velocity is characterized by the use of a Stokes number, St, which is the ratio of the stopping distance of the particle to a characteristic dimension of the flow, such as a dimension of the obstacle causing the flow disturbance, d_c:

$$\mathrm{St} = \frac{x_s}{d_c} = \frac{\tau U_0}{d_c} \tag{4.28}$$

Here U_0 is interpreted as the undisturbed flow (and therefore particle) velocity at infinite distance upstream of the flow disturbance. Thus, the Stokes number can be interpreted as the ratio of the effects of particle inertia to fluid drag, a measure of the tenacity with which particles hold to their course rather than following the streamlines. Small values of St indicate good "tracking" of the streamlines, while large values indicate that the particle will "resist" changes in direction or velocity.

An example of the use of the Stokes number in designing cyclones is given in Chapter 5.

In practice, for an assembly of particles the settling velocity is typically much lower than the single-particle terminal velocity. This reduction in settling velocity arises from two complementary effects:

1. The displacement of fluid by the settling particles causes an upflow in the spaces between the particles;
2. The drag on each individual particle is increased by the effect of the neighboring particles on the velocity profile in the interstitial fluid.

Provided that direct particle–particle interactions such as electrostatic forces are negligible, the combined effect is described by the well-known Richardson–Zaki correlation (see Kay and Nedderman, 1985):

$$U'_t - U_t\varepsilon^n \tag{4.29}$$

where U'_t and U_t are, respectively, the "hindered settling velocity" and the single-particle terminal velocity, and ε is the void fraction, i.e., $(1 - C_V)$, where C_V is the volumetric concentration of particles. The index n depends on the single-particle Reynolds number, and values are given in Table 4.2. Note that the effect of solids concentration is strongest at the lowest Reynolds number. Equation (4.29) has been shown to describe the expansion of fluidized beds in the nonbubbling mode, as well as sedimentation.

Table 4.2 Values for the Richardson–Zaki Index

Re_p at Terminal Velocity	Value of n
$Re_p \leq 0.2$	4.6
$0.2 < Re_p < 1$	$4.4\ Re_p^{-0.03}$
$1 \leq Re_p < 500$	$4.4\ Re_p^{-0.1}$
$500 \leq Re_p$	2.4

REFERENCES

Clift, R., Grace, J.R., Weber, M.E., 1978. Bubbles, Drops and Particles. Academic Press, New York (reprinted 2005 by Dover Publications, New York).

Grace, J.R., 1986. Contacting modes and behaviour classification of gas—solid and other two-phase suspensions. Canadian Journal of Chemical Engineering 64, 353−363.

Hill, R., Power, G., 1956. Extremum principles for slow viscous flow and the approximate calculation of drag. Quarterly Journal of Mechanics and Applied Mathematics 9, 313−319.

Hinds, W.C., 1982. Aerosol Technology. Wiley-Interscience, New York.

Kay, J.M., Nedderman, R.M., 1985. Fluid Mechanics and Transfer Processes. Cambridge University Press, Cambridge.

Seville, J.P.K., Tüzün, U., Clift, R., 1997. Processing of Particulate Solids. Chapman & Hall, London.

Turton, R., Levenspiel, O., 1986. A short note on the drag correlation for spheres. Powder Technology 47, 83−86.

CHAPTER

Gas–Solid Systems

5

Contact between phases is a very common operation in chemical and process engineering. This chapter concerns the contact between multiple solid particles and a continuous gas phase. It draws upon the content of Chapter 4, on the interaction between single particles and a fluid.

Multiphase systems containing particles present inherent difficulties which do not exist in simple multiphase systems, particularly related to the treatment of particle size distributions, as discussed in Chapter 3. For example, in many practical calculations engineers will use an "average" particle size. In practice, process behavior often depends to some extent on the shape of the particle size distribution, particularly on the fraction of "fines" (smaller particles) and not just on the average. These matters are considered in Chapter 3.

The range of practical devices and equipment which are used to enable contact between particles and a gas is extremely wide. In this chapter we consider a few of the most frequently encountered.

5.1 GAS–SOLID CONTACT REGIMES—THE WHOLE PICTURE

Consider a heap of particles sitting in a column and supported by some form of porous plate, through which gas can pass (Fig. 5.1).

What happens as the gas velocity is increased? Initially, the particles do not move; they remain stationary and the gas flows around them. This is a *settled or "packed" bed*, shown at point A. As the gas velocity increases further, the particles may begin to move due to the forces exerted on them by the gas, the particle packing loosens, and the bed may expand and/or bubble. Its appearance then resembles a liquid with gas bubbles rising within it. This is represented by point B and is called *fluidization*. As the gas velocity is increased still further, the drag forces on the particles can become large enough to carry them completely out of the column. This is the beginning of *pneumatic transport*.

Consider now what happens to the pressure drop across the bed, represented at the bottom of the diagram. It increases initially as the gas flow increases through the packed bed but then levels off in the fluidization regime, before starting to increase again as pneumatic transport is reached. The details of this behavior are considered further below.

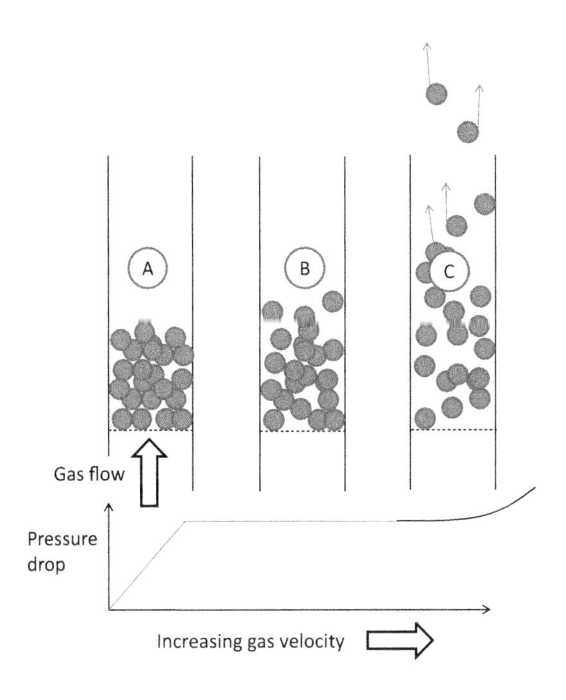

FIGURE 5.1

Regimes of gas–solid contact.

Figure 5.2 shows the practical experimental setup necessary to carry out the experiment described above. In this case the column is shown as cylindrical, although it could be of any shape, and of cross-sectional area A. There is a device for pushing the gas through the bed, commonly known as a blower, and a means of measuring the total volumetric gas flow, which is here denoted as Q (and understood to be under the pressure and temperature conditions in the bed), so that the

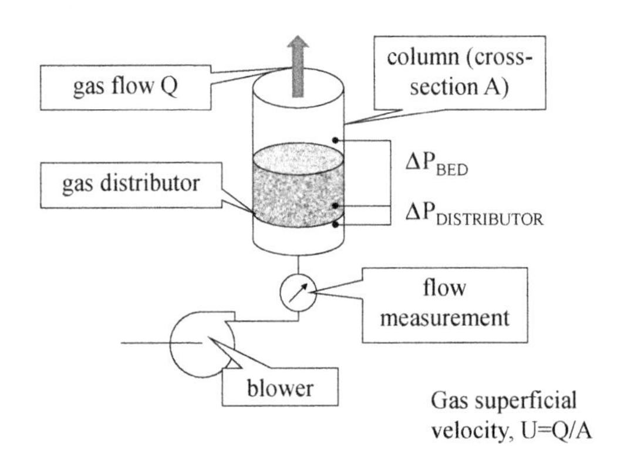

FIGURE 5.2

Gas–solid contact: experimental details.

superficial gas velocity in the column is $U = Q/A$. The pressure drop across the bed, ΔP_{BED}, is measured, and it is also useful to measure the pressure drop across the gas distributor, $\Delta P_{DISTRIBUTOR}$.

5.2 FLOW THROUGH A PACKED BED

When a fluid passes through a packed bed of particles, a pressure drop results. It is best to describe this in terms of the *manometric* pressure drop; the manometric pressure difference between two points is the total pressure difference minus the hydrostatic pressure difference arising when a stationary fluid is present between the two points. In other words, the manometric pressure difference is the pressure difference which results solely from the motion of the fluid. The distinction between total and manometric pressure difference is only of practical importance if the density of the fluid is significant, i.e., when the fluid is a liquid, as in Chapter 6, but not usually when it is a gas.

The most commonly used approach to estimate pressure drop in packed beds is that due to Ergun (1952)[1]:

$$\frac{\Delta P}{H} = 150\frac{(1-\varepsilon)^2}{\varepsilon^3}\frac{\mu U}{d^2} + 1.75\frac{(1-\varepsilon)}{\varepsilon^3}\frac{\rho U^2}{d} \tag{5.1}$$

where ΔP is the manometric pressure difference between two points in the bed a distance H apart in the direction of flow, and U is the superficial fluid velocity. The *void fraction* in the bed is denoted by ε. This includes interstitial voids (i.e., voids between the particles) but not interparticle voids (i.e., voids *within* the particles). A typical value of ε for a narrow size distribution of sand at the point of minimum fluidization might be 0.41. d is the particle diameter, ρ is the fluid density, and μ is its viscosity.

The Ergun equation is semiempirical in nature, that is to say that the form of each of the two terms has a good theoretical justification, but the numerical coefficients have been obtained by fitting to experimental results. Nevertheless, this equation has been found to be quite reliable in a wide range of situations, from packed columns to filters.

As introduced in Chapter 4, fluid flow is often described in terms of dimensionless groups, the most common of which is the Reynolds number, $\rho Ud/\mu$. The value of the Reynolds number provides a simple indication of whether flow behavior is dominated by fluid viscosity or density—that is by viscous or inertial effects.

[1]At low Reynolds numbers, the second term in the Ergun equation disappears, and the equation then becomes virtually the same as the well-known "Carman–Kozeny equation." The simple result that the pressure gradient is proportional to the flow rate (which is only true at low Reynolds numbers) is generally credited to Darcy (1856). For a more extensive explanation of the basics of particles in fluids, see Seville et al. (1997), Chapters 2 and 6.

In the context of flow through beds of particles, the form of the Reynolds number to be used is the *particle* Reynolds number, $\rho U d_p/\mu$, where d_p is the particle diameter.

The first term on the right of Eq. (5.1) dominates in creeping flow, i.e., when the particle Reynolds number, Re_p, is small so that drag is dominated by fluid viscosity and not affected by its density; thus $\Delta P \propto U$, as shown in Fig. 5.1. The second term dominates at relatively high Re_p, i.e., when drag is dominated by the inertia of the fluid and therefore affected by ρ but not μ; thus, at high Re_p, $\Delta P \propto U^2$.

5.3 FLUIDIZATION

As shown in Fig. 5.1, as the gas velocity through the packed bed is increased, the pressure drop across the bed also increases until it equals the weight per unit area of the bed. At this point (the point of incipient or minimum fluidization, U_{mf}), the bed is said to be fluidized. At gas velocities in excess of the minimum fluidization velocity[2], some of the fluidizing gas passes through the bed in the form of bubbles, which resemble (in some respects) bubbles in a viscous liquid (Fig. 5.3).

A fluidized bed exhibits the following properties, which make it useful in many chemical and process engineering applications:

1. It behaves like a liquid of the same bulk density—particles can be added or withdrawn freely, the pressure varies linearly with depth, heavy objects will sink and light ones float;
2. Particle motion is rapid, leading to good solids mixing—hence little or no variation in bed temperature with position;
3. A very large surface area is available for reaction/mass and heat transfer.

There are, however, some disadvantages, which should be considered:

1. Particles can be carried out of the column by the gas, especially with wide size distributions. This is called *entrainment*. In practice, entrainment can be controlled by increasing the height of the column and/or by increasing the diameter of the column at the top and so reducing the exit gas velocity. Particles which are entrained can be separated in a cyclone or filter and, if required, returned to the column. Circulating fluidized beds operate on exactly this principle.
2. Particles can be damaged by collisions with each other and the wall of the column (termed *attrition*), and the walls can be damaged by such collisions (termed *erosion*). These effects can be controlled by good design and appropriate selection of operating parameters.

An apparent disadvantage for a gas–solid contactor such as a catalytic reactor is that bubbling provides a means whereby gas can bypass the bed without contacting

[2]This is a simplification, as explained later.

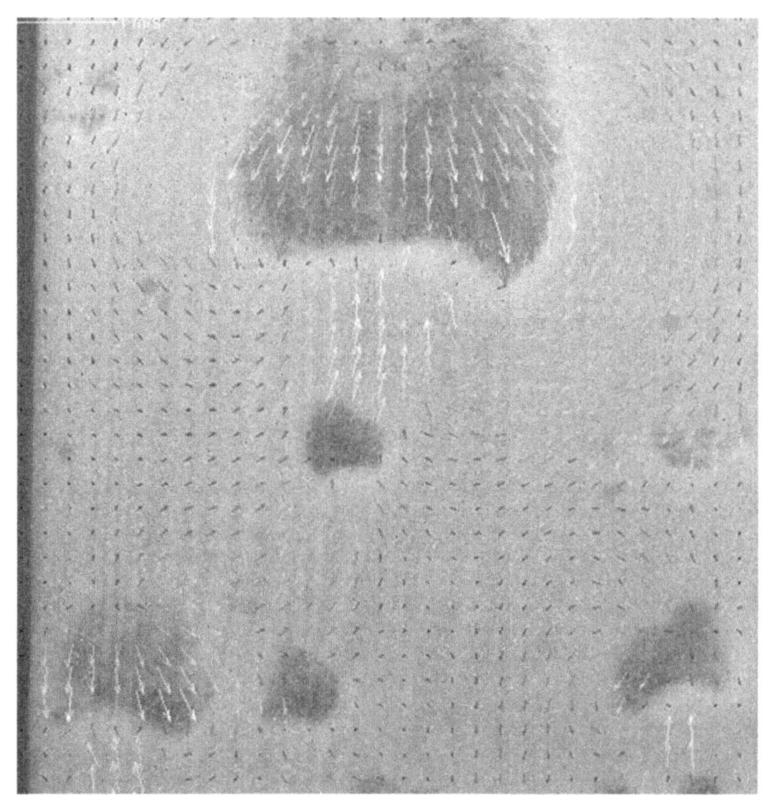

FIGURE 5.3

Bubbles in a fluidized bed. (From optical investigation of a "two-dimensional" bed; courtesy of Niels Deen and Hans Kuipers, University of Eindhoven. Arrows show gas and solid velocities.)

the solids. However, bubbles in fluidized beds differ from those in liquids in that there is a strong throughflow of gas in bubbles in fluidized beds, which is not the case in a bubble in a liquid. In effect, the bubble is a "void" in the bed—a "shortcut" for gas to take, as shown in Fig. 5.3.

The favorable properties listed above have given rise to many applications of fluidized beds in industry, some of which are listed in Table 5.1.

5.3.1 MINIMUM FLUIDIZATION VELOCITY

When a fluid passes upward through a packed bed, the manometric pressure gradient increases as U increases. When the pressure drop is just sufficient to support the immersed weight of the particles, then the particles are supported by the fluid and not by resting on neighboring particles. Therefore, at this point, the particles become free to move around in the fluid and are said to be "fluidized." At this point (see Fig. 5.4),

$$\Delta p/H = (1 - \varepsilon_{\mathrm{mf}})(\rho_{\mathrm{p}} - \rho)g \tag{5.2}$$

Table 5.1 Classification of Fluidized Bed Applications According to Predominating Mechanisms (Geldart, 1986)

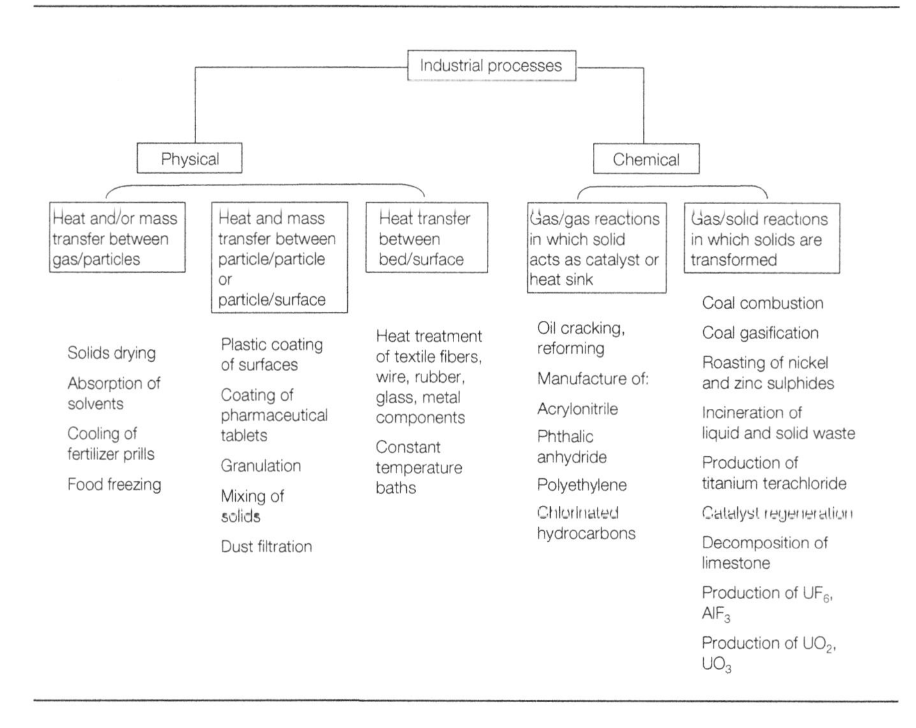

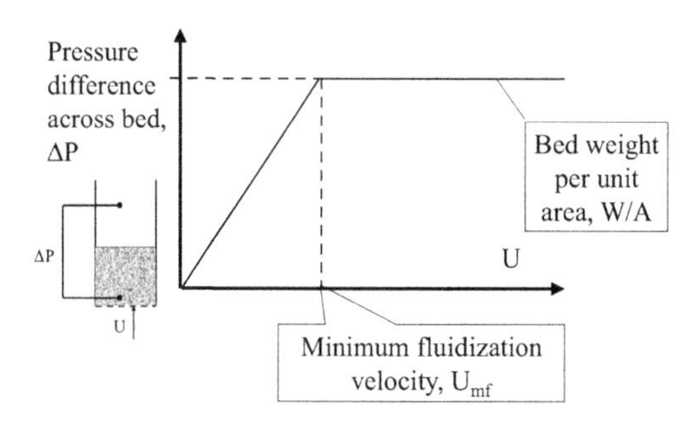

FIGURE 5.4

Determination of the minimum fluidization velocity (variants on this behaviour are considered in Box 5.1).

BOX 5.1 THE TRANSITION TO FLUIDIZATION

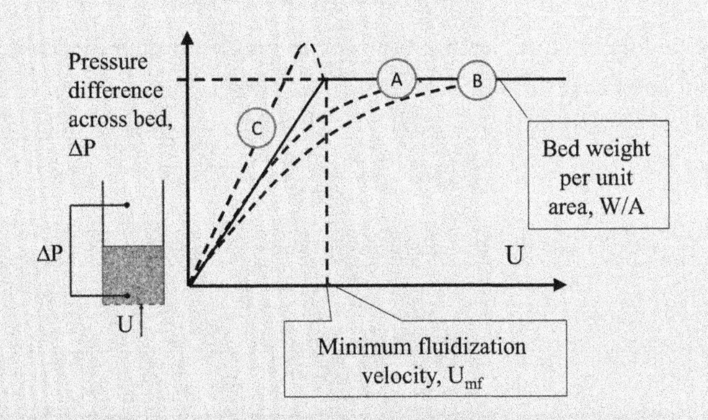

Figure 5.4 represents ideal behavior in an experiment in which the gas velocity is gradually increased. In practice, variants on this shape of pressure drop versus gas velocity curve can often be seen, some of which are shown above. Curves A and B represent the behavior to be expected for the same mean particle size as the width of the particle size distribution increases, B representing a wider distribution than A. Curve C shows a steeper packed bed gradient than for the ideal case and an "overshoot," so that the pressure difference across the bed initially exceeds the bed weight per unit area and then drops back. This is the behavior to be expected if the initial bed is tightly packed into the container. The bed void fraction is then reduced (consider the effect of this on Eq. (5.1)) and the forces within the bed are transmitted through friction to the wall. Eventually particles will slip at the wall and rearrangement will occur, so that the void fraction increases slightly and the pressure difference drops back to equal W/A.

where the subscript "mf" is used to denote minimum fluidization conditions. Using Eq. (5.1) to evaluate $\Delta P/H$ leads to an equation for the minimum fluidization velocity, U_{mf}, which rearranges to

$$\frac{\rho d^3(\rho_p - \rho)g}{\mu^2} = \frac{150(1-\varepsilon_{mf})}{\varepsilon_{mf}^3}\frac{\rho d}{\mu}U_{mf} + \frac{1.75}{\varepsilon_{mf}^3}\frac{\rho^2 d^2}{\mu^2}U_{mf}^2 \tag{5.3}$$

Each individual term in Eq. (5.3) is dimensionless. It is therefore convenient to rewrite it in terms of a dimensionless diameter and the particle Reynolds number at minimum fluidization:

$$d^* = d\left[\rho\frac{\rho_p - \rho}{\mu^2}g\right]^{1/3} \tag{5.4A}$$

$$\mathrm{Re}_{mf} = \rho U_{mf} d/\mu \tag{5.4B}$$

In these terms, and combining the numerical constants with the voidage terms as suggested by Wen and Yu (1966), Eq. (5.3) becomes

$$(d^*)^3 = 1650\text{Re}_{\text{mf}} + 24.5\text{Re}_{\text{mf}}^2 \tag{5.5}$$

which is very widely used for estimation of minimum fluidization velocities.

For low d^*, such that the viscous term in Eq. (5.5) dominates,

$$U_{\text{mf}} = \frac{d^2(\rho_{\text{p}} - \rho)g}{1650\mu} \tag{5.6}$$

For high d^*, where the inertial term dominates,

$$U_{\text{mf}} = \left[\frac{d(\rho_{\text{p}} - \rho)g}{24.5\rho}\right]^{1/2} \tag{5.7}$$

The different dependencies on particle size and fluid properties should be noted. Figure 5.5 shows some numerical values, calculated from Eq. (5.5), to illustrate these effects.

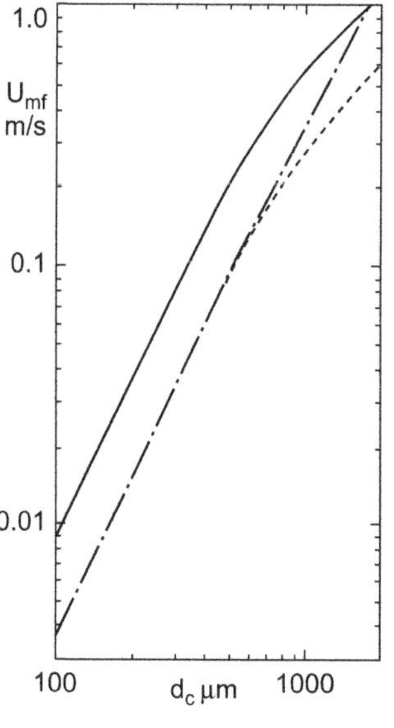

FIGURE 5.5

Superficial gas velocity of air at minimum fluidization, for spherical particles of density 2500 kg/m^3 (Seville et al., 1997).

In Fig. 5.5, the continuous line represents values of U_{mf} at normal atmospheric conditions (300 K and 1 bar). Note the linearity of this log–log plot for small particle sizes, where U_{mf} is proportional to the square of diameter (Eq. (5.6)). Increase in temperature to 1100 K, shown by the dashed lines, increases the gas viscosity, which affects U_{mf} directly (see Eq. (5.6) again). Change in pressure has little effect for small particles but is important for larger ones, because it affects the gas density.

As regards fluidization behavior, the most important particle properties are density, size, and size distribution. The density of interest is what is sometimes called the "piece density" or "envelope density" (see Box 5.2). The average diameter to use is the surface-volume mean (see Chapter 3). This is the particle size whose surface area per unit mass or per unit solid volume is the average value for the whole particulate. It is therefore the best single measure of particle size for processes controlled by the interfacial area between gas and solids; this includes mass transfer processes and, to a first approximation, fluid/particle drag at *low* particle Reynolds numbers.

The relevant properties of the gas in a fluidized bed are its density ρ and viscosity μ. For virtually all practical purposes, the density of a gas or gas mixture can be estimated from the ideal gas laws; it is proportional to absolute pressure and inversely

BOX 5.2 WHICH DENSITY?

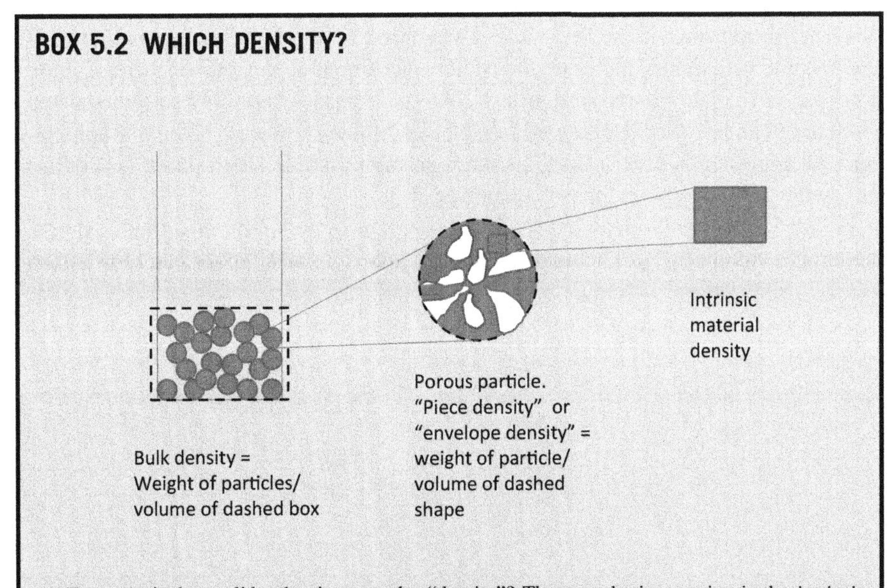

For a particulate solid, what is meant by "density"? The most basic meaning is the *intrinsic density* of the fully dense solid, which is a material property found in reference books. However, particles of that solid may have internal porosity, so the *particle density* may be much lower than the material density. The particles are then stored or transported in bulk, which introduces the idea of *bulk density,* which is lower still. For example, alumina has an intrinsic density of about 4000 kg/m^3. As a porous catalyst carrier, this may be reduced to 2000 kg/m^3 and the bulk density of the as-delivered material may be only 1000 kg/m^3. Particle density is difficult to measure. This is normally done by a displacement method. More details are given in Chapter 2.

proportional to absolute temperature. To a good first approximation, the viscosity of a gas or gas mixture is independent of pressure but increases with increasing temperature; the variation is as $T^{1/2}$ according to elementary kinetic theory and is usually somewhat stronger in practice. Temperature can have an effect on the particles too, if it is high enough to cause sintering (time-dependent bonding) or melting.

5.3.2 BUBBLES

Most gas-fluidized beds operate in the *bubbling* regime. To a first approximation, in a bubbling fluidized bed only sufficient gas to support the particles flows in the small spaces ("interstices") between the particles; this is termed the *interstitial flow*. Any excess passes through the bed as distinct bubbles. Taking the analogy with bubbles in liquids further, it is possible to distinguish between *the dense phase* (also known as the "particulate phase" or "emulsion phase"), consisting of the bed particles fluidized by the interstitial gas, and *the lean phase*, consisting of rising bubbles virtually free of bed particles.

A bubbling bed can be regarded as a bed in which the bubble phase is dispersed, and the particulate phase is continuous—as in a bubbling liquid. At higher gas velocity, the proportion of the bed volume occupied by the bubbles, ε_b, increases. It may become sufficiently high that the bed can no longer be described as "lean phase dispersed/particulate phase continuous." The two "phases" are now so interspersed that neither can be described as continuous, sometimes known as "turbulent fluidization." At higher velocities still, ε_b becomes so high that the "lean phase" is continuous, with thc "particulatc phasc" dispersed in it.

Figure 5.6 shows an idealized section through a single bubble in a fluidized bed. The bubble volume is V_b. The upper surface is approximately spherical, with radius

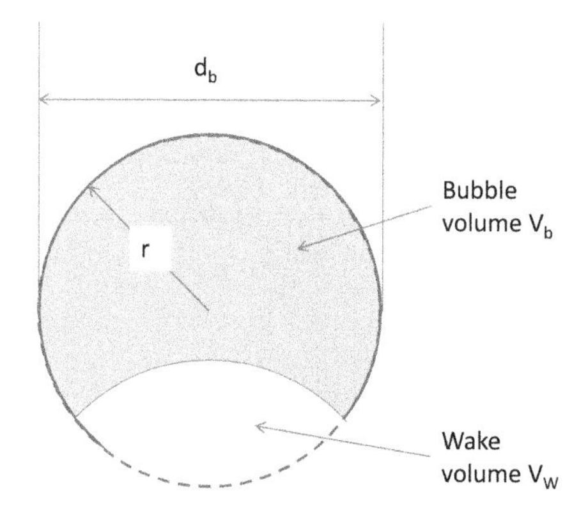

FIGURE 5.6

Spherical cap bubble.

of curvature r. The base is typically indented. The volume filling in the sphere is called the *wake* and, to a good approximation, this volume V_W consists of dense phase rising with the bubble. Bubble shapes vary according to the properties of the bed particles, particularly their size; in general, the smaller the particles, the larger the ratio of wake volume to bubble volume.

The wake volume has an important effect on particle motion. The rising sphere, corresponding to the bubble plus its wake, displaces the surrounding dense phase; the effect is roughly equivalent to dragging up a volume equal to $1/2(V_b + V_W)$, by the process known as drift transport. If $V_W = V_b/3$, then the drift volume is roughly equal to $2V_b/3$. Therefore, the total transport of dense phase, by drift and in the bubble wake, is

$$2V_b/3 + V_W \approx 2V_b/3 + V_b/3 = V_b \tag{5.8}$$

That is, to a first approximation, a bubble rising through a fluidized bed transports its own volume of dense phase. It is this rapid turnover of the bed particles which gives fluidized beds many of their important properties, such as good temperature uniformity.

The description above refers to isolated bubbles. Bubbles whose dimensions approach those of the bed behave rather differently and are known as *slugs*. Slugging is generally undesirable because slugs are not as effective at mixing the bed and they cause extreme pressure fluctuations, which may ultimately damage the equipment.

Theoretically, a large isolated bubble in a fluid of relatively low viscosity rises at velocity u_b (Davies and Taylor, 1950), where:

$$u_b = 2\sqrt{gr}/3 \tag{5.9}$$

Bubble velocities in fluidized beds vary erratically, but Eq. (5.9) seems to give a good estimate for mean rise velocity. In a freely bubbling bed, the rise velocity is greater than the value given by Eq. (5.9), and this is discussed further below.

Usually the radius of curvature, r, is not known. The characteristic dimension frequently used is the volume-equivalent sphere diameter, d_v, i.e., the diameter of a sphere whose volume is equal to the bubble volume V_b. The rise velocity of an isolated bubble is then given by

$$u_b = K\sqrt{gd_v} \tag{5.10}$$

where K depends on the relationship between r and d_v. Commonly, it is assumed that $K = 0.71$, as for bubbles in water, although values of 0.5–0.6 may be more realistic for smaller particles (below say 100 μm).

In a real bubbling fluidized bed, bubbles are seen to undergo both combination ("coalescence") and splitting. Bubbles in fluidized beds break up by the process shown schematically in Fig. 5.7. An indentation forms on the upper surface of the bubble and grows as it is swept around the periphery by the particle motion. If the indentation grows sufficiently to reach the base of the bubble before being swept away, the bubble divides. Splitting dominates in beds of smaller particles (below say 100 μm) and also tends to become more frequent at higher pressures.

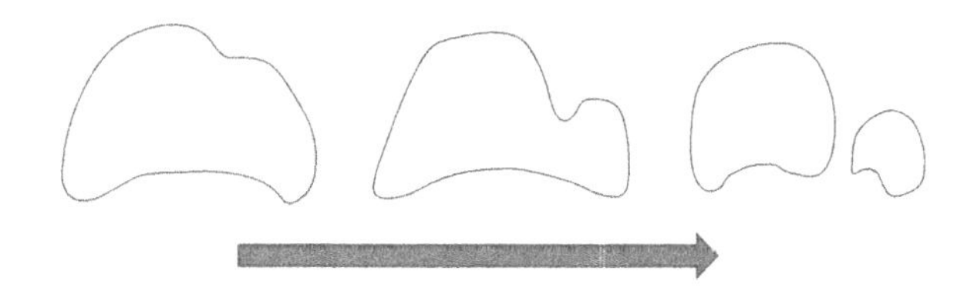

FIGURE 5.7

Bubble splitting.

After Clift (1986).

Splitting leads to the idea of there being a maximum stable bubble size, which effectively limits the size to which they can grow. This size seems to be of order 1 cm for particles of size ~60 μm, increasing to more than 1 m for particles of 250 μm, and over 10 m for particles of 1 mm in size. So in practice, splitting is only relevant for smaller particles.

When a bed of particles is fluidized at a gas velocity above the minimum bubbling point, bubbles form continuously and rise through the bed, which is said to be "freely bubbling." Bubbles coalesce as they rise, so that the average bubble size increases with distance above the distributor (Fig. 5.3) until the bubbles approach the maximum stable size. For smaller particles, thereafter, splitting and recoalescence cause the average bubble size to equilibrate at a value close to the maximum stable value.

Bubble coalescence can also have an influence on circulation of the dense phase. The effect is shown schematically in Fig. 5.8. Bubbles usually coalesce by overtaking a bubble in front (Fig. 5.8(b)(i)) and may move sideways into the track of another bubble (Fig. 5.8(b)(ii)). Thus coalescence can cause lateral motion of bubbles. Bubbles near a bed wall can only move inward, because bubbles surrounding them are only on the side away from the wall, while bubbles well away from the walls are equally likely to move in any horizontal direction. As a result of this preferential migration of bubbles away from the wall, an "active" zone of enhanced bubble flow forms a small distance from the wall. In this zone, coalescence is more frequent so that the bubbles become larger than at other positions on the same horizontal plane. Because the region between the "active" zone and the wall is depleted of bubbles, coalescence continues to cause preferential migration toward the bed axis. Eventually the "active" zone comes together to form a "bubble track" along which the lean phase rises as a stream of large bubbles. Large beds may divide into several "cells," with several preferential bubble tracks in the bed. Because of the transport of dense phase by the bubbles, the solids tend to move up in regions of high bubble activity and down elsewhere. In the upper levels, the motion is up near the bubble tracks and down near the walls. At lower levels, the particle motion is down near the axis and outward across the distributor; this motion can in turn enhance bubble activity near the walls close to the distributor.

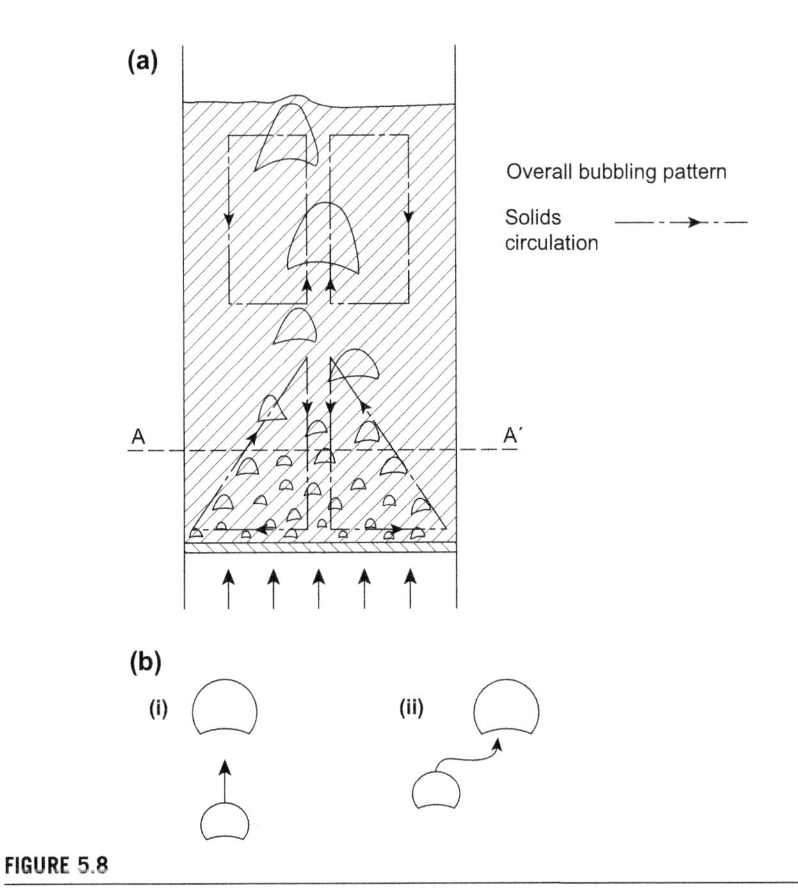

FIGURE 5.8

(a) Bubbles and solids flow pattern and (b) coalescence behavior: (i) bubble overtaking; (ii) bubble sideways motion (Clift, 1986).

In design calculations it is necessary to predict bubble size and rise velocity, for which there are a number of empirical and semiempirical correlations. The approach due to Darton et al. (1977) has been shown to be reasonably reliable and is relatively convenient to use. The mean bubble size formed at the distributor $d_{v,o}$ is first estimated by an expression whose form has a fundamental theoretical basis:

$$d_{v,o} = 1.63(A_o(U - U_{mf}))^{0.4} g^{-0.2} \tag{5.11}$$

where A_o is the distributor area associated with one gas inlet orifice or nozzle. The effect of coalescence on bubble growth above the distributor is then given by

$$d_v = 0.54(U - U_{mf})^{0.4}\left(z + 4\sqrt{A_o}\right)^{0.8} g^{-0.2} \tag{5.12}$$

where z is the distance above the distributor. It is to be remembered, however, that for smaller particles the maximum bubble size is often reached quite close to the distributor, and this approach does not account for that effect.

The rise velocity of bubbles in a freely bubbling bed can be predicted approximately by the approach of Davidson and Harrison (1963), which gives the mean bubble rise velocity, u_A, as

$$u_A = (U - U_{mf}) + u_b \tag{5.13}$$

where u_b is the rise velocity of a single isolated bubble (given by Eq. (5.10)). Instantaneous bubble velocities vary widely about this result and the mean velocity for a given bubble diameter varies with position in the bed, being highest in the "active" bubbling zones where coalescence is frequent.

5.3.3 TYPES OF FLUIDIZATION

Early in the development of fluidization it was realized that the kind of fluidization behavior which can be achieved depends on the particle properties. For example, particles larger than about 1 mm tend to fluidize in a jerky, explosive sort of way while smaller particles fluidize much more smoothly. However, very small particles (say below 10 μm in size) do not fluidize at all, or only under special circumstances, apparently because their cohesive forces are so large compared with their particle weight (see Chapter 8).

There have been several attempts to devise theoretical and empirical classifications of fluidization behavior. Of these, the most widely used is the empirical classification of Geldart (1973), who divided fluidization behavior according to mean particle size and density difference between the solids and the fluidizing gas (Fig. 5.9). Geldart recognized four behavioral groups, designated

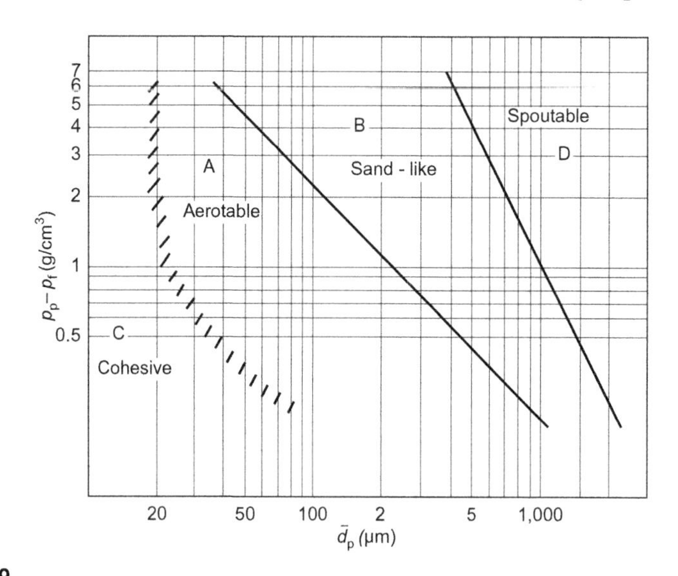

FIGURE 5.9

Simplified diagram for classifying powders according to their fluidization behavior in air at ambient conditions (Geldart, 1973).

A, B, C, and D. Typical fluidization behavior of groups A, B, and C is illustrated in Fig. 5.10.

Group B (for "bubbling") particles fluidize easily, with bubbles forming at or only slightly above the minimum fluidization velocity.

Group C (for "cohesive") particles are cohesive and tend to lift as a plug or channel badly; conventional fluidization is usually difficult or impossible to achieve.

Group A particles are intermediate in particle size and in behavior between groups B and C and are distinguished from group B by the fact that appreciable (apparently homogeneous) bed expansion occurs above the minimum fluidization velocity but before bubbling is observed. There is now much experimental evidence that group A particles are also intermediate in cohesiveness between groups B and C, their interparticle cohesive forces being of the same order as the single particle weight.

Group D particles are those which are "large" and/or abnormally dense. Such particles fluidize poorly, but can be made to "spout," rather than fluidize. A "spouted bed" is shown schematically in Fig. 5.11; its main features are a conical–cylindrical shape and a very pronounced solids circulation—up in the central lean "spout" and down in the dense annular ring around it. Gas is introduced through a small diameter entry at the base. This is not true fluidization because the gas does not support all the particles.

Typical properties associated with Geldart's groups are summarized in Table 5.2.

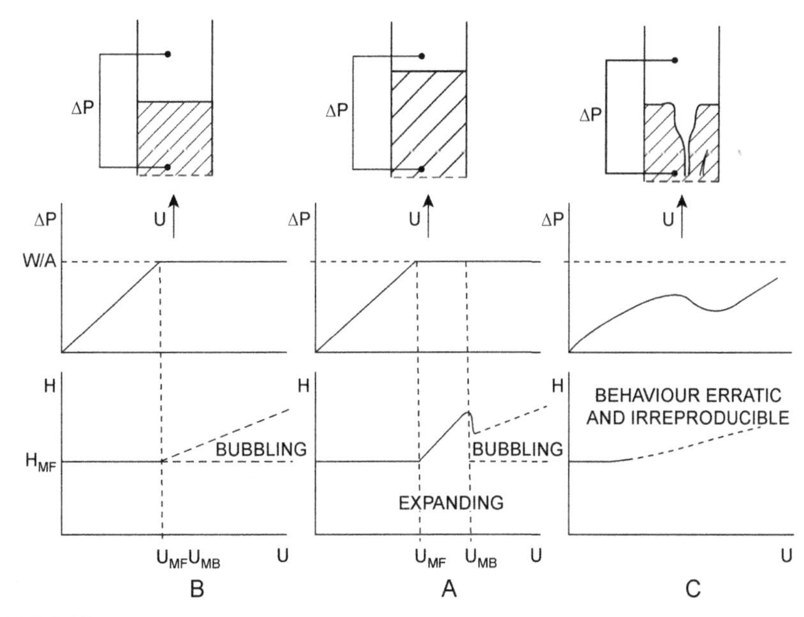

FIGURE 5.10

Typical fluidization behavior by Geldart groups B, A, and C (from left to right). Note that the scales are different for each group.

After Seville (1987).

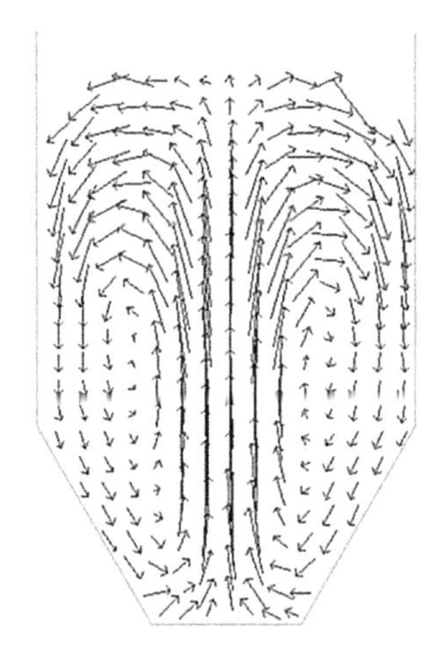

FIGURE 5.11

Solids circulation in a spouted bed (The length of the arrows is proportional to the average speed of the particles).

Courtesy of Christian Seiler.

Table 5.2 Characteristic Features of Geldart's (1973) Classification of Fluidization Behavior (after Geldart, 1986)

Group	C	A	B	D
Typical examples	Flour, cement	Cracking catalyst	Building sand, table salt	Crushed limestone, coffee beans
Property				
1. Bed expansion	Low when bed channels; can be high when fluidized	High	Moderate	Low
2. Deaeration rate	Can be very slow	Slow	Fast	Fast
3. Bubble properties	Channels	Splitting and coalescence predominate. Maximum size. Large wake.	No limit on size	No known upper size; small wake.
4. Solids mixing	Very low	High	Moderate	Low
5. Spouting	No, except in very shallow beds	Shallow beds only	Shallow beds only	Yes, even in deep beds

A different kind of classification of fluidization behavior due to Grace (1986) is shown in Fig. 5.12. This uses a dimensionless gas velocity, U^*, and particle diameter, d_p^* as in Section 5.3.1, to define the behavioral groups, where

$$d_p^* = d_p \left[\frac{\rho(\rho_p - \rho)g}{u^2} \right]^{1/3} \tag{5.14}$$

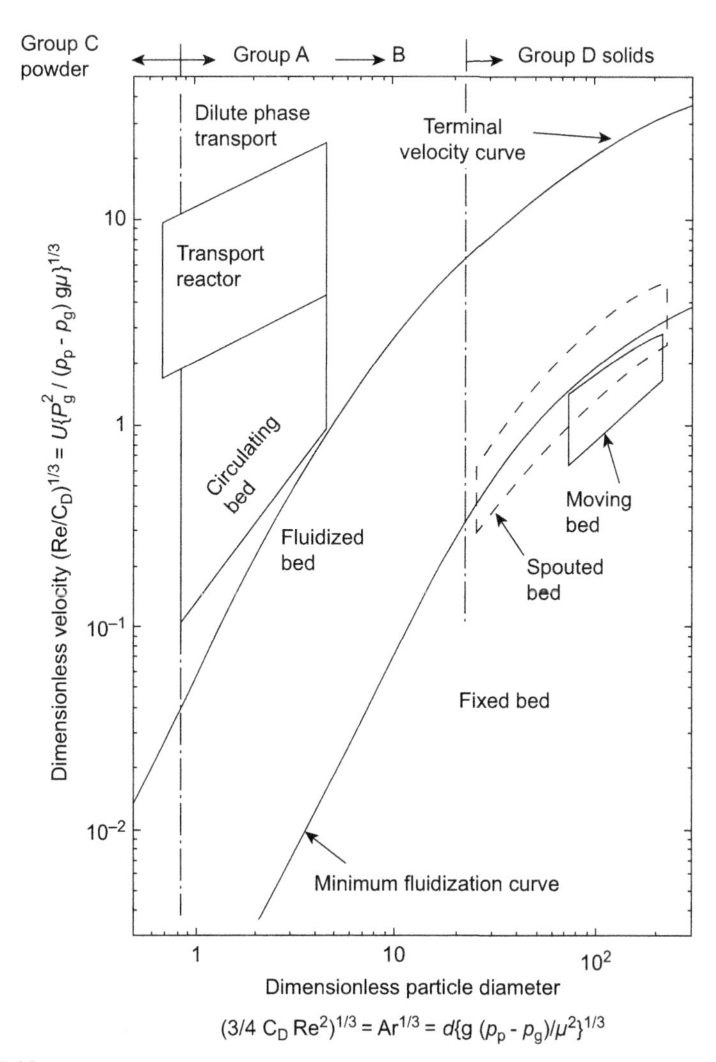

FIGURE 5.12

Regime/processing mode diagram, grouping systems according to type of powder and gas velocity used.

After Grace (1986).

and

$$U^* = U\left[\frac{\rho^2}{\mu(\rho_p - \rho)g}\right]^{1/3} \tag{5.15}$$

Figure 5.12 also shows the various processing options which might be considered for particles of various sizes and gases of different properties. Grace's classification successfully accounts for the effects of variation in gas properties due to operation at elevated temperature and pressure but there is, as yet, no satisfactory classification which also takes into account interparticle forces (see Chapter 8), which in many practical situations may be of considerable importance.

5.4 PNEUMATIC CONVEYING

As shown in Fig. 5.1, if the gas velocity is sufficient to carry the particles out of a fluidized bed, this is, in effect, a form of conveying, termed pneumatic conveying. In Fig. 5.1, vertical pneumatic conveying is shown. However, pneumatic conveying also operates horizontally, and at an angle to the vertical, although the last of these is not common.

In general, solid particles can be moved from place to place in bags, using forklift trucks and using mechanical devices such as screw conveyors and bucket elevators. Pneumatic conveying has advantages and disadvantages by comparison with these alternative methods. It requires low-operating labor input (runs unattended) and very little space. It also maintains the solids in a totally enclosed pipe so that contamination (both of the product and from the product) is prevented. Dispersed solids are an explosion risk (see, for example, Barton, 2002), but an inert gas can be used to convey the powder, which mitigates this risk. A disadvantage of pneumatic conveying is possible damage to the solids being conveyed, and/or erosion of the conveying pipe material, possibly causing contamination to the product.

Pneumatic conveying systems fall into different categories, as shown in Table 5.3.

First, it is possible to convey solids in a fully dispersed state or as a more-or-less coherent dense plug. These approaches are termed "lean" and "dense" phase, respectively, where the dividing concentration is about 15 kg solid/kg gas. A further distinction lies in whether the conveying pipe is vertical or horizontal. Many systems

Table 5.3 Categorizing Pneumatic Conveying Systems

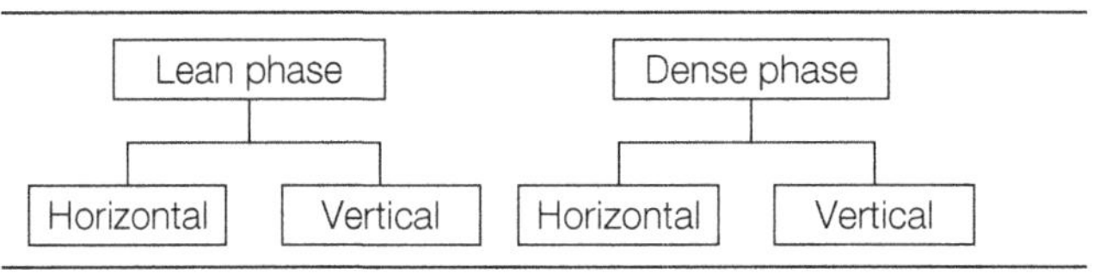

contain both, of course, but the particle behavior is very different in the two directions, as discussed later. Finally, it is possible to blow powder along a pipe or up a column by exerting an excess pressure, or to suck a powder along a pipe under negative pressure (with respect to the atmosphere), as in a domestic vacuum cleaner. It is also possible to have systems which employ a combination of "blow" and "suck." Examples of such systems are given below.

The design process for pneumatic conveying can be summarized as:

- Choose a conveying system type
- Choose a gas velocity which is high enough to transport the solids but low enough not to be too costly (or too damaging to the solids)

One guide to selection of pneumatic conveying systems (from the company Neu Engineering) recommends the following:

1. System duty
   a. Choose low pressure systems for conveying duties up to 10 t/h and up to 150 m in distance.
   b. Choose high pressure systems for conveying duties up to 50 t/h and up to 500 m in distance.
2. Process layout
   a. If there are multiple sources of solids to be considered, choose a suction system
   b. If there are multiple destinations for solids to be considered, choose a blowing system
   c. If there are multiple sources and destinations to be considered, choose a combination suck–blow system

In addition to these selection criteria there are obviously also product considerations, some of which are summarized in Table 5.4.

Figures 5.13–5.15 show three examples of pneumatic conveying systems.

Figure 5.13 shows a typical pressurized system in which blowers are used to pressurize a transport tanker and convey particles—in this case flour—from the tanker to two storage silos, from which they can be further transported to a sieving unit.

Table 5.4 Product Considerations in Pneumatic Conveying

Product	System choice
Toxic	Suction usually
Explosive	Inert gas/positive pressure
Large particle size	Low pressure
Fragile	Low velocity/dense phase
Abrasive	Low velocity/dense phase
Cohesive/damp	Slug phase
Temperature-sensitive	Suction
Hygroscopic	Positive pressure

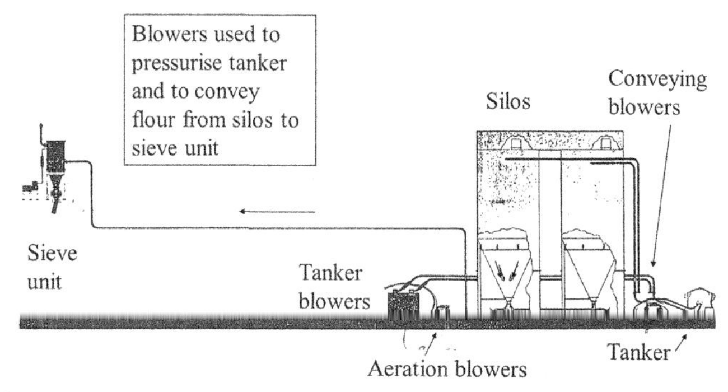

FIGURE 5.13

Blow-type flour delivery system, with aerated silos.

Courtesy of Neu Engineering.

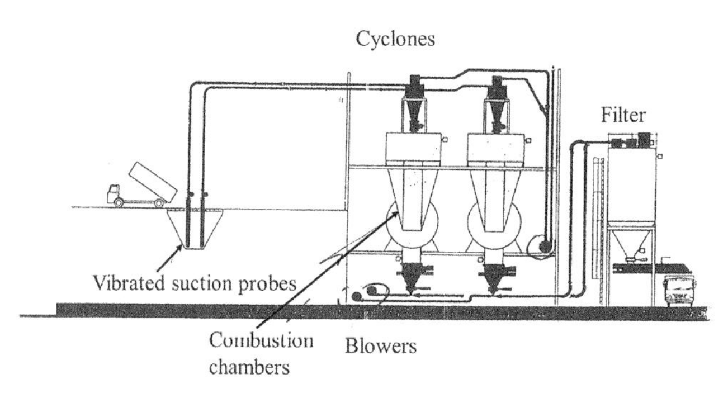

FIGURE 5.14

Solid fuel/ash handling system.

Courtesy of Neu Engineering.

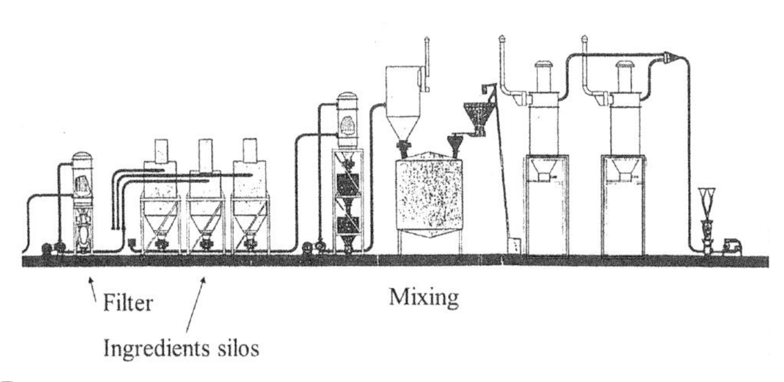

FIGURE 5.15

Multiple source/multiple destination food mix system.

Courtesy of Neu Engineering.

Figure 5.14 shows a typical suction system in which a pelletized solid fuel is sucked up from a storage pit into two cyclones situated above combustion chambers, where the solids are separated into the gas and fed into the chambers. In a separate system, ash falls under gravity from underneath the combustion chambers, where it is conveyed under pressure to a filter situated above a storage silo, from which the ash can be taken away by truck.

Figure 5.15 shows a very flexible system in which solids can be extracted under suction from a combination of ingredient silos, shown on the left, into a filter situated above a metering silo from which the particles are conveyed under pressure to a mixing vessel (also shown being fed with a small mass ingredient using a separate screw conveyor).

It will be apparent from the designs shown here that there are a number of auxiliary devices needed to make up a pneumatic conveying system, including devices to introduce the solids to the gas, such as eductors, rotary valves, and blow tanks (see Fig. 5.16), and devices to separate the solids from the gas, such as cyclones and filters (considered in Section 5.5).

The pressure drop versus gas velocity relationships for pneumatic conveying are complex. In general, the pressure drop in a system has six constituent terms, which are due to:

- Acceleration of the gas (proportional to ρU^2)—usually small
- Acceleration of the particles (proportional to $\rho_s U_s^2$)—could be large
- Gas-to-pipe friction—usually small
- Solids-to-pipe friction—usually large
- Static head of gas (ρHg)—usually small
- Static head of solids ($\rho_s U(1-\varepsilon)Hg$)—large if vertical

where ρ and ρ_s are the gas and particle densities, U and U_s are the gas and particle velocities, ε is the void fraction, H is the vertical conveying height, and g is the acceleration due to gravity.

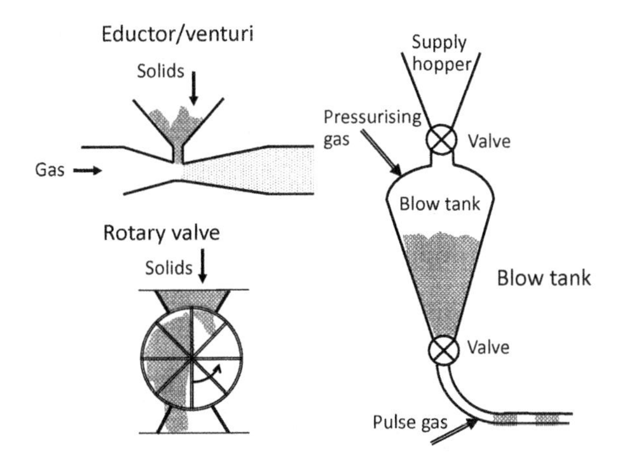

FIGURE 5.16

Devices for introducing particulate solids to a gas stream.

From this comparison it is possible to say that the largest contributions to pressure drop are likely to be due to the solids—their initial acceleration as they are introduced to the flowing gas by a rotary valve or alternative device, their pipe friction due to repeated collisions with the wall and their consequent loss of momentum, and their static head contribution in vertical conveying. Figure 5.17 shows the first two of these effects schematically.

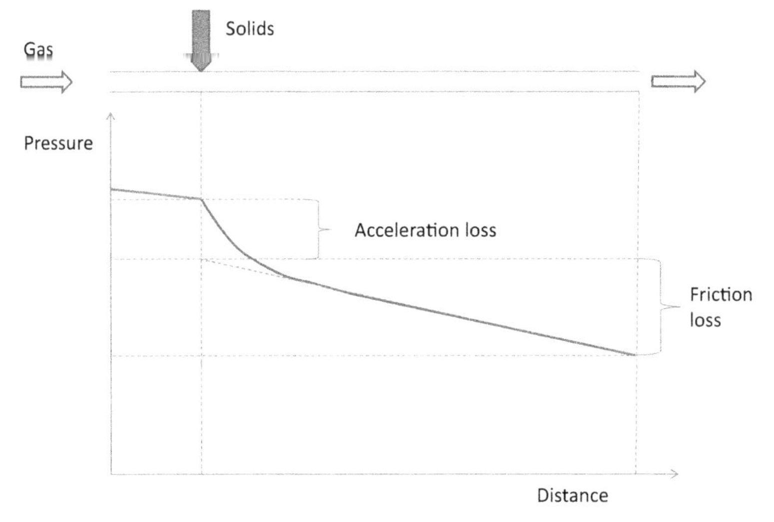

FIGURE 5.17

Pressure drop in horizontal conveying—acceleration loss and friction loss.

After Seville et al. (1997).

Figure 5.18 shows the pressure drop versus gas velocity behavior to be expected in vertical conveying. The curves represent constant values of the solids mass flow rate, with the lowest one representing gas flow only. Consider what happens as the gas velocity is reduced with the solids flow rate kept constant: the *in situ* particle concentration goes up so that the "hydrostatic" contribution to the pressure gradient goes up. This contribution eventually outweighs the decrease in the frictional pressure gradient, so that the total gradient passes through a minimum and then increases. Eventually it rises more sharply, to a condition in which the riser pipe contains a *slugging* (or, more rarely, *bubbling*) fluidized bed. This condition is called *choking*. A "choked" vertical conveying line is typically characterized by pressure fluctuations, associated with the rise and eruption of slugs. As for horizontal conveying, the transition velocity depends on the solids flow rate.

Figure 5.19 shows the different modes which can be observed in horizontal conveying, while Fig. 5.20 shows the pressure drop versus gas velocity behavior. In horizontal flow, the transition from lean- to dense-phase flow is usually termed the *saltation velocity*. At constant solids flow rate, G, and starting with a high superficial high gas velocity, U, the mixture is in lean-phase flow. If the gas flow

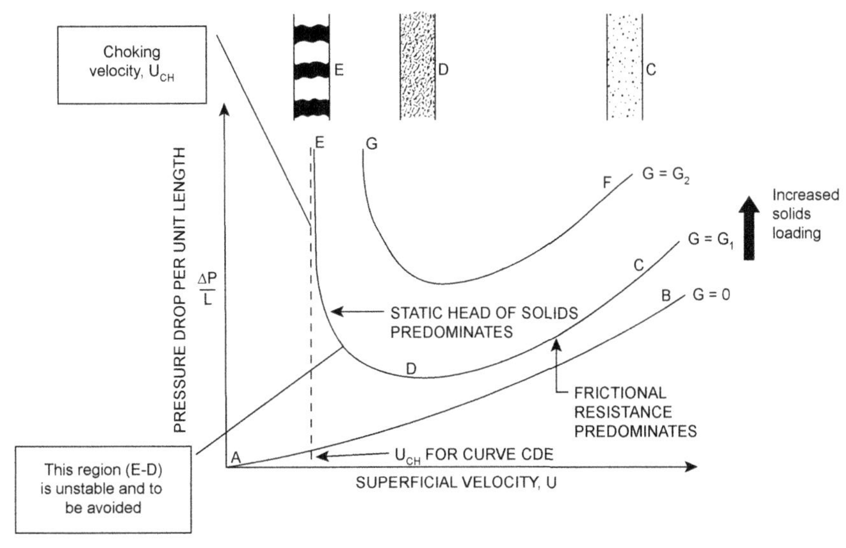

FIGURE 5.18

Phase diagram for vertical conveying.

After Knowlton (1986).

is reduced, the pressure gradient at first decreases as the frictional contribution is reduced. When the saltation velocity (U_{salt}) is reached, the pressure gradient increases sharply, which indicates a transition to dense-phase flow. Further reduction in gas flow causes the pressure gradient to increase further. The reverse transition occurs on increasing flow, without appreciable hysteresis if the flow is allowed to reach steady state. The saltation velocity increases when the solids

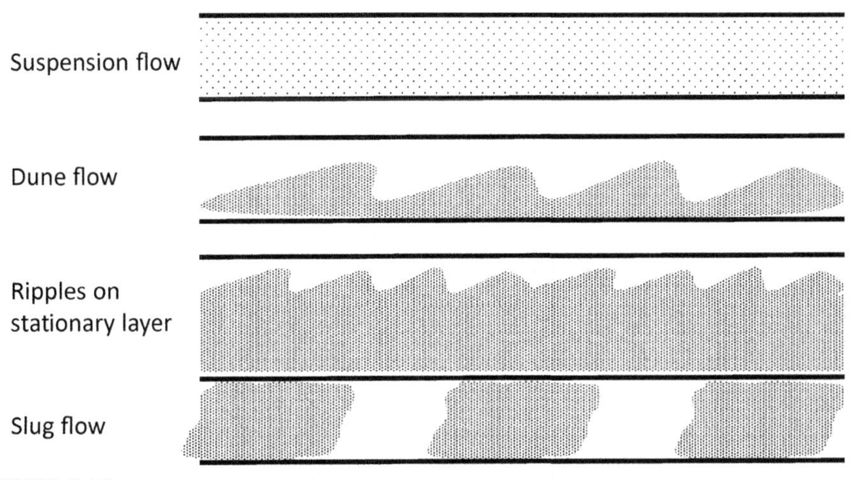

FIGURE 5.19

Horizontal pneumatic conveying regimes.

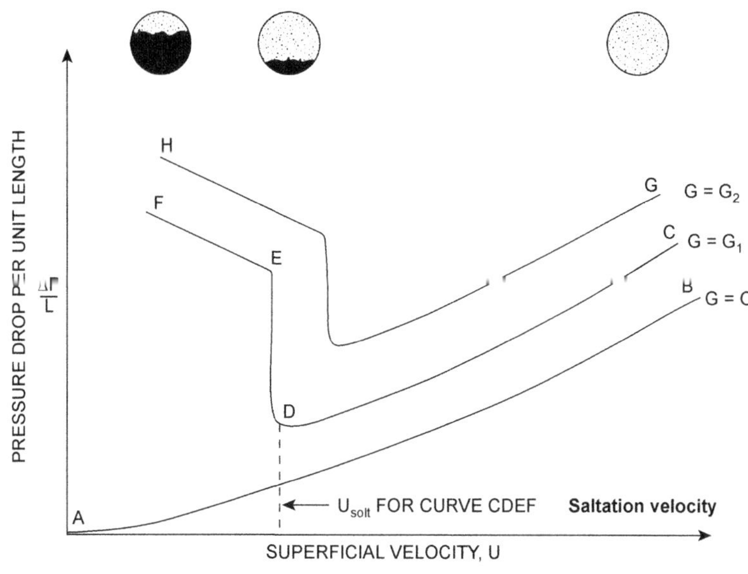

FIGURE 5.20

Phase diagram for horizontal pneumatic conveying.

After Knowlton (1986).

flow rate is increased. Therefore, the transition from lean to dense flow may be triggered by an increase in solids flow rate, if the line is operating close to the saltation point.

Pneumatic systems often contain bends, which can have a substantial effect on the pressure drop because the solids can be knocked out of the flow and have to be resuspended. Figure 5.21 shows a schematic illustration of this.

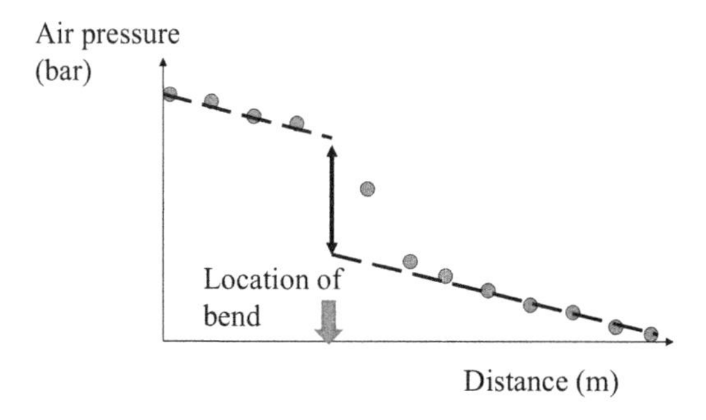

FIGURE 5.21

Schematic of pressure drop at a bend in pneumatic conveying.

5.5 GAS–SOLID SEPARATION*

The separation of particles from gases is an important technical problem in engineering and has applications in many industries, both for environmental protection and for prevention of fouling and damage to equipment. The devices which have been proposed to achieve this separation are astonishingly varied in design, in the physical principles upon which they rely, and in their effectiveness. For example, such devices include

- Cyclones, in which the particles are separated by aerodynamic effects
- Filters, in which particles are mechanically separated from the gas, which is forced to pass through some sort of porous material
- Wet scrubbers, in which the particles are separated by being captured in droplets or films of liquid
- Electrostatic precipitators, in which the particles are charged and then caused to migrate in an electric field to collection plates

This chapter concerns only the first two of these devices—the cyclone and the filter—which are by far the most commonly used.

5.5.1 CYCLONES

The cyclone is an example of a generic device termed an *inertial separator.* Inertial separators concentrate or collect particles by changing the direction of motion of the flowing gas, in such a way that the particle trajectories cross over the gas steamlines so that the particles are either concentrated into a small part of the gas flow or are separated by making contact with a surface. In a cyclone the gas undergoes some kind of vortex motion so that the gas acceleration is centripetal, and the particles therefore move centrifugally towards the outside of the cyclone.

Figure 5.22 shows the most important features of a typical *reverse-flow cyclone*, where the vortex motion is induced by introducing the gas tangentially to a cylindrical section called the *cyclone barrel*. The gas exits through an axial pipe sometimes called the *vortex finder.* The end of the vortex finder extends beyond the gas outlet, so that the gas executes a helical *outer vortex* in the barrel and tapered section or *cone*, before moving into the much narrow *inner vortex* and leaving through the vortex finder. The particles are flung out to the wall while, in a properly designed cyclone, the helical motion in the outer vortex pushes them down toward the apex of the cone. Most commonly, cyclones are mounted vertically for easier discharge of particles from the cone, but they can be (and sometimes are) used in different orientations. A *disengagement hopper* is sometimes provided to control particle discharge, as shown in Fig. 5.22. However, in some applications where the particles are retained within the process, a *dip-leg* is used instead, in the form of a pipe down which the disengaged particles move in

*Unmodified from Seville et al. 1997 (p. 264)

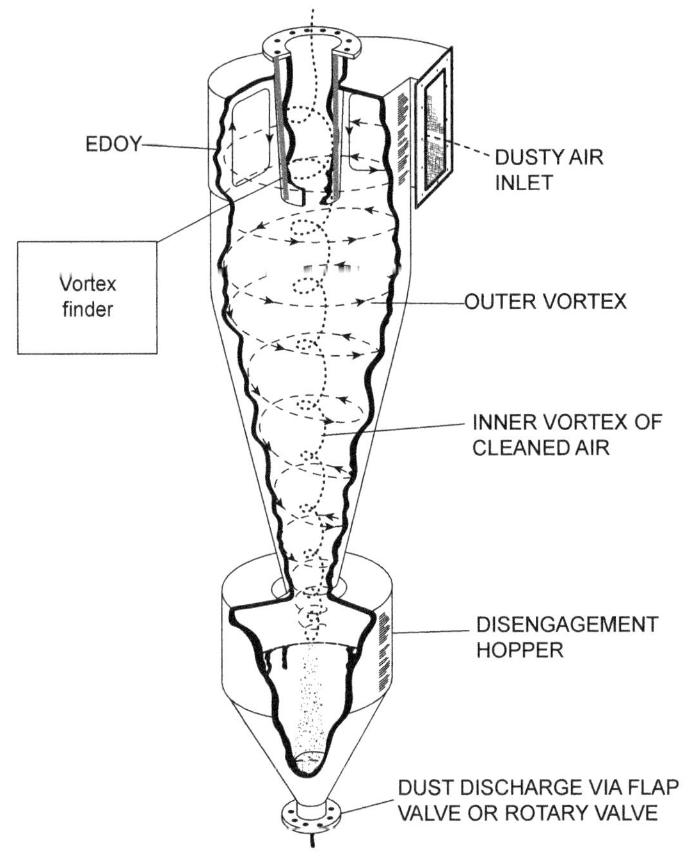

FIGURE 5.22

A typical reverse-flow cyclone, showing gas motion.

After Strauss (1975).

dense-phase flow. This is common practice in collecting particles from the gases leaving fluidized bed reactors or combustors, for example, and the dip-leg then usually extends down below the surface of the bed.

The gas entry to the cyclone is critical. It is preferable to ensure a sufficient length of straight pipe—at least several pipe diameters—before entering the cyclone. The different gas entry configurations are exemplified by the archetypal designs of Stairmand[3] (e.g., 1951, 1955) shown in Fig. 5.23, which are almost "industry standards." A *simple tangential entry*, shown in Fig. 5.23(a), is cheaper to construct but may give higher pressure drop across the cyclone (see below);

[3]The Stairmand cyclone designs originate in the UK. In practice, the commonly used North American designs by Lapple and others are similar and are in turn similar to designs originating in the Soviet bloc in the 1960s.

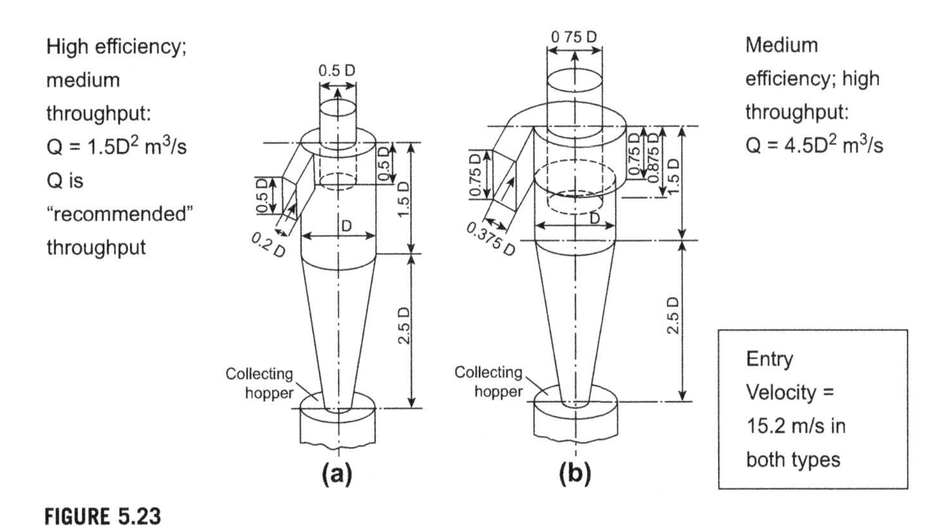

FIGURE 5.23

Standard cyclone designs.

Strauss (1975); after Stairmand (1951).

Stairmand used the tangential entry for his "high efficiency, medium throughput" design. For lower pressure drop (but increased construction cost), a *volute* (or "scroll" or "wrap-around") entry can be used, as in Stairmand's "high throughput, medium efficiency" design shown in Fig. 5.23(b).

Other types of inertial separator have been suggested and some, such as axial-entry types with swirl vanes for air-intakes, have become popular in some specialized applications, but types based on the two basic designs shown in Fig. 5.23 remain commonly used.

It is possible to analyze particle motion in cyclones using various mechanistic and computational approaches, but for design work it is much easier to resort to the common practice of dimensional analysis. For most purposes, this is quite adequate, because reputable cyclone suppliers have good data on performance which can readily be scaled or extrapolated using appropriate dimensionless groups.

5.5.1.1 Dimensional Analysis of Cyclone Performance

When particles are separated from a gas this is termed *collection*. The efficiency of particle collection, η, by an inertial separator, operated at given gas properties and throughput, depends on particle diameter, d, and density, ρ_p. The relationship between η and d for given ρ_p is called the *grade efficiency curve*. Grade efficiency curves for the two Stairmand designs of Fig. 5.23 are shown in Fig. 5.24. These curves refer to low inlet particle loading in the inlet gas (mass of solids per mass of gas); the effect of increasing particle loading is to increase efficiency, so designs using this approach should be conservative. A simple scaling approach can now be used to apply measured grade efficiency curves to different conditions and different sizes of cyclones.

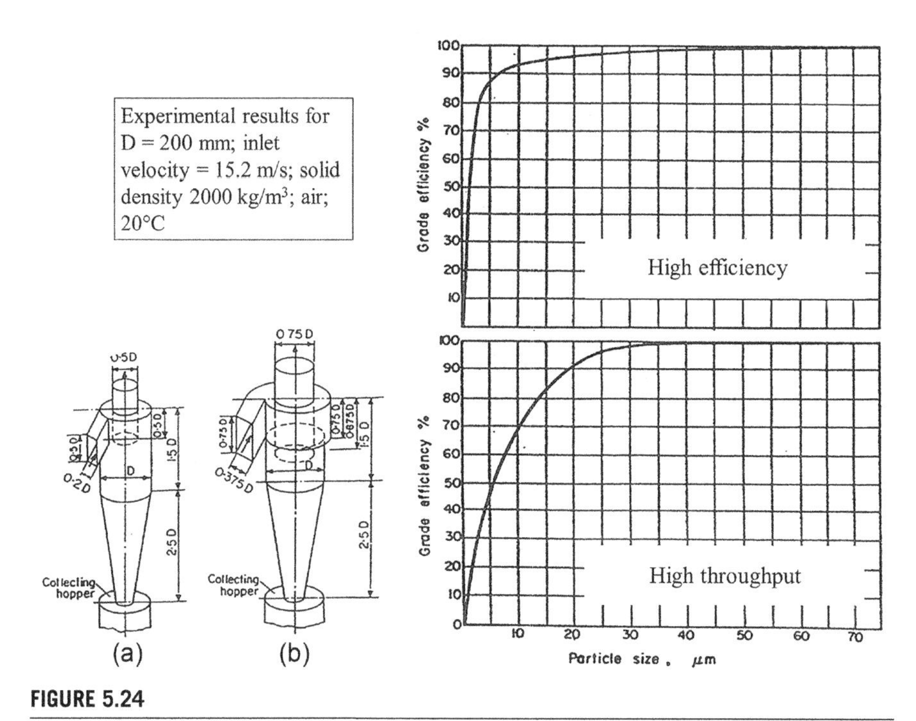

FIGURE 5.24

Grade efficiencies for standard cyclones.

Strauss (1975); after Stairmand (1951).

The collection efficiency for a particle of diameter d depends on the following variables:

$$\eta = f[d, \rho_p, \rho, \mu, D, Q] \tag{5.16}$$

where D is the characteristic cyclone dimension, typically the barrel diameter, and Q is the gas volumetric flow rate. Proceeding as above, the problem can be generalized by writing it in terms of six minus three, i.e., three dimensionless groups. (Collection efficiency is already dimensionless.) Hence we can write:

$$\eta = f[\mathrm{St}, \mathrm{Re}, d/D] \tag{5.17}$$

where St is the Stokes number derived from the ratio of the stopping distance (see Chapter 4) to cyclone diameter:

$$\mathrm{St} = \frac{\rho_p d^2 U_e}{18\mu D} = \frac{2}{9\pi}\frac{\rho_p d^2 Q}{\mu D^3} \tag{5.18}$$

However, Eq. (5.17) can be simplified. In practice, cyclones operate at very high Reynolds number so that, as for many devices which operate in the fully turbulent flow range, the effect of changes in Reynolds number can be neglected. Similarly the size ratio, d/D, is usually so small that its effect can be ignored. This leaves:

$$\eta = f[\mathrm{St}] \tag{5.19}$$

Equation (5.19) shows that the grade efficiency can be put into a general nondimensional form by expressing η as a function of St rather than d. In particular, the particle size which can be collected by a cyclone is commonly expressed in terms of the *cut size*, d_{50}, i.e., the particle size which is collected (at low particle loading) with 50% efficiency, $\eta = 0.5$. From Eqs (5.18) and (5.19), for a given geometric cyclone design,

$$\frac{\rho_p d_{50}^2 Q}{\mu D^3} = \text{constant} \tag{5.20}$$

or

$$d_{50} \propto \sqrt{(\mu D^3 / \rho_p Q)} \tag{5.21}$$

Equation (5.21) shows that the cut size decreases (i.e., the cyclone efficiency improves) if the throughput increases, and also that the performance of small cyclones is better than that of large cyclones, all other things being equal (Box 5.3).

A similar dimensional approach can be taken to prediction of the *pressure drop* across the cyclone, ΔP, which depends on the cyclone dimensions, the gas volumetric flow rate, Q, and the gas density and viscosity, ρ and μ:

$$\Delta P = f[D, Q, \rho, \mu] \tag{5.22}$$

BOX 5.3 CYCLONE SCALING

Equation (5.20) is extremely useful for design purposes, since it means that the performance of any size of cyclone under any conditions of gas flow can be predicted from that of a geometrically similar cyclone for which experimental data are available. So, for the two cyclones shown below,

$$\mathrm{St}_{50} = \left[d_{50}^2 \rho_p U / 18\mu D\right]_A = \left[d_{50}^2 \rho_p U / 18\mu D\right]_B$$

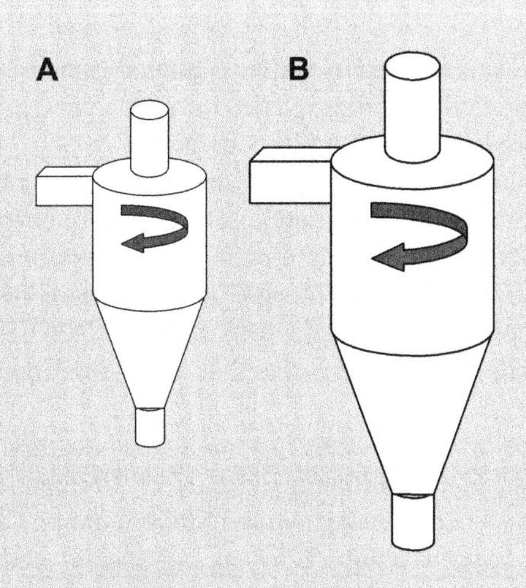

Equation (5.22) applies for the gas flow alone; particle loading effects are discussed below. Given the five parameters in Eq. (22) and that the system has the usual three dimensions (mass, length, and time), the problem can be expressed in terms of two (i.e., five minus three) dimensionless groups. These are, most conveniently, the *pressure coefficient*:

$$\Pi = \Delta P / \left(\rho U_e^2 / 2 \right) = \frac{\pi^2 D^4 \Delta P}{8 Q^2 \rho} \tag{5.23}$$

and a *cyclone Reynolds number*:

$$\mathrm{Re_c} = \frac{\rho U_e D}{\mu} = \frac{4 \rho Q}{\pi \mu D} \tag{5.24}$$

so that, for a given cyclone geometry,

$$\Pi = f[\mathrm{Re_c}] \tag{5.25}$$

In these equations, U_e represents a superficial velocity through the cyclone:

$$U_e = Q / \left(\pi D^2 / 4 \right) \tag{5.26}$$

and is used here as a measure of gas velocity. The inlet velocity is typically of order $10U_e$, the precise value depending on the geometric design of the cyclone.

The effective value of Π depends on the cyclone design; for geometrically similar cyclones over the normal velocity range, Π is approximately constant.

In practice, cyclones are often operated in combination, both in parallel (with the flow split between several cyclones) and in series (so that the larger particles are collected by the first cyclone, with progressively smaller particles collected by subsequent cyclones designed to have a smaller cut size). A discussion on multi-cyclone systems is presented in Box 5.4.

5.5.2 FILTERS

Filtration is the commonest method of particle removal from a fluid and refers to the process by which the particle-laden gas passes through a permeable membrane or an array of fibers, commonly termed the filter "medium."

Commonly-used filter media include woven and nonwoven fabrics, paper, ceramics, metals in various forms, and granular materials. An example of a nonwoven filter material is shown in Fig. 5.25. In industrial use, filters often consist of arrays of vertically hung cylindrical elements ("bags" in the case of flexible fabric filters; "candles" in the case of rigid media) as shown in Fig. 5.26. The flow is from the outside inward and exits via a plenum chamber at the top. Many other geometries are known.

A distinction can be drawn between two main types of filtration behavior: *depth* filtration and *surface* or *barrier* filtration (Fig. 5.27). In depth filtration, collection of particles from the gas occurs throughout the filter medium. The two main mechanisms by which this happens (Fig. 5.28) are *diffusional* collection and *inertial*

BOX 5.4 MULTI-CYCLONE SYSTEMS

For given gas and particle properties (i.e., ρ, μ, and ρ_p fixed) from Eq. (5.23)

$$\Delta P \propto Q^2/D^4 \tag{5.27}$$

while, from Eq. (5.21),

$$d_{50} \propto (D^3/Q)^{1/2} \propto D^{1/2} \Delta P^{-1/4} \tag{5.28}$$

so that lower d_{50} (i.e., improved collection performance) can also be achieved at the expense of higher pressure drop.

The analysis above also explains the use of a number of identical cyclones in parallel. For N cyclones handling a total gas flow rate Q, the flow per cyclone should be

$$q = Q/N \tag{5.29}$$

so that, from Eq. (5.21)

$$\Delta P \propto q^2/D^4 \propto Q^2/N^2 D^4 \tag{5.30}$$

To maintain collection performance, from Eq. (5.27) D^3/q must be kept constant; i.e., ND^3/Q must be constant, implying that

$$D^3 \propto N^{-1} \tag{5.31}$$

for given total flow rate Q. From Eqs (5.30) and (5.31)

$$\Delta P \propto Q^2 N^{-2/3} \tag{5.32}$$

That is, for given collection performance, the pressure drop can be reduced by using a number of cyclones in parallel. In practice, the number of parallel cyclones is limited by the difficulty of ensuring uniform gas distribution between them, although this general result explains the interest in devices using batteries of small cyclones in parallel, usually of an axial flow design.

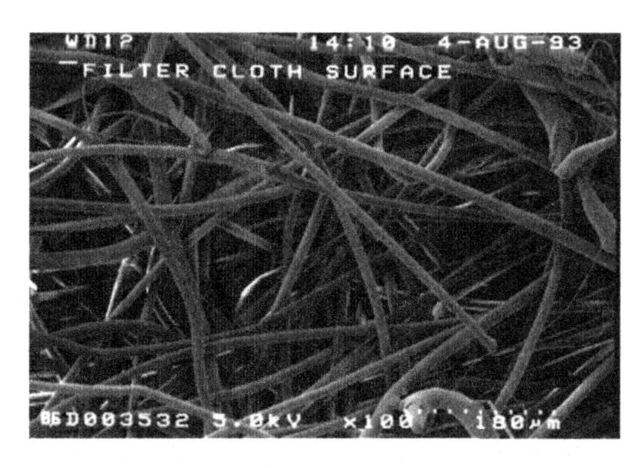

FIGURE 5.25

Scanning electron microscope picture of a nonwoven polyester filter cloth.

Courtesy of Freudenberg.

collection. In diffusional collection, particles deviate from the gas streamlines in a "random walk," due to molecular impacts; in inertial collection, particles deviate from the streamlines because of their inertia. Diffusion decreases as the particle size increases, while inertia increases with size. The combination of the two effects

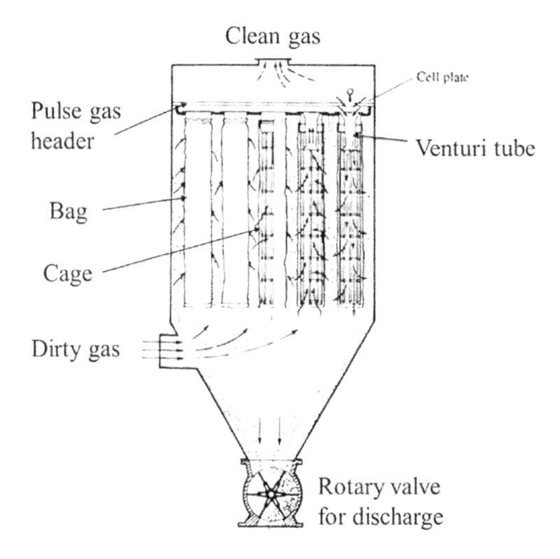

FIGURE 5.26

A typical bag filter arrangement (Barton, 2002).

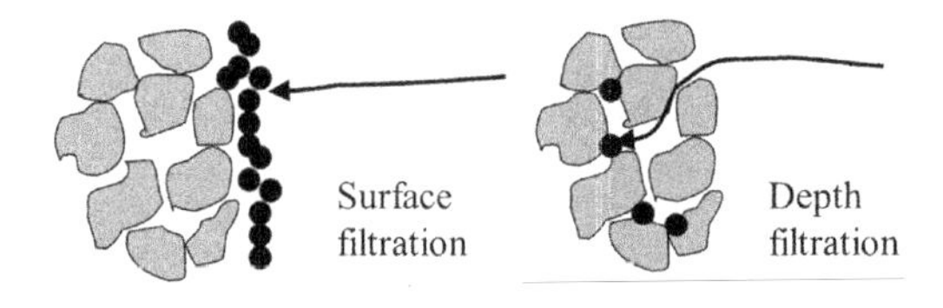

FIGURE 5.27

Surface and depth filtration—schematic.

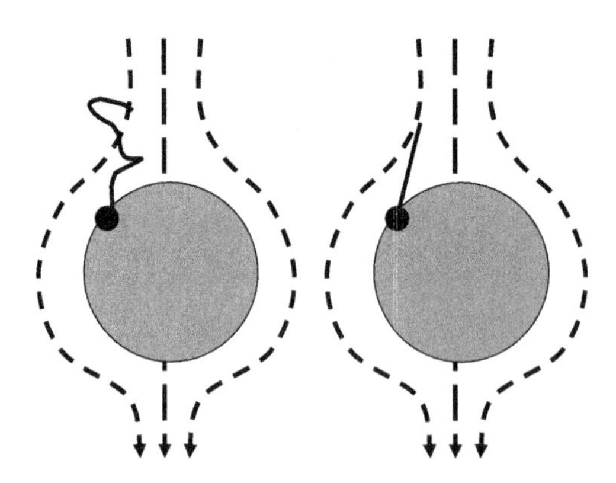

FIGURE 5.28

Diffusional collection (left) and inertial collection (right). The larger sphere represents an element of the filter medium.

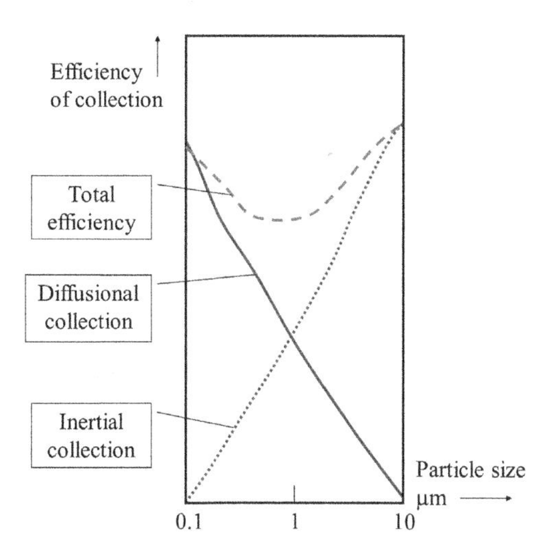

FIGURE 5.29

Typical grade efficiency for depth filtration in a fibrous medium (both scales are logarithmic).

is shown in Fig. 5.29. Large and small particles are collected with high efficiency, but there is a minimum efficiency at the "most penetrating size," typically about 1 μm.

In surface filtration, the medium acts as a barrier to the solids so that a dust "cake" is built up on the upstream surface (Fig. 5.30), with no penetration into the medium itself. In practice, filtration behavior depends on the properties of both the dust and the medium, not only the relative size of the pores in the medium

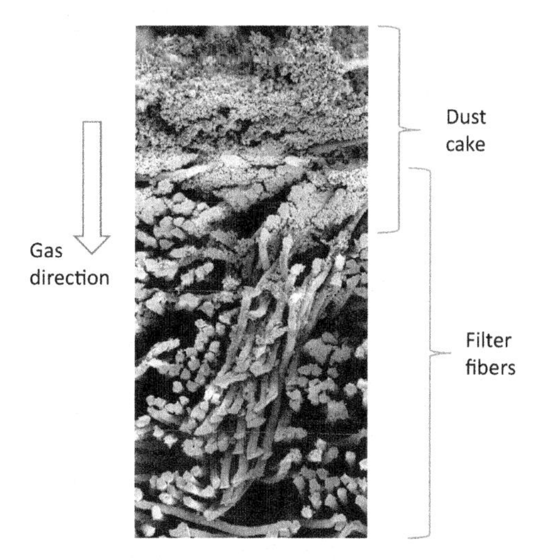

FIGURE 5.30

Dust cake build up on a fibrous filter. (A section through a polyester needle felt filter of fiber diameter about 13 μm, with a limestone dust; courtesy of Eberhard Schmidt, Institute of Particle Technology, University of Wuppertal.)

but also the surface properties (such as adhesion) of both. In general, gas cleaning using membranes and the finer grades of fibrous and granular filtration media approximates to surface filtration. Perfect surface filtration is rare, and with a new "surface filter" it is usual for there to be a short period of penetration into the surface layers of the medium before cake formation begins. During this short period the filtration efficiency may be lower than in steady operation.

In depth filtration it is not usually possible to remove particles effectively from the filter after they have been captured, certainly not "on line" (i.e., without shutting down the filter and removing the filtration medium). Barrier filters, however, are usually cleaned *in situ*, sometimes by shaking or other mechanical action but more commonly by gas flow in the opposite direction to that of the filtration flow.

In general, barrier filters are operated cyclically (Fig. 5.31). During filtration, dust builds up on the filter. After a prescribed time, or when the resistance to flow reaches a prescribed level, the medium is cleaned. The usual cleaning action is a reverse pulse of gas, applied to the clean side of the filter while it is on line. This detaches the cake of deposited particles, which then falls into a collecting hopper at the base of the unit, and the cycle is restarted. During the first few cycles, the "residual pressure drop" (after cleaning) increases, but after some time in a well-designed filter it usually levels off, as shown.

The most important design relationship for filters is that between pressure drop and gas velocity. In a filter the usual velocity which is chosen is the *face velocity* or the total volumetric flow at filtration conditions divided by the surface area of the filter which is available for flow. For a cylindrical bag, for example, this surface is the sum of the outer surface of the cylinder plus the end cap.

The total pressure drop of a surface filter is due to the resistance of the filter medium plus the resistance of the dust cake; the latter may be much larger than the former. To complicate matters further, even after pulse cleaning, the resistance of the medium does not return to its original value, as shown in Fig. 5.31. This effect is very system-specific.

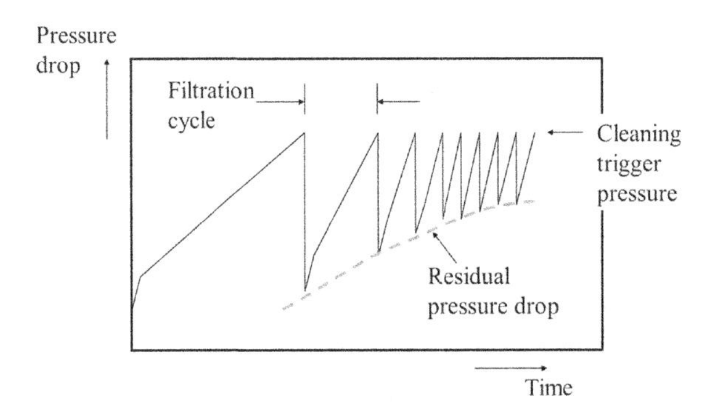

FIGURE 5.31

Cyclic behavior of a pulse-cleaned barrier filter.

Nevertheless, it is of interest to determine the pressure drop of the clean filter medium for comparison purposes. In general, the pressure drop through a planar porous medium can be represented (see Eq. (5.1)) as

$$-\frac{dP}{dz} = k_1 \mu U + k_2 \rho U^2 \tag{5.33}$$

where $(-dP/dz)$ is the pressure gradient in the direction of flow, and U is the superficial fluid velocity, i.e., the actual volumetric flow rate divided by the area available for flow. In the case of the media considered here, the Reynolds number $(U\rho d_p/\mu)$ is much less than unity, so that the second term in Eq. (5.33) can be neglected, and k_1 can be replaced by the Kozeny expression provided that the void fraction ε is not too high (Kyan et al., 1970):

$$k_1 = k_k (1 - \varepsilon)^2 \varepsilon^{-3} S_o^2 \tag{5.34}$$

where ε is the void fraction, S_o the specific surface area of the medium, and K_k is the Kozeny parameter, which depends on the geometrical structure. Equation (5.34) can be used to "design" a medium with the desired resistance characteristics. In cases where $\varepsilon > 0.95$, such as in many fibrous media, the prediction of pressure drop is much more complex; reviews are presented by Strauss (1975) and Brown (1993).

REFERENCES

Barton, J., 2002. Dust Explosion Prevention and Protection. IChemE, Rugby.

Brown, R., 1993. Air Filtration. Pergamon, Oxford.

Clift, R., 1986. In: Geldart, D. (Ed.), Gas Fluidisation Technology. John Wiley & Sons, New York.

Darcy, H.P.G., 1856. Les Fontaines Publiques de la Ville de Dijon. Victor Dalmont, Paris.

Darton, R.C., LaNauze, R.D., Davidson, J.F., Harrison, D., 1977. Bubble-growth due to coalescence in fluidised-beds. Transactions of the Institution of Chemical Engineers 55, 274–280.

Davidson, J.F., Harrison, D., 1963. Fluidised Particles. Cambridge University Press, Cambridge.

Davies, R.M., Taylor, G.I., 1950. The mechanics of large bubbles rising through extended liquids and through liquids in tubes. Proceedings of the Royal Society of London A 200, 375–390.

Ergun, S., 1952. Fluid flow through packed columns. Chemical Engineering Progress 48, 89–94.

Geldart, D., 1973. Types of gas fluidisation. Powder Technology 7, 285–292.

Geldart, D., 1986. Gas Fluidisation Technology. John Wiley & Sons, New York.

Grace, J.R., 1986. Contacting modes and behaviour classification of gas—solid and other two-phase suspensions. The Canadian Journal of Chemical Engineering 64, 353–363.

Knowlton, T.M., 1986. In: Geldart, D. (Ed.), Gas Fluidization Technology. John Wiley & Sons, New York.

Kyan, C.P., Wasan, D.T., Kinter, R.C., 1970. Flow of single-phase fluids through fibrous beds. Industrial and Engineering Chemistry Fundamentals 9, 596–603.

Seville, J.P.K., 1987. Particle cohesion in gas/solid fluidisation. In: Briscoe, B.J., Adams, M.J. (Eds.), Tribology in Particle Technology. Adam Hilger, Bristol, pp. 173–190.

Seville, J.P.K., Tüzün, U., Clift, R., 1997. Processing of Particulate Solids. Chapman and Hall, London.

Stairmand, C.J., 1951. The design and performance of cyclone separators. Transactions of the Institution of Chemical Engineers 29, 356–383.

Strauss, W., 1975. Industrial Gas Cleaning, second ed. Pergamon, Oxford.

Wen, C.Y., Yu, Y.H., 1966. A generalized method for predicting the minimum fluidization velocity. AIChE Journal 12, 610–612.

CHAPTER

6 Liquid–Solid Systems

Chapter 5 is concerned with multiphase systems where the continuous phase is a gas; this chapter concerns the case where the continuous phase is a liquid. The most obvious difference is that in liquid–solid systems the difference in densities between the phases is much lower than in a gas–solid system, so that the particle settling velocity is also much reduced. In practice, suspensions of solid particles in liquids are divided into "settling" and "nonsettling." Roughly speaking, particles above about 100 μm in size in water show noticeable settling behavior, while those below this size will remain in suspension quite easily. An example of a "nonsettling" system is a pharmaceutical oral suspension of a nonsoluble drug, where the particles are typically in the size range 5–50 μm. From the settling velocity equations (section 4.3), it is apparent that the tendency toward settling is reduced by (1) decreasing the particle size and (2) increasing the liquid viscosity. Even so, bottles of such products often carry the instruction "shake before use"! Other products which are sold as suspensions of particles in a continuous liquid phase include paint, ink, abrasive cleaning products, and foods. If the particle size is below about 1 μm, the dispersion is known as *colloidal.*

The behavior of liquid–solid systems depends very much on the ratio of phase volumes, as shown in Fig. 6.1. The phase volume, ϕ, is the fraction of the total volume occupied by the phase in question. On the left is shown a dilute suspension, with a solid-phase volume less than 10%, for which the flow properties are dominated by the liquid phase. In the center is a dense suspension, where the particles are approaching a dense packing. In this case the flow properties are strongly influenced by the presence of the solids and the interactions between them. On the right is a loose aggregate of particles with liquid in the pore spaces between them. Each of these situations is considered in more detail below.

As the particle size is reduced, not only is settling less evident, but surface forces between the particles increase due to electrostatic interactions, van der Waals' forces, and steric interactions between adsorbed molecules. At the same time, the random movement of the particles due to molecular impacts (Brownian diffusion) increases. The combination of these effects can lead to particle aggregation, which can cause settling and phase separation. These issues and the surface chemistry associated with them are considered in detail by Shaw (1992).

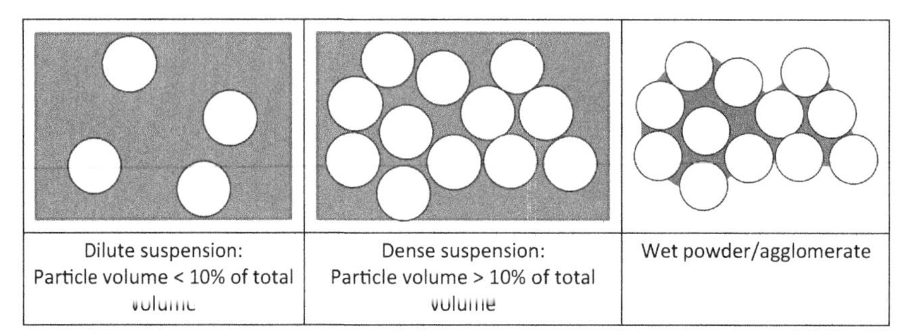

FIGURE 6.1

Liquid–solid systems: effects of phase volume.

6.1 RHEOLOGY OF SUSPENSIONS

Rheology is the study of flow; rheology of suspensions is of great importance in the development of suspension products, which are usually designed to flow in particular ways. For example, rheology is an important contributor to "mouth feel" in food and drink products.

Introduction of particles allows the rheological behavior to be manipulated to give the desired product properties. For example, paint needs to flow well at high shear rates (under the brush) but be resistant to flow when applied (not running or dripping). Suspensions which are to be pumped or sprayed need to have low viscosity under shear.

It is to be expected that the addition of particles to the liquid will have a fluid mechanics effect, in general causing an increase in viscosity. However, if the particles carry charge, which is invariably the case in colloidal systems, they can form structures ranging from aggregated *flocs*, which can trap some of the continuous phase within them, to *gel*-like materials which behave in an elastic way and can immobilize the system at low shear rates.

Figure 6.2 shows a liquid, with and without suspended particles, in simple shear.

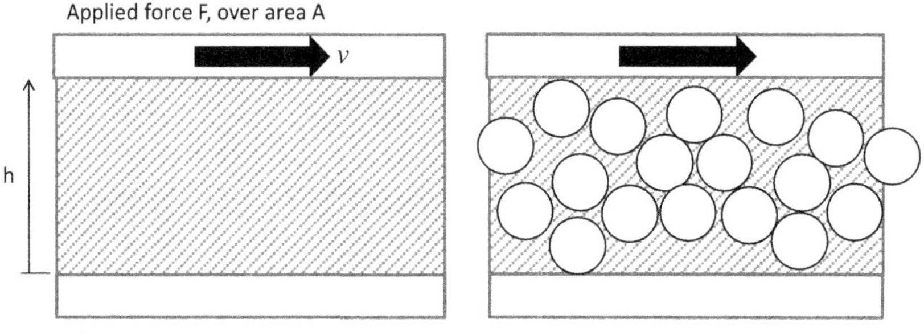

FIGURE 6.2

Simple shear of a continuous liquid and a suspension.

Viscosity is the constant of proportionality between the force per unit area (the shear stress) and the resulting velocity gradient in the direction perpendicular to the flow direction (the shear rate):

$$\tau = \eta \dot{\gamma} \tag{6.1}$$

where $\tau = F/A$ is the shear stress, η is the fluid viscosity, and $\dot{\gamma}$ ($=v/h$) is the shear rate, where v is the velocity at height h. Here A is the surface on which the shear force F is acting and h is the height of the fluid element being sheared.

If the fluid obeys this linear relationship, then it is termed *Newtonian*, which means that its viscosity is independent of shear rate; water is the best-known example. Newtonian liquids show no variation in viscosity with time of shearing, and when shearing stops, the shear stress returns immediately to zero.

In general, adding particles to a Newtonian liquid increases its viscosity, but it may also cause it to behave in a non-Newtonian way, showing viscosity which depends on shear rate, time-dependent behavior, or even the occurrence of an apparent yield stress.

Figure 6.2 also shows the shear of a suspension, which was studied by Einstein (see Barnes et al., 1989), who showed theoretically that the addition of particles increased the viscosity according to

$$\eta = \eta_s(1 + 2.5\phi) \tag{6.2}$$

where η is the viscosity of the suspension, η_s is the viscosity of the suspension liquid, and ϕ is the volume fraction of the solid phase, as before. The derivation of this equation considers the particles as isolated spheres, and so the result is only valid at small phase volumes, $\phi < 10\%$. At higher phase volumes, results of the form $\eta = \eta_s(1 + a\phi + b\phi^2 + \ldots)$ are available, but these are of limited applicability.

Clearly, there is a maximum solid-phase volume which corresponds to the maximum packing fraction ϕ_m. This is readily calculable for equally sized spheres.

Arrangement	Maximum Packing Fraction
Simple cubic	0.52
Body-centered cubic	0.68
Face-centered cubic/ hexagonal close packed	0.74

Other approaches to the problem of predicting the viscosity of suspensions, such as averaging techniques, result in equations such as the widely used Krieger and Dougherty expression:

$$\eta = \eta_s(1 - \phi/\phi_m)^{-[\eta]\phi_m} \tag{6.3}$$

where $[\eta]$ is the (dimensionless) "intrinsic viscosity," which takes values between 2.5 and about 10, depending on particle size, size distribution, and shape. In practice, ϕ_m and $[\eta]$ are fitted to experimental results, as shown by Barnes et al. (1989). At low ϕ, Eq. (6.3) gives results equivalent to Eq. (6.2).

So far we have considered only Newtonian liquids and suspensions, i.e., those for which the viscosity is independent of shear *rate*, as shown in Fig. 6.3(a).

More generally, suspensions show a range of non-Newtonian behavior, in which the viscosity is dependent on shear rate, as in Fig. 6.3(c) and (d) and Fig. 6.4, and/or shows a yield stress, as in Fig. 6.3(b). Nonlinear behavior such as that shown in Fig. 6.3(c) and (d) is often characterized as that of a "power law" fluid:

$$\tau = k\gamma^n \tag{6.4}$$

where k is a constant and $n < 1$ for shear thinning and $n > 1$ for shear thickening.

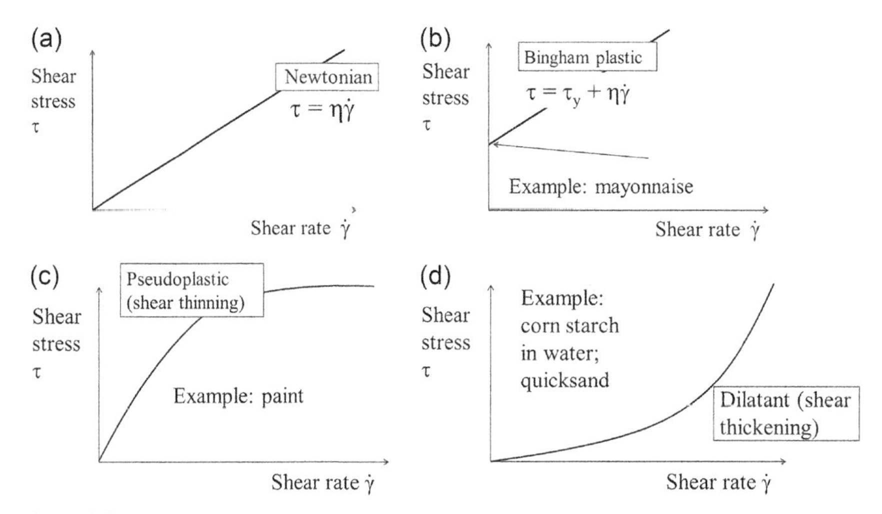

FIGURE 6.3

Typical shear stress versus shear rate curves—(a) Newtonian; (b) Bingham plastic, showing a yield stress; (c) pseudoplastic (shear thinning); (d) dilatant (shear thickening).

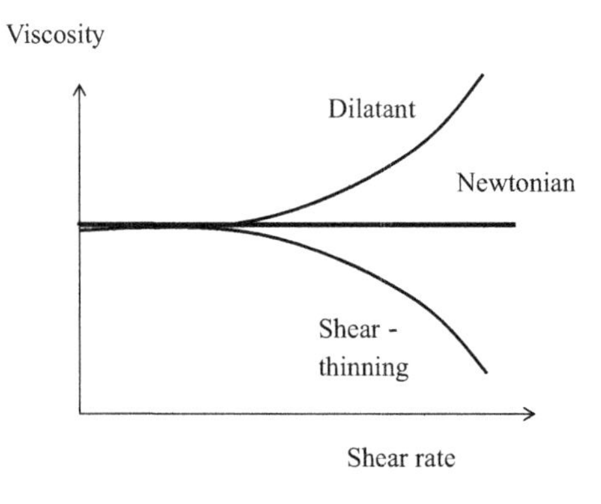

FIGURE 6.4

Viscosity versus shear rate, for Newtonian, dilatant and shear-thinning suspensions.

If the suspension shows both a yield stress and power-law type behavior, it may obey the widely used Herschel–Bulkley equation:

$$\tau = \tau_y + k\dot{\gamma}^n \tag{6.5}$$

In pipe flow, both Bingham and Herschel–Bulkley fluids show a region of plug flow near the centerline of the pipe and high strain rate near the wall.

The same suspension can show both shear-thinning and shear-thickening behavior. Typically, the former is associated with alignment of the particles into shear bands or two-dimensional structures which slide easily over one another, while the latter occurs when such structures are disrupted at higher shear rates. Suspension rheology becomes extremely sensitive to shear rate, particle size, and particle shape at values of solid-phase volume around 0.5, and such suspensions can show a "critical shear rate" at which the viscosity increases by as much as two orders of magnitude (Barnes et al., 1989). This jump in viscosity can be reduced by widening the particle size distribution.

An additional complication in rheology of suspensions is time-dependent behavior. For example, *thixotropy* is the time-dependent reduction in viscosity on shearing, caused by breakdown in the microstructure and normally followed by full or partial recovery of properties. Many clays, such as bentonite, show this behavior.

If one extreme of suspension behavior is Newtonian liquidlike behavior, the other is solidlike linear-elastic behavior, associated with Hooke's law:

$$\tau = G\gamma \tag{6.6}$$

where G is the shear modulus (independent of the stress and time constant) and γ is the shear strain. Suspensions can show this form of behavior if they are strongly cross-linked into a solid gel structure. More commonly, gels show a combination of liquidlike and solidlike behavior known as *viscoelasticity*. For a viscoelastic material which is held at constant strain, the shear stress relaxes over time. This kind of behavior can be investigated in an oscillatory shear device and the material response characterized in terms of both magnitude and phase.

Figure 6.5 shows an example of how the behavior of a suspension can be manipulated from predominantly viscous to highly elastic by making a small change in composition.

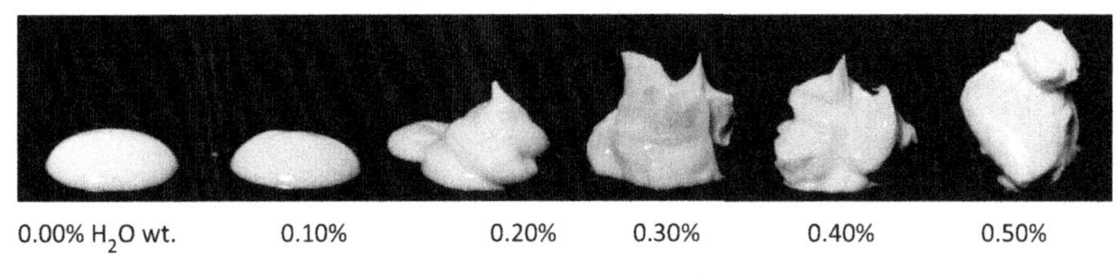

FIGURE 6.5

Transition from weakly elastic, predominantly viscous to highly elastic, gel-like behavior (Willenbacher and Georgieva, 2013). The effect is produced by adding 0.5% wt of water to a suspension of hydrophobically modified $CaCO_3$ (diameter = 800 nm, $\phi = 0.11$) in diisononyl phthalate. The water forms capillary bridges between the particles.

6.2 PASTES

Pastes are highly filled suspensions with a solid-phase volume approaching the maximum packing fraction. Apart from well-known domestic examples such as toothpaste and some food products, pastes represent a very important stage in a number of manufacturing processes, particularly for extruded products such as catalysts, foods, pharmaceuticals, and ceramics. Pastes can be divided into "stiff" and "soft" (Wilson and Rough, 2012), according to the relative magnitude of the two terms on the right hand side of Eq. (6.5). If τ_y is larger, the paste is "stiff" and shows predominantly plastic behavior, as in many ceramic pastes, which are required to maintain their shape after extrusion while firing into a final form. An example of a "soft" paste is molten chocolate, which shows plastic behavior at comparatively low temperatures.

Just as in the tabletting process (Chapter 10), wall slip is a common feature of paste extrusion, so that paste may move in plug flow with deformation only at the wall.

The most commonly applied method of characterizing pastes is due to Benbow and Bridgwater (1993) and makes use of the simple flow geometry shown in Fig. 6.6. The essential idea behind this simple experimental approach is that the overall extrusion pressure, P_{ex}, is made up of the sum of P_1, the pressure required to achieve the deformation required at the die entry, and P_2, the pressure required to push the paste plug along the die land:

$$P_{ex} = P_1 + P_2 = 2\sigma_Y \ln(D_0/D) + 4\tau_W \cdot (L/D) \tag{6.7}$$

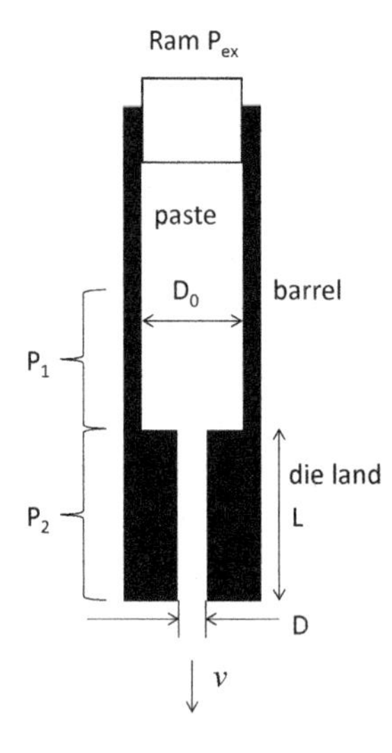

FIGURE 6.6

Schematic of Benbow and Bridgwater ram extruder configuration; the ram forces material from the barrel (diameter D_0) through the die land of diameter D and length L, at a mean extrudate velocity, v.

Equation (6.7) assumes homogeneous deformation of a perfect plastic undergoing a change in cross-sectional area from πD_0^2 to πD^2 (Fig. 6.7), where σ_Y is the uniaxial bulk yield stress of the paste and τ_W is the wall shear stress in the die land.

In general, both σ_Y and τ_W can be functions of the strain rate, so can be written in terms of the extrusion velocity, v:

$$\sigma_Y = \sigma_0 + \alpha v^m \tag{6.8}$$

and

$$\tau_W = \tau_0 + \beta v^n \tag{6.9}$$

A series of experiments is then carried out in which v and L/D are varied, from which the four (if $m = n = 1$) or six fitting parameters are obtained. This approach has been used successfully for a wide range of paste materials. Its usefulness and comparison with more rigorous analyses are considered by Wilson and Rough (2012).

Two particular problems associated with paste extrusion should be mentioned: post-extrusion fracture and liquid-phase migration. Just as tablets ejected from dies show stress relaxation which can lead to fracture, paste extrudates leaving the die land can show similar behavior, including the distinctive circumferential fracture shown in Fig. 6.8.

Liquid-phase migration occurs when the pore pressure in the liquid causes it to move relative to the solids, so causing local differences in phase volume; low liquid

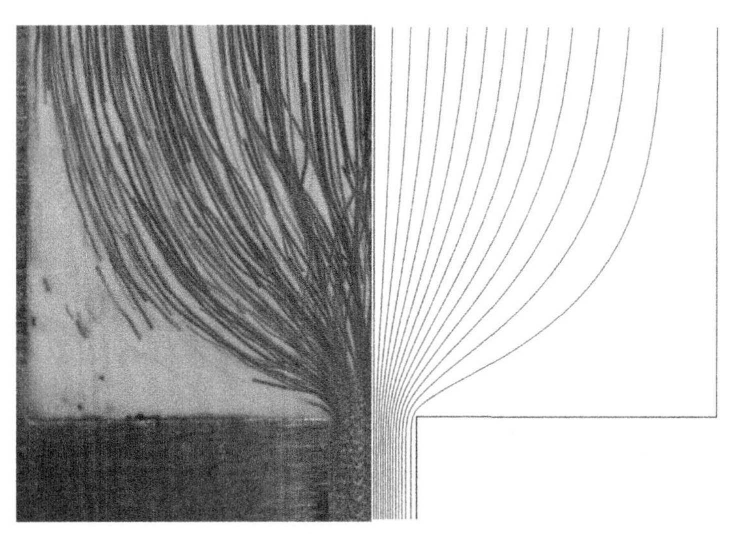

FIGURE 6.7

Flow visualization of a microcrystalline cellulose/water paste extruding (top to bottom) through a semicylindrical die: streak photograph of streamlines (left); fluid dynamics simulation assuming Bingham plastic behavior with Navier-type wall slip: streamlines (right).

Courtesy of Matthew Bryan, Sarah Rough and Ian Wilson, University of Cambridge.

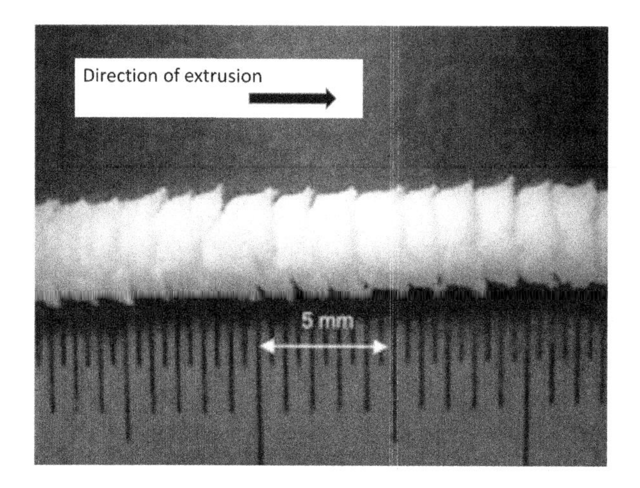

FIGURE 6.8

Circumferential fracture of paste extrudate (Russell et al., 2006).

content gives rise to increased bulk yield stress and can cause blockages and poor product quality. Liquid-phase migration depends on the permeability of the solids phase and the properties of the liquid, as described in Section 5.2, and therefore puts limits on the paste formulation. Prediction is difficult, however, because of consolidation of the solid during extrusion.

6.3 AGGLOMERATION

Many commonly used products such as foods, beverages, and pharmaceuticals are sold as aggregated or agglomerated structures, or pass through such a form in their manufacture, and there are many products of the chemical industry, such as catalysts and solid fuel pellets, which are similarly manufactured. A common method consists of mixing the fine constituent powder(s) mechanically or by means of fluidization, for example, while spraying a liquid binder into or onto the agitated powder bed. This is known as wet agglomeration, as distinct from dry processes such as roll pressing (Wu et al., 2010). Examples of wet agglomeration devices are shown in Fig. 6.9.

Although the devices used for agglomeration vary in design, the processes which occur within them have some similarity. Figure 6.10 shows these processes schematically: (i) wetting and nucleation, (ii) consolidation and coalescence, and (iii) attrition and breakage. These are not usually sequential but occur simultaneously, especially in batch agglomerators.

As the binder liquid is sprayed onto the powder bed, a process of nucleation of potential agglomerates occurs. This is a rapid process which requires only capillary action. Unlike nucleation in crystallization, there is no energy barrier to be overcome. Although fine sprays are often used, in practice the liquid is never uniformly

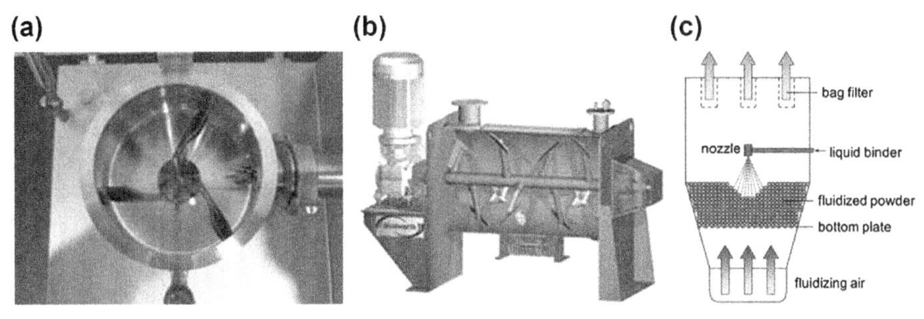

FIGURE 6.9

Common types of batch industrial agglomeration devices: from left to right—a high-shear mixer-granulator, a ploughshare mixer, and a fluidized bed granulator. [Images reproduced with permission (all rights reserved): (a) Surplus Solutions, LLC, http://www.ssllc.com/; (b) Winkworth Machinery Ltd., http://www.mixer.co.uk; (c) Fries et al., 2013].

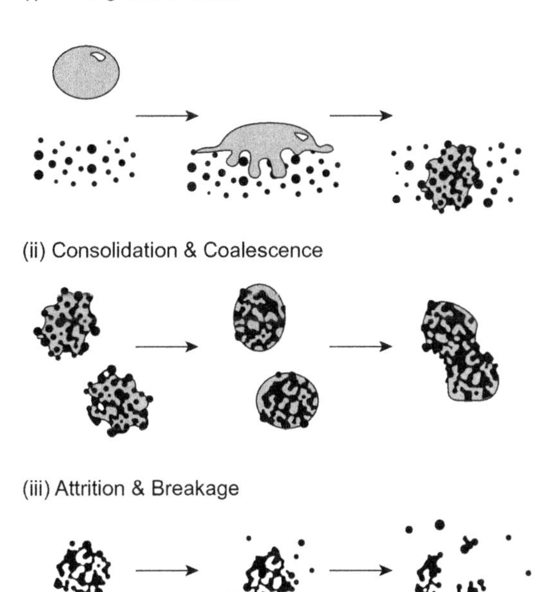

FIGURE 6.10

Processes occurring in agglomeration (Iveson et al., 2001).

distributed, the nonuniformity becoming more pronounced with increasing scale of operation. It is normal to form a bimodal agglomerate size distribution at this stage, which may persist throughout the process, as shown in Fig. 6.11.

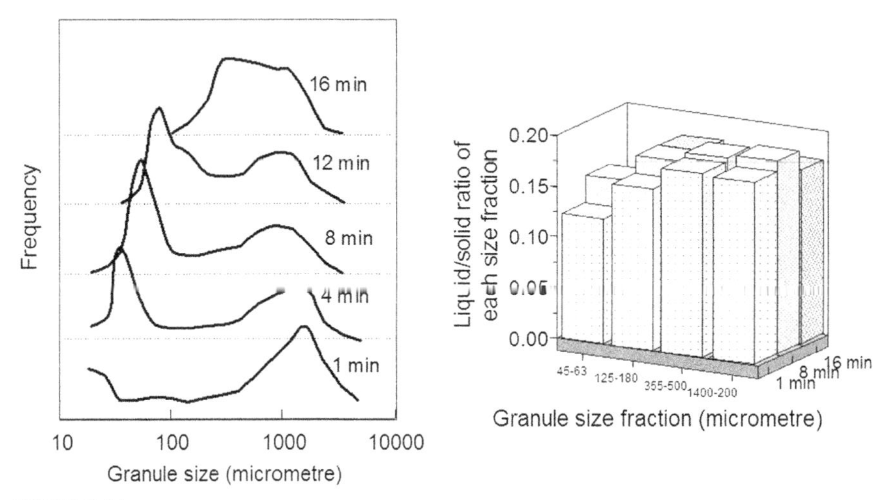

FIGURE 6.11

Agglomerate size and liquid distributions at different times after liquid addition in a high-shear mixer-granulator (4 μm solid, liquid/solid ratio = 0.17) (Knight et al., 1998).

The size of the nuclei depends on the size of the sprayed liquid drops, if this is the method of addition. If the liquid is poured into the mixer rather than sprayed, then the size of the nuclei depends on the mixing intensity.

This initial nuclei formation is considered further by Litster et al. (2001), for a spray of droplets of size d, of total volumetric flow rate V', arriving at a surface being renewed at a rate A' per square meter per second, as shown in Fig. 6.12.

The number of drops per unit time, n', is given by:

$$n' = V'/\left(\pi d^3/6\right) \tag{6.10}$$

and the cross-sectional area per drop, a, is $\pi d^2/4$, so that the new area covered by drops per unit time is given by

$$an' = 3V'/2d \tag{6.11}$$

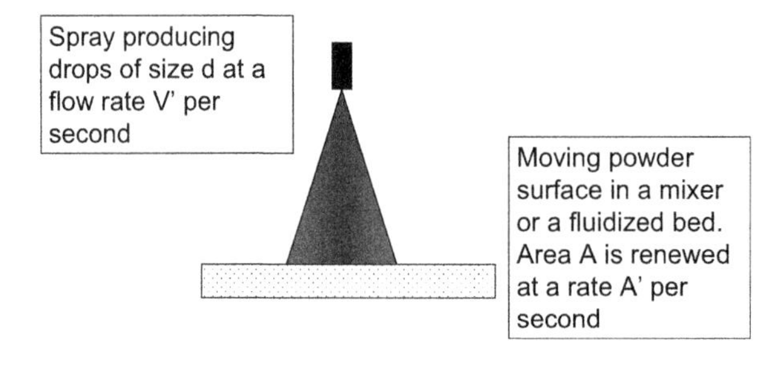

FIGURE 6.12

Spraying onto the surface of a powder.

It is then possible to write a dimensionless spray flux (the wetted area per unit time/area of powder renewed per unit time):

$$\Psi = a'n/A' = 3V'/2A'd \tag{6.12}$$

Litster et al. (2001) demonstrated experimentally a clear dependence of the nucleated agglomerate number on the dimensionless spray flux, as shown schematically in Fig. 6.13.

The most important parameter in wet agglomeration processes is saturation, S, which is defined as the volume of voids filled with liquid divided by the total volume of voids. The total void volume depends on the packing of the particles, which can change over time. In general, therefore, saturation increases not only with the amount of liquid added but also as the agglomerates consolidate in response to the agitating action. Figure 6.14 shows agglomerates with saturations from zero (loose powder) through partially saturated (weak agglomerate) to fully saturated (paste-like). Note that if the agglomerate is surface-wet, the saturation may exceed 1. The picture is further complicated by the fact that drying may occur simultaneously or sequentially with liquid addition.

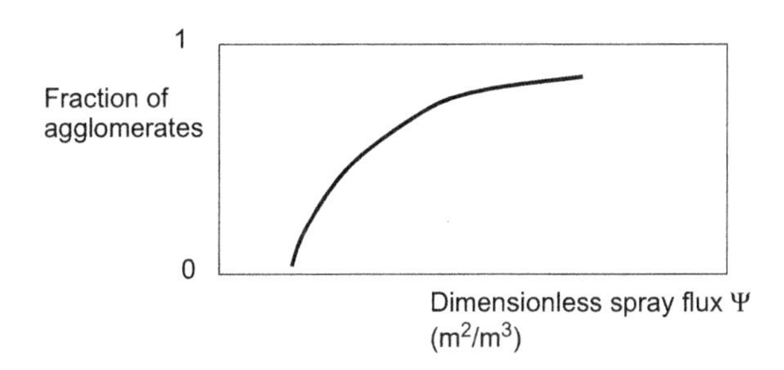

FIGURE 6.13

Dependence of agglomerate number on dimensionless spray flux.

After Litster et al. (2001).

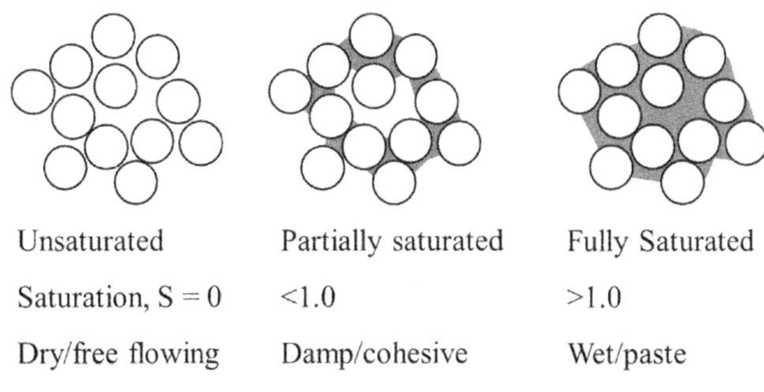

FIGURE 6.14

Agglomerates with different degrees of saturation.

Figure 6.15 shows the consolidation process occurring during agitation and mixing. As mentioned above, consolidation reduces the pore volume and so increases the saturation. It also removes air-filled pores. Visually, very little is seen as the process continues until the surface of the initially-formed agglomerates becomes wet, which may occur quite suddenly. When the surface becomes wet, the initially-formed small agglomerates may aggregate with each other, and rapid size enlargement then occurs. This effect is usually much more marked with fine powders (below about 10 μm), because the initial void fraction of such fine powders in the dry state is usually high (~0.8), as shown in Fig. 6.16. This makes it difficult to control agglomeration processes for fine powders, and changing the liquid-to-solid ratio around the saturation point can have a rapid effect on growth, as shown in Fig. 6.17. All of these phenomena will be familiar to those who are used to mixing liquids into powders in a domestic setting—in making pastry or mixing cement, for example.

The problem of modeling the outcome of collisions between wet agglomerates (as in Fig. 6.18) has been considered by many researchers, because this is the key to constructing a mathematical model for agglomeration processes. Collision between a particle and a wet surface is considered in Chapter 8. However, the approach developed there applies in the case of elastic particles which show no plastic deformation, so that all of the energy dissipated in collision is taken up in viscous dissipation in the liquid layer. The case shown in Fig. 6.18 is clearly more complex, since energy can be dissipated here in plastic deformation of the agglomerates,

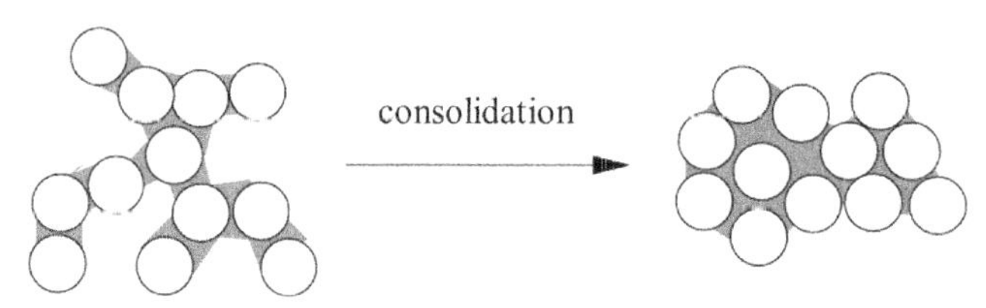

FIGURE 6.15

Consolidation of a wet agglomerate.

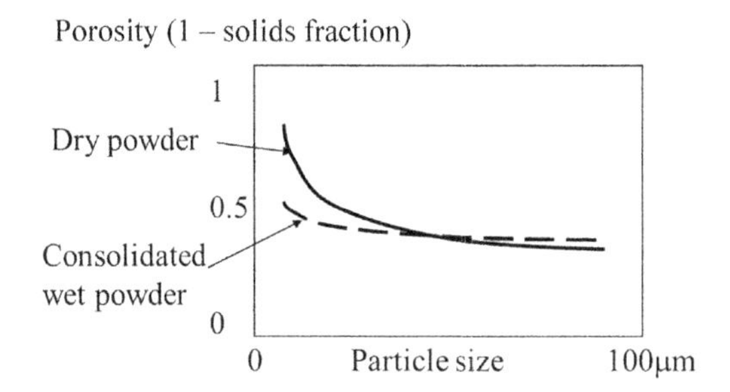

FIGURE 6.16

Change in porosity of agglomerates during mixing, wet and dry, as a function of initial particle size (Knight, 2004).

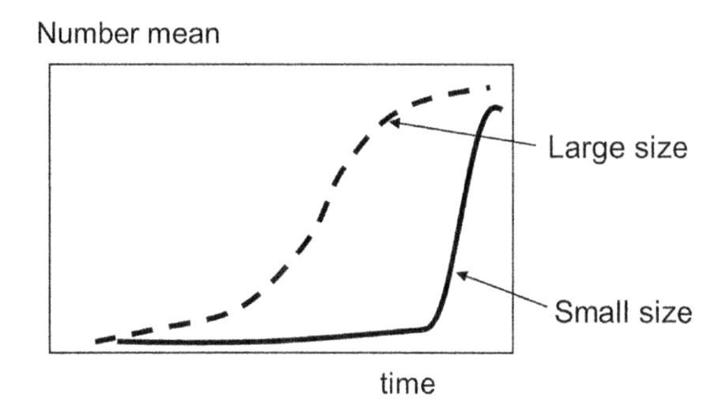

FIGURE 6.17

Change in mean size of agglomerates with time, for two initial particle sizes (large = ~50 μm; small = ~5 μm) (Knight, 2004).

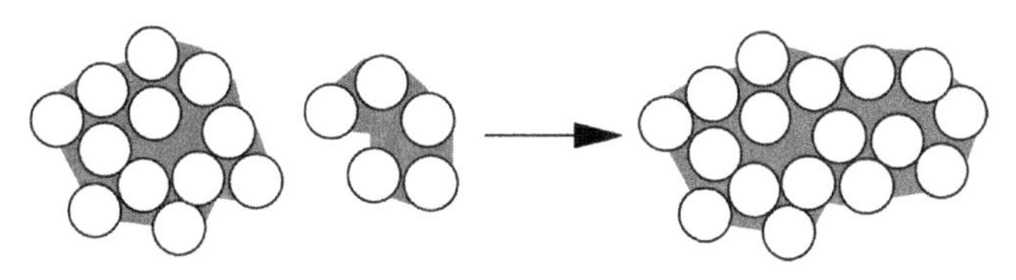

FIGURE 6.18

Collision between two wet agglomerates to produce a larger one.

accompanied by liquid migration, friction between the constituent particles and possibly breaking of liquid bridges. A different but related approach was therefore developed by Iveson, Litster, and coworkers (Iveson and Litster, 1998; Iveson et al., 2001). They defined a deformation number,

$$D_e = \rho U^2/2Y \tag{6.13}$$

where ρ is the agglomerate density, U is a characteristic collision velocity, and Y is a dynamic yield stress for the agglomerates. This approach treats the impacts as fully plastic. They were able to measure Y by a compression method, concluding that it depends on the binder liquid viscosity, the liquid surface tension (with low viscosity binders), and the constituent particle size. As the deformation number is increased, the extent of breakage also increases, until a critical value of D_e, above which "crumb" is formed instead of granules.

Using these concepts, Iveson et al. (2001) presented a regime map for wet agglomeration processes, as shown in Fig. 6.19. Considering first the limits to behavior, small proportions of added liquid will result in nucleation only, and if the deformation number is large (strong agitation), a free-flowing powder is still seen. At the other extreme, if the proportion of liquid is high (saturation near or above 100%) a paste or slurry results. Between these two extremes, various forms

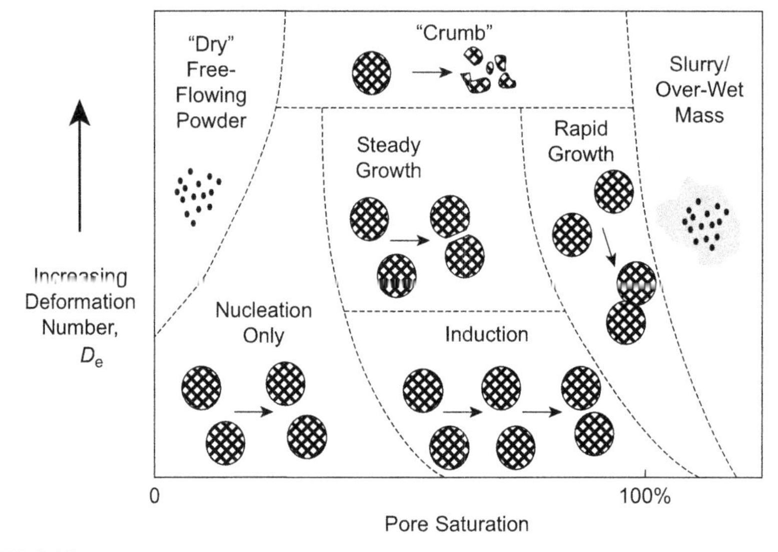

FIGURE 6.19

Agglomeration regime diagram (Iveson et al., 2001).

of growth are possible, unless the deformation number exceeds a critical value, in which case agglomerates are destroyed and crumb results. Approaching 100% saturation, growth is very rapid, as discussed earlier.

This diagram is clearly approximate in nature and there are difficulties in assigning numerical values to the components of the deformation number. However, it gives a useful conceptual framework for understanding wet agglomeration processes.

REFERENCES

Barnes, H.A., Hutton, J.F., Walters, K., 1989. An Introduction to Rheology. Elsevier.

Benbow, J., Bridgwater, J., 1993. Paste Flow and Extrusion, Clarendon Press, Oxford.

Fries, L., Antonyuk, S., Heinrich, S., Dopfer, D., Palzer, S., 2013. Collision dynamics in fluidised bed granulators: A DEM-CFD study. Chemical Engineering Science 86, 108–123.

Iveson, S.M., Litster, J.D., 1998. Growth regime map for liquid-bound granules. AIChE Journal 44, 1510–1518.

Iveson, S.M., Litster, J.D., Hapgood, K., Ennis, B.J., 2001. Nucleation, growth and breakage phenomena in agitated wet granulation processes: a review. Powder Technology 117, 3–39.

Knight, P.C., 2004. Personal Communication.

Knight, P.C., Instone, T., Pearson, J.M.K., Hounslow, M.J., 1998. An investigation into the kinetics of liquid distribution and growth in high shear mixer agglomeration. Powder Technology 97, 246–257.

Litster, J.D., Hapgood, K.P., Michaels, J.N., Sims, A., Roberts, M., Kameneni, S.K., Hsu, T., 2001. Liquid distribution in wet granulation: dimensionless spray flux. Powder Technology 114, 32–39.

Russell, B.D., Blackburn, S., Wilson, D.I., May 2006. A study of surface fracture in paste extrusion using signal processing. Journal of Materials Science 41 (10), 2895–2906.

Shaw, D.J., 1992. Introduction to Colloid & Surface Chemistry, fourth ed. Elsevier.

Wilson, D.I., Rough, S.L., 2012. Paste engineering: multi-phase materials and multi-phase flows. The Canadian Journal of Chemical Engineering 90, 277–289.

Willenbacher, N., Georgieva, K., 2013. Rheology of Disperse Systems. In: Bröckel, U., Meier, W., Wagner, G. (Eds.), Product Design and Engineering: Formulation of Gels and Pastes. Wiley-VCH Verlag.

Wu, C.Y., Hung, W.-L., Miguélez-Morán, A.M., Gururajan, B., Seville, J.P.K., 2010. Roll compaction of moist pharmaceutical powders. International Journal of Pharmaceutics 391 (1–2), 90–97.

CHAPTER

Mechanics of Bulk Solids

7

Powders are unique as a material form, but they can manifest gas-like, solid-like, and liquid-like properties at different stress states. When powders are highly agitated, as in a high-speed bladed mixer or the rotating mixer which produces the results of some national lotteries, collision between particles dominates and the powder can fill the entire volume of the chamber. In some respects this behavior mimics the collision of molecules of a gas that similarly occupies the entire container in which it is held. Under different conditions the same powder can also behave like a single block of a solid material, especially when consolidated under a high compression pressure, as in compressed tablets and "green" (prefired) compacts of metal or ceramic powders. Such compacts exhibit the mechanical strength and brittle fracture/plastic deformation of a conventional continuous solid. They also show a small degree of elasticity, as a consequence of the elasticity of the single particle–particle contacts.

When powders are subjected to a high shear stress at high shear rates, as in some forms of mixing equipment and in flow down an inclined chute, they can behave like a liquid. An interesting example of liquid-like behavior can be observed when a projectile is dropped into a powder bed, as shown in Fig. 7.1. As the projectile penetrates into the bed, it produces a void behind it, which is rapidly re-filled by mobile powder, causing a vertical jet to emerge as shown. This is similar to the

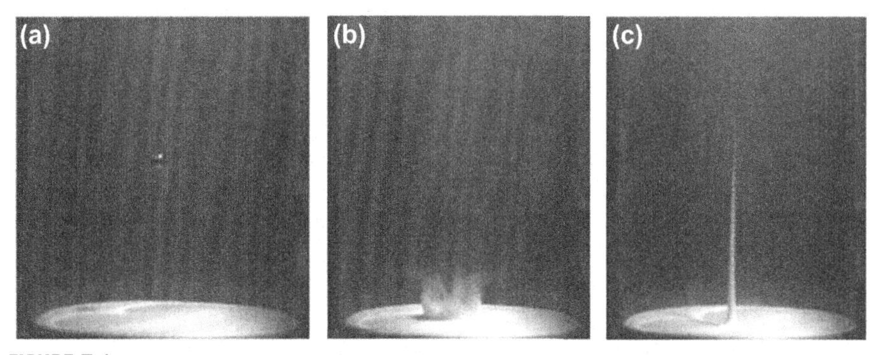

FIGURE 7.1

Formation of a granular jet during the impact of a 10 mm sphere with a 70-mm-deep powder bed of 78 μm particles; times from impact -20, 50, 150 ms (Marston et al., 2008).

Particle Technology and Engineering.

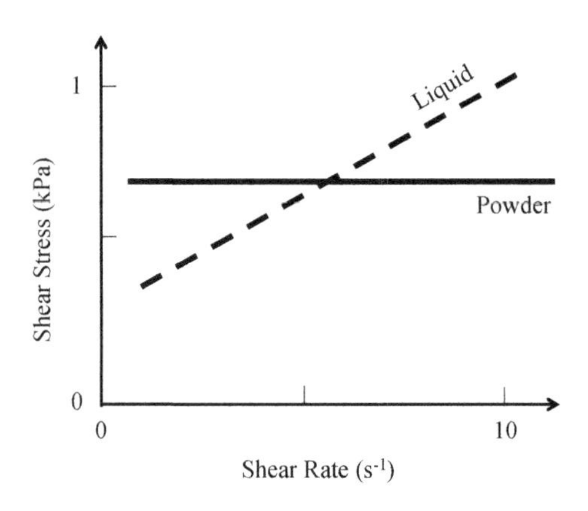

FIGURE 7.2

The variation of shear stress with shear rate for a powder and a liquid.

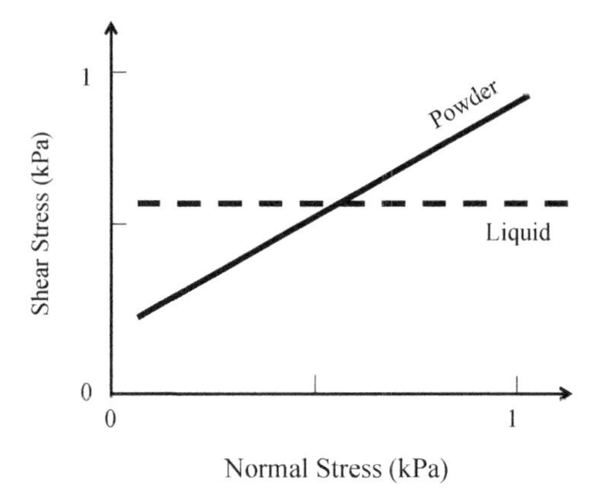

FIGURE 7.3

The variation of shear stress with normal stress for a powder and a liquid.

formation of the water jet when a pebble is thrown into water, implying that a powder can behave like a liquid in such impact processes.

Despite the similarities discussed above, the mechanical behavior of powders is complicated and distinctive from that of gases, solids, and liquids. The most distinct aspects of powders are that they are frictional and their mechanical behavior is state-dependent and determined by particle–particle interactions. Two fundamental differences between powders and liquids are illustrated in Figs 7.2 and 7.3. In frictional flow of powders, the shear stress varies little with shear *rate* (Fig. 7.2), whereas for a liquid, the shear stress increases as the shear rate increases; there is a linear relationship if the liquid is Newtonian, for which the proportionality is the liquid viscosity. On the other hand, powder flow is determined not only by the shear stress but also by the normal stress (see Fig. 7.3), whereas the viscosity of a liquid is almost independent of the pressure applied.

A powder is composed of millions of individual solid particles, and each particle has certain physical properties: elasticity, plasticity (when the yield stress is reached), and failure stress (at which the solid fractures), together with surface properties such as surface energy and roughness. When particles with these properties are assembled into a powder, the bulk properties of the powder will depend on those individual particle properties. Prediction of bulk properties from individual particle properties is a challenging scientific problem and has attracted increasing intention. In particular, the discrete element method (DEM) has proved to be a useful numerical tool for this purpose and will be introduced in Chapter 9.

7.1 FRICTION AND THE COULOMB MODEL

For a powder to flow, relative displacements must occur between particles and particles, and between particles and walls, or more commonly both. From the point of view of engineering mechanics, there is a material *failure* when a powder flows, as the temporary connections between particles are broken. When a powder flows, it must overcome the resistance arising from friction, either internally (between particles along the failure surface, see Fig. 7.4(a)) or at the interface with the container wall (see Fig. 7.4(b)). This is often referred to as frictional failure or shear. It is therefore important to know the conditions for frictional failure both internally and at the wall. These are governed by the force balances at the internal failure surface (i.e., internal friction) and at the wall (i.e., wall friction). Generally, the force required to initiate sliding is greater than that required to maintain continuous sliding, under the same normal force. Hence there are two types of friction: static friction and dynamic friction. Static friction occurs between two objects (e.g., particles, walls) which are not in relative movement, while dynamic friction occurs between two objects moving relative to each other (e.g., particle–particle and particle–wall interactions during continuous sliding). Static friction must be overcome in order to initiate relative movement. Both forms of friction are characterized by a coefficient of friction, μ, which is defined as the ratio of the shear force, F_S, to the normal force, F_N, as shown in Fig. 7.4.

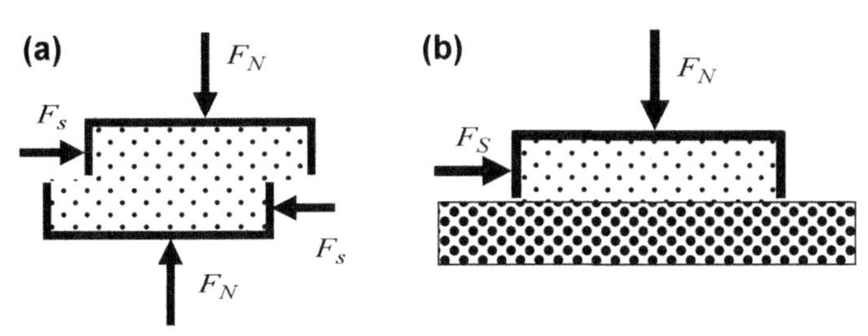

FIGURE 7.4

Illustration of (a) friction internally and (b) friction at a wall.

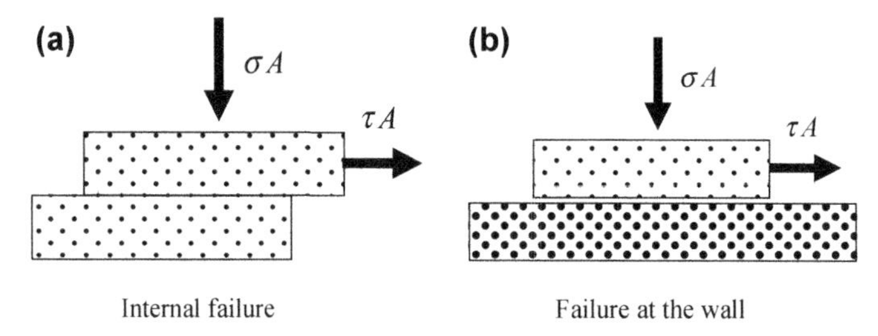

FIGURE 7.5

Schematic illustration of (a) internal and (b) wall frictional failure (A is the area of the failure surface).

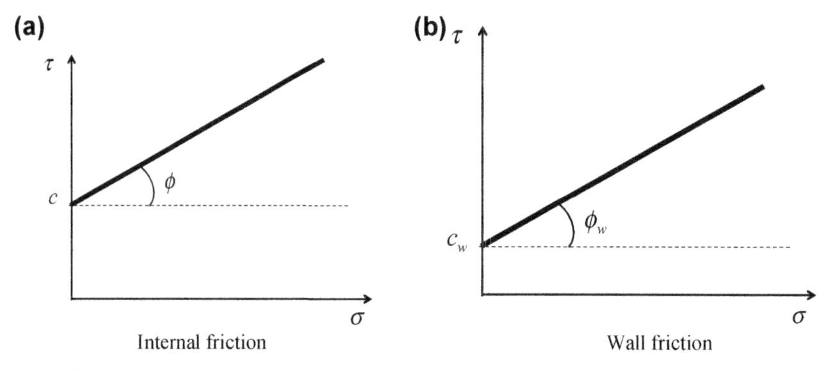

FIGURE 7.6

Graphical representation of (a) internal and (b) wall frictional failure.

For a powder to start to shear, Coulomb and others (see Seville et al. (1997)) proposed a linear relationship between the shear stress τ and the normal stress σ:

$$\tau = \mu\sigma + c \tag{7.1}$$

where μ is the internal friction coefficient and c is the cohesive shear stress or "cohesion," i.e., the shear stress of the material under zero applied load. Cohesion may have many causes (e.g., mechanical interlocking, interparticle cohesive forces induced by van der Waals interaction, and liquid bridges), and can lead to "caking" and serious flow problems.

Similarly, for initiation of failure at the wall, the following condition must be satisfied:

$$\tau = \mu_W\sigma + c_W \tag{7.2}$$

where μ_W is the wall friction coefficient and c_W is the wall adhesive shear stress. These two situations are represented schematically in Fig. 7.5 and graphically in Fig. 7.6.

The coefficient of friction is also commonly represented as the tangent of an angle, ϕ, the angle of internal friction, or ϕ_w, the wall friction angle, i.e.,

$$\mu = \tan \phi \tag{7.3a}$$

$$\mu_w = \tan \phi_w \tag{7.3b}$$

Typically, μ is in the range 0.45−1 ($\phi = 25°-45°$) and μ_w is in the range 0.25−0.7 ($\phi_w = 15°-35°$). Plots such as those shown in Fig. 7.6 can be obtained from a range of commercially available shear cell testers (see, e.g., Box 2.4 in Chapter 2).

7.2 STRESS ANALYSIS IN STORAGE VESSELS

Bulk solids, such as cements, agricultural grains, and food powders, are stored in a wide variety of vessels, including *hoppers, silos* and *bins*. (These terms are often used interchangeably but silos are normally larger and of cylindrical shape. All may have a converging discharge section.). Predicting the stresses and flow patterns of bulk solids in storage vessels is necessary for proper design and safe operation of these containers. Improper storage vessel design and usage due to poor understanding of the stresses in bulk solids and their flow patterns may lead to disastrous consequences, as exemplified in Fig. 7.7, which shows the

FIGURE 7.7

Silo collapse due to the development of a different flow pattern from that envisaged in the original design.

Courtesy of Jenike & Johanson, Inc. Reproduced with permission.

collapse of a silo due to the development of a different flow pattern (see Section 7.4.1) from the one for which it was initially designed.

Due to the frictional nature of bulk solids, stress analysis based on the analogies with liquids can be very misleading. As illustrated Fig. 7.8(a), the stress in a liquid is hydrostatic and hence varies linearly with depth, irrespective of the shape of the container; the stress acting on the bottom of the container is then equal to ρgh (where ρ is the liquid density, h is the depth of the liquid in the container, and g is the gravitational acceleration). However, for the storage of a bulk solid with the same depth h (Fig. 7.8(b)), some of the weight is supported by friction at the walls, so that the stress within the powder is generally less than the hydrostatic value, and the stress on the bottom of the container will be less than the hydrostatic pressure $\rho_b gh$ (where ρ_b is the bulk density).

Furthermore, at the microscopic level, the stress in a bulk solid is often very far from uniform. Figure 7.9 shows the force network of a packed powder bed obtained using DEM (Guo et al., 2009), in which the widths of the lines indicate the magnitude of the local force (i.e., the contact force between two particles): the thicker the line, the larger the contact force. It can be seen that the contact forces form a network of heterogeneous chains. Some particles within these chains are supporting a large load and others, between the chains, may be taking no load at all. These factors are important in analyzing force transmission in bulk solids during storage and compression.

In engineering practice, it is often necessary to estimate the stress in a storage vessel at the macroscopic level. Let us consider the storage of a bulk solid in a vertical-sided cylindrical container of diameter D, as shown in Fig. 7.10. At depth z, the stress in the vertical direction is σ_{zz} and is assumed to be a function only of the depth and the shear stress at the wall, τ_w. A first-order estimate of stresses inside the storage container was obtained using the method of differential slices by Janssen in 1895, and is based upon the following assumptions:

1. the powder is cohesionless;
2. the stresses are uniformly constant across the cross-section of the storage vessel;

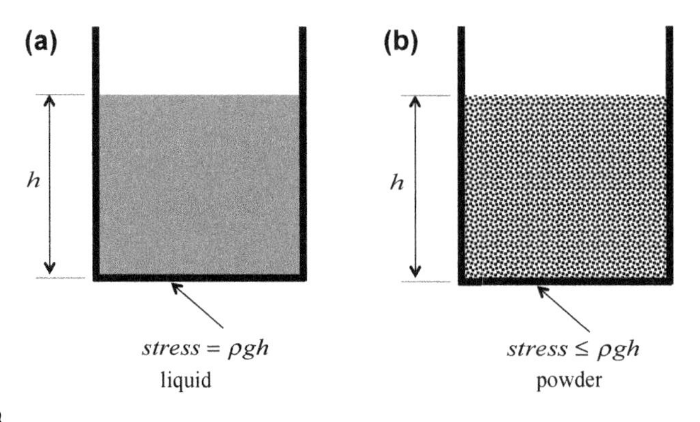

FIGURE 7.8

Illustration of the stresses in (a) a liquid and (b) in a powder during storage.

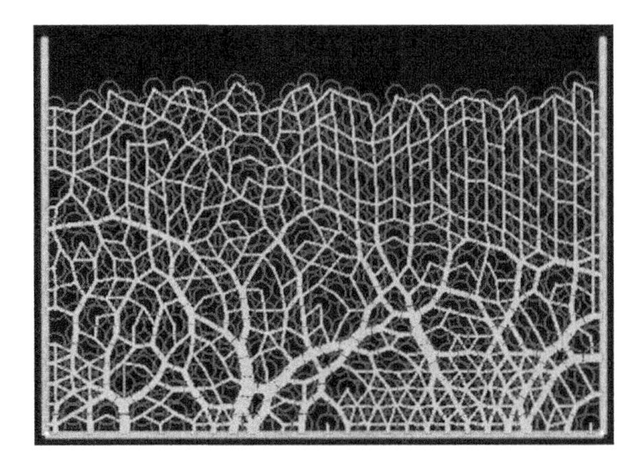

FIGURE 7.9

Stress transmission in a packed powder bed of monosized particles obtained from discrete element method (DEM) simulations.

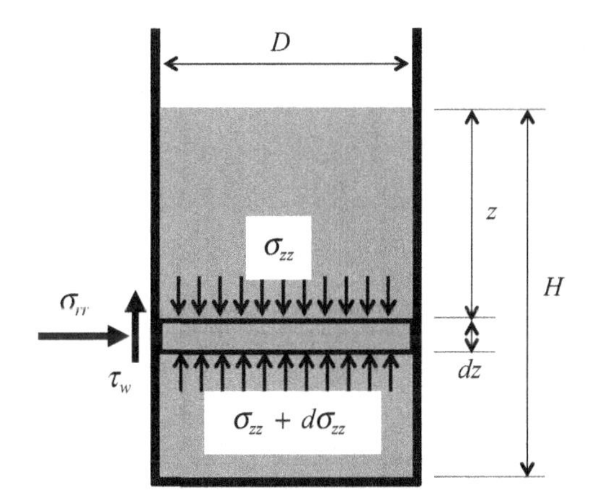

FIGURE 7.10

Force balance on an infinitesimal slice at depth z below the surface of a stored bulk solid (the total height of the bulk solid in the container is H).

3. the ratio of vertical stress to horizontal (radial) stress is a constant;
4. the container cross-section is circular.

Using these assumptions, a force balance on an infinitesimal elemental slice with a thickness dz at depth z is performed: the downward force on the element is $A\sigma_{zz}$, which is opposed by an upward force $A(\sigma_{zz} + \mathrm{d}\sigma_{zz})$ and an upward shear stress at the wall $\pi D \mathrm{d}\tau_\mathrm{w}$. The weight of the elemental slice is $A\rho_\mathrm{b} g \mathrm{d}z$, where ρ_b is the bulk density (assumed constant throughout) and g is the gravitational acceleration.

The force balance in the vertical direction is then given as:

$$\frac{\pi D^2}{4}\sigma_{zz} + \frac{\pi D^2}{4}\rho_b g dz = \frac{\pi D^2}{4}(\sigma_{zz} + d\sigma_{zz}) + \pi D dz \tau_w \tag{7.4}$$

Re-arranging Eq. (7.4), we have

$$\frac{d\sigma_{zz}}{dz} + \frac{4\tau_w}{D} = \rho_b g \tag{7.5}$$

Based on Janssen's assumption that the radial stress acting normal to the wall, σ_{rr}, is proportional to the vertical stress σ_{zz}, we have

$$K = \frac{\sigma_{rr}}{\sigma_{zz}} \tag{7.6}$$

where K is the stress transmission ratio and is also known as the "Janssen constant."

Since the bulk solid is assumed to be cohesionless, using Eq. (7.2), the wall shear stress, τ_w, can be related to the normal stress on the wall σ_{rr} by

$$\tau_w = \mu_w \sigma_{rr} \tag{7.7}$$

Finally combining Eqs (7.5)–(7.7):

$$\frac{d\sigma_{zz}}{dz} + \frac{4\mu_w K}{D}\sigma_{zz} = \rho_b g \tag{7.8}$$

Equation (7.8) is a first-order differential equation of standard form, with the solution:

$$\sigma_{zz} = \frac{\rho_b g D}{4\mu_w K} + \overline{A}\exp\left(-\frac{4\mu_w K z}{D}\right) \tag{7.9}$$

where $\overline{A}$ is an arbitrary constant that can be determined using the boundary condition: $\sigma_{zz} = 0$ on $z = 0$, as the top surface of the powder bed is stress-free. Hence the solution of Eq. (7.8) is given as

$$\sigma_{zz} = \frac{\rho_b g D}{4\mu_w K}\left[1 - \exp\left(-\frac{4\mu_w K z}{D}\right)\right] \tag{7.10}$$

and

$$\sigma_{rr} = \frac{\rho_b g D}{4\mu_w}\left[1 - \exp\left(-\frac{4\mu_w K z}{D}\right)\right] \tag{7.11}$$

$$\tau_w = \frac{\rho_b g D}{4}\left[1 - \exp\left(-\frac{4\mu_w K z}{D}\right)\right] \tag{7.12}$$

Figure 7.11 shows the resulting variation of σ_{zz} with depth z. It is clear that σ_{zz} increases sharply with depth until a plateau (i.e., a limiting stress) is reached; this is known as the "great depth" limit (although, as seen here, the depth at which it is effectively reached may be a small number of diameters). At great depths, i.e., $z \to \infty$, the limiting stresses can be determined from Eqs (7.10)–(7.12) as

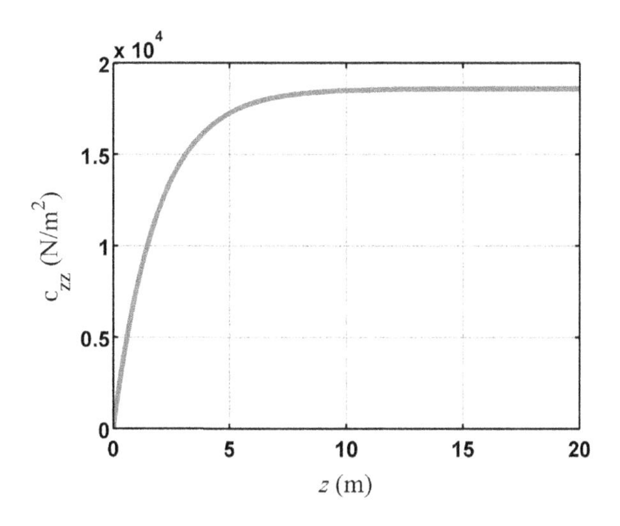

FIGURE 7.11

The variation of vertical stress σ_{zz} with depth z in a bulk solid ($\rho_b = 1000\ \text{kg/m}^3$; $\mu_w = 0.4$; $K = 0.33$) stored in a cylindrical container with $D = 1$ m.

$$\sigma_{zz} = \frac{\rho_b g D}{4\mu_w K} \tag{7.13}$$

$$\sigma_{rr} = \frac{\rho_b g D}{4\mu_w} \tag{7.14}$$

$$\tau_w = \frac{\rho_b g D}{4} \tag{7.15}$$

It may be noted that these limiting stresses can also be determined directly from the differential equation (Eq. (7.8)) by setting $\frac{d\sigma_{zz}}{dz} = 0$. In industrial applications and design, it is of interest to estimate the load supported by the base of the container. This is discussed in detail in Box 7.1.

Much more sophisticated analysis methods than Janssen's are now available, but the general result shown here remains true: for even moderately deep containers much of the weight is supported by friction at the wall and the internal stresses are therefore less than hydrostatic.

7.3 STRESS ANALYSIS FOR COMPRESSION OF A POWDER BED

In a number of industries, such as pharmaceutical, agrochemical, fine chemicals, food, automotive, and powder metallurgy, powders are compressed under high pressures to form coherent powder compacts. Knowing the stresses in the bulk solid under compression is of importance in process optimization and product quality

BOX 7.1 CALCULATION OF THE LOAD SUPPORTED BY THE BASE OF A CONTAINER

Taking the total height of the bulk solid in the container as H, as shown in Fig. 7.10, the force acting on the container base, F, can be determined from Eq. (7.10); i.e., setting $z = H$ gives

$$F = A\,\sigma_{zz}\big|_{z=H} = \frac{\pi D^2}{4}\left\{\frac{\rho_b g D}{4\mu_w K}\left[1 - \exp\left(-\frac{4\mu_w KH}{D}\right)\right]\right\} \tag{7.16}$$

or

$$F = \frac{\pi \rho_b g D^3}{16\mu_w K}\left[1 - \exp\left(-\frac{4\mu_w KH}{D}\right)\right] \tag{7.17}$$

The total weight of the bulk solid in the container W is

$$W = \frac{\pi \rho_b g D^2 H}{4} \tag{7.18}$$

Using Eqs (7.17) and (7.18), the fraction of the weight supported by the base of the container can be obtained as

$$\frac{F}{W} = \frac{1}{4\mu_w K}\frac{D}{H}\left[1 - \exp\left(-4\mu_w K\frac{H}{D}\right)\right] \tag{7.19}$$

It is interesting to notice from Eq. (7.19) that the fraction of the weight of bulk solids supported by the container base is independent of the bulk density ρ_b. Figure 7.12 shows the typical variation of F/W with H/D. It can be seen that the value of F/W decreases sharply as the ratio H/D increases, indicating that the percentage of the weight of the bulk solid supported by the base reduces dramatically with increase in the height/diameter ratio. For this particular bulk solid considered in Fig. 7.12, less than half of its weight is supported by the base when $H/D > 3$. In other words, most of its weight is supported by friction at the wall.

The calculation of wall stresses (both compressive σ_{rr} and in shear τ_w) is of particular importance in the design of large (>500 ton) silos. The Janssen method is adopted by most design codes for

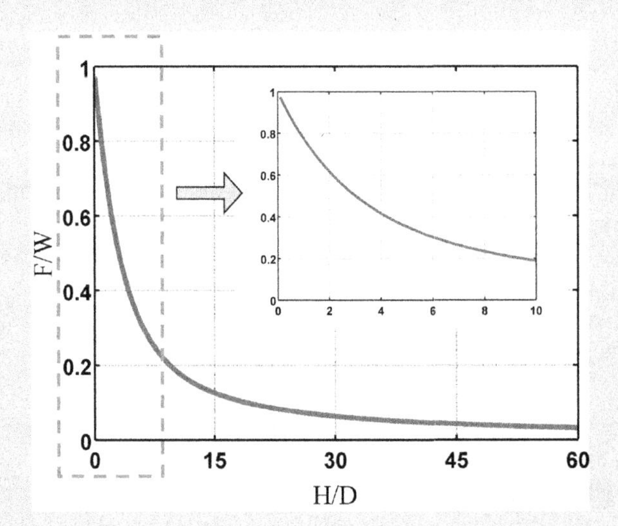

FIGURE 7.12

The variation of F/W with H/D in a bulk solid ($\mu_w = 0.4$; $K = 0.33$) stored in a cylindrical container.

BOX 7.1 CALCULATION OF THE LOAD SUPPORTED BY THE BASE OF A CONTAINER—cont'd

silos, with empirically derived corrections and safety factors. Improvements to the method by relaxing some of the assumptions are available, but at the cost of greater complexity. The principal uncertainty in the model is the value of K, which can be derived from the internal frictional angle ϕ (see Seville et al., 1997).

$$K = \frac{1 - sin\phi}{1 + sin\phi} \tag{7.20}$$

assurance. A stress analysis very similar to that of Janssen's analysis for bulk solids in a storage vessel (see Section 7.2) can be performed. As the powder mass used in compaction is generally very small, the total gravity force is negligible compared to the compression forces used. Moreover, the cohesive force is also very small compared with the compression force. It is hence reasonable to neglect the weight and the cohesion of the powder (i.e., it can be treated as cohesionless).

Let us consider the compaction of a powder in a cylindrical die of diameter D, as illustrated in Fig. 7.13. Ignoring the powder weight, a force balance for an arbitrary infinitesimal elemental slice dz at a distance z below the punch gives:

$$\frac{\pi D^2}{4}\sigma_{zz} = \frac{\pi D^2}{4}(\sigma_{zz} + \mathrm{d}\sigma_{zz}) + \tau_{\mathrm{w}}\pi D\mathrm{d}z \tag{7.21a}$$

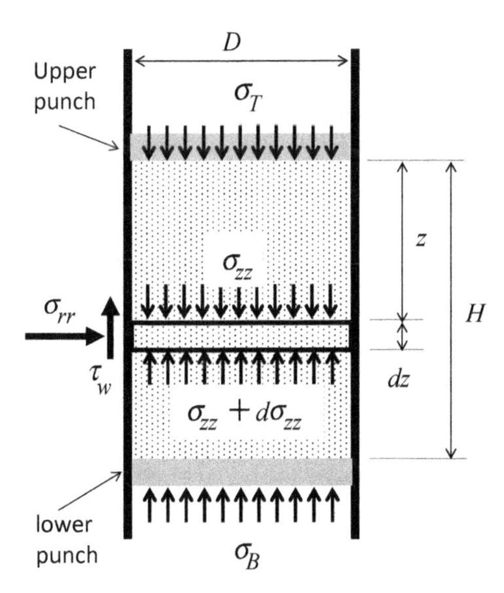

FIGURE 7.13

Force balance on an infinitesimal slice a distance z below the punch for a powder under compression.

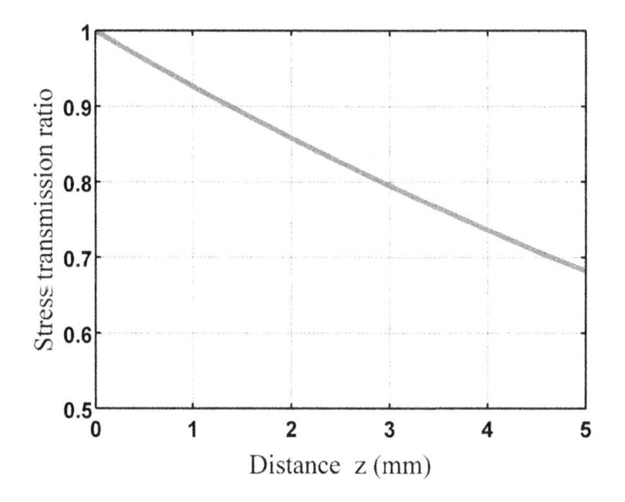

FIGURE 7.14

Stress transmission ratio as a function of distance below the punch, z ($D = 10$ mm; material properties: $\mu_w = 0.4$; $K = 0.33$).

Equation (7.21a) can be simplified to

$$\frac{d\sigma_{zz}}{dz} + \frac{4\tau_w}{D} = 0 \tag{7.21b}$$

Using Eqs (7.6) and (7.7) and substituting in terms of K and μ_w, we have

$$\frac{d\sigma_{zz}}{dz} + \frac{4\mu_w K}{D}\sigma_{zz} = 0 \tag{7.22}$$

Given that at $z = 0$, the applied stress (F/A) is σ_T, solving Eq. (7.22) gives:

$$\sigma_{zz} = \sigma_T\left[\exp\left(-\frac{4\mu_w K z}{D}\right)\right] \tag{7.23}$$

Figure 7.14 shows a typical result for the stress transmission ratio, σ_{zz}/σ_T, as a function of distance z below the top punch.

Equation (7.23) indicates that the stress decreases exponentially with distance from the punch. A practical consequence of this is that, when only a single punch is used to compress the powder, there is a limit to the thickness of a tablet that can be pressed so that the powder is able to form a coherent compact. This explains why double-ended compression is normally used in practice (i.e., powders are compressed from above and below by both upper and lower punches simultaneously) Using the solution given in Eq. (7.23), the stress state of the powder during compression can be well defined. Furthermore, the above analysis can be further extended to determine the die-wall friction, as shown in Box 7.2.

BOX 7.2 DETERMINATION OF THE STRESS STATE AND DIE-WALL FRICTION DURING COMPACTION

Eq. (7.23) can be rewritten as

$$\ln\left(\frac{\sigma_{zz}}{\sigma_T}\right) = -\frac{4\mu_w Kz}{D} \tag{7.24}$$

When $z = H$, $\sigma_{zz} = \sigma_B$. We have

$$\ln\left(\frac{\sigma_B}{\sigma_T}\right) = -\frac{4\mu_w KH}{D} \tag{7.25}$$

From Eqs (7.24) and (7.25), the stress σ_{zz} can be expressed in terms of the axial stress at the top and bottom surfaces of the powder bed as follows:

$$\sigma_{zz} = \sigma_T\left(\frac{\sigma_B}{\sigma_T}\right)^{\frac{z}{H}} \tag{7.26}$$

Equation (7.26) can be used to estimate the axial stress in the powder bed as a function of z; substituting Eq. (7.26) into Eqs (7.6) and (7.7), the radial and shear stresses as a distance z from the top surface can also be determined, i.e.,

$$\sigma_{rr} = K\sigma_T\left(\frac{\sigma_B}{\sigma_T}\right)^{\frac{z}{H}} \tag{7.27}$$

$$\tau_w = K\mu_w\sigma_T\left(\frac{\sigma_B}{\sigma_T}\right)^{\frac{z}{H}} \tag{7.28}$$

Using Eqs (7.6), (7.25), and (7.26), the wall friction coefficient μ_w can be determined as:

$$\mu_w = \frac{D}{4H}\frac{\sigma_T}{\sigma_{rr}}\left(\frac{\sigma_B}{\sigma_T}\right)^{\frac{z}{H}}\ln\left(\frac{\sigma_T}{\sigma_B}\right) \tag{7.29}$$

Equation (7.29) indicates that the wall friction coefficient μ_w can be calculated using the measurements of the radial stress σ_{rr}, the axial stresses applied by the upper and lower punches, σ_T and σ_B. This offers a robust method to determine wall friction with special applications to die compaction and roll compaction, as it derives the wall friction using the data collected from the actual process conditions. The method is therefore representative of the interaction behavior of the powder bed during compaction and the stress level which the powder experiences during the process (Sinka et al., 2003; Wu et al., 2005). A list of die-wall coefficients for compression of various powders in a steel die obtained using this method is given in Table 7.1.

Die-wall friction plays an important role in determining the powder behavior during compaction. It is the primary cause of nonuniform density distributions in tablets, as pointed out by Train (1957). His pioneering work on the typical density distribution of tablets, as illustrated in Fig. 7.15, showed that there is a dense core and upper annulus (next to the upper punch), but a low-density lower annulus and a less dense region in the upper tablet. This has been confirmed repeatedly by finite element analysis (see also Fig. 10.20, which will be discussed in detail in Chapter 10).

Continued

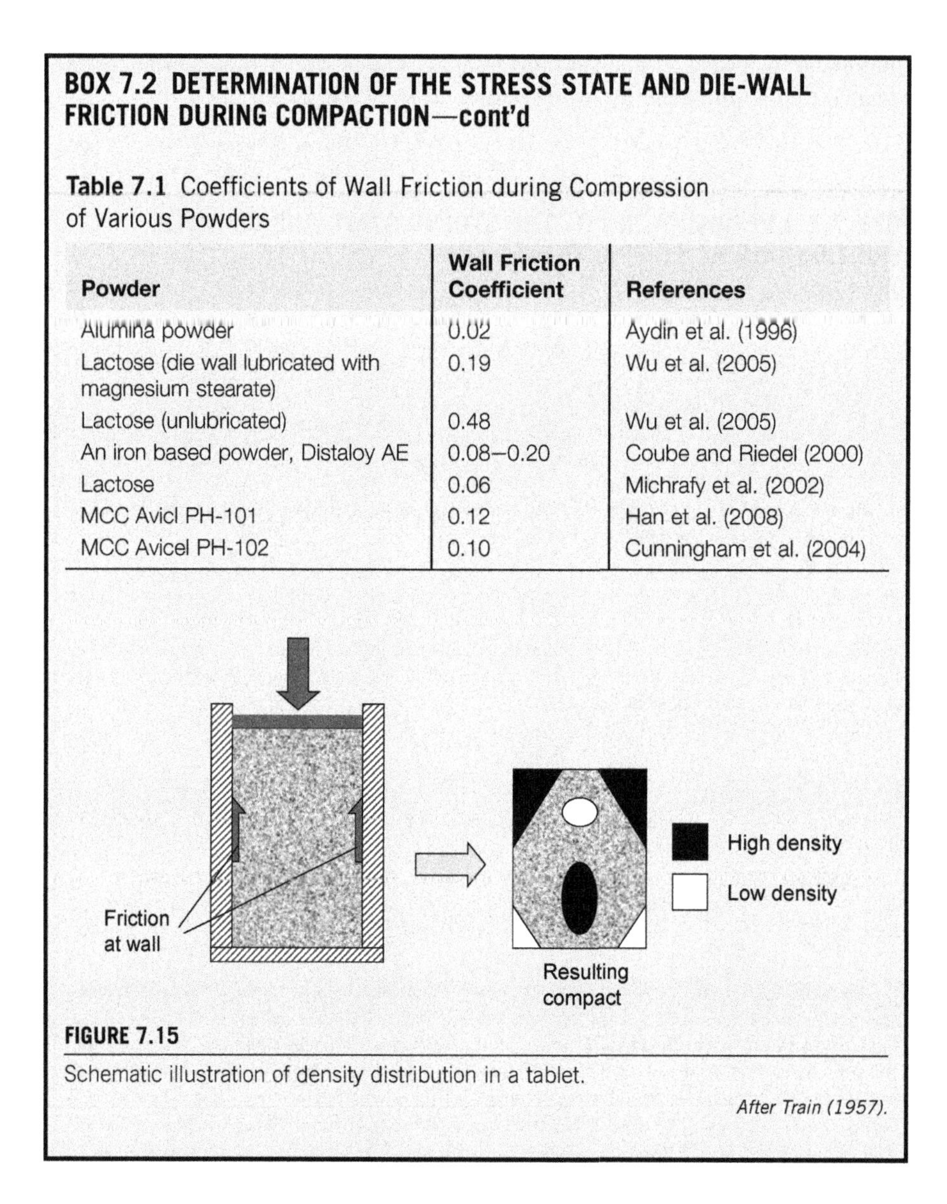

BOX 7.2 DETERMINATION OF THE STRESS STATE AND DIE-WALL FRICTION DURING COMPACTION—cont'd

Table 7.1 Coefficients of Wall Friction during Compression of Various Powders

Powder	Wall Friction Coefficient	References
Alumina powder	0.02	Aydin et al. (1996)
Lactose (die wall lubricated with magnesium stearate)	0.19	Wu et al. (2005)
Lactose (unlubricated)	0.48	Wu et al. (2005)
An iron based powder, Distaloy AE	0.08–0.20	Coube and Riedel (2000)
Lactose	0.06	Michrafy et al. (2002)
MCC Avicl PH-101	0.12	Han et al. (2008)
MCC Avicel PH-102	0.10	Cunningham et al. (2004)

FIGURE 7.15

Schematic illustration of density distribution in a tablet.

After Train (1957).

7.4 DISCHARGE DYNAMICS—HOPPER FLOW

7.4.1 FLOW PATTERNS

Many industrial processes involve discharge of bulk solids from storage hoppers, silos or bins. The flow pattern during discharge of bulk solids from these vessels depends on the vessel geometry, wall friction, and the material properties of the bulk solid, especially the internal friction and cohesion. Two distinct flow patterns have been widely observed during the discharge of bulk solids from these vessels: (1) *mass flow* and (2) *funnel flow*, also known as *core flow*, as illustrated in Fig. 7.16.

In mass flow, all bulk solids within the vessel are moving during discharge, even though particles may move at different speeds (Fig. 7.16(a)). No distinct flow zone is identifiable and the vessel walls are also the boundaries of the entire mass flow region. The materials fed into the vessel first will also be discharged first, i.e., "first in, first out." All bulk solids stored in the vessel can be completely discharged. Mass flow is generally observed when discharging free-flowing bulk solids from steep, smooth, conical, and wedge-shaped vessels.

In funnel flow, there are two distinct regions in the vessel: a central flowing region and stagnant regions near the walls, as shown in Fig. 7.16(b). Only particles near the center of the vessel or above the discharge orifice (i.e., the flowing region) are moving during discharge, while the particles close to the vessel walls are stationary, i.e., they are in the stagnant region. The size of the stagnant zone is determined by the physical properties of the bulk solids, such as the particle size, size distribution and cohesion, and the vessel design, especially the wall angle and the nature of the surface. The wall angle is also called the hopper angle, i.e., the angle of the walls with respect to the horizontal. The stagnant zone generally decreases, even disappears, if the hopper angle is increased sufficiently. The stagnant zones can reach the top surface of the powder bed. As the bulk solid is discharged through the central flowing region, a significant depression at the top surface is induced (Fig. 7.16(b)). Once the slope of the free surface exceeds a certain angle that is related to the angle of repose, the material initially in the stagnant zone cascades toward the center. The material in the stagnant zone may never be discharged from the vessel or can

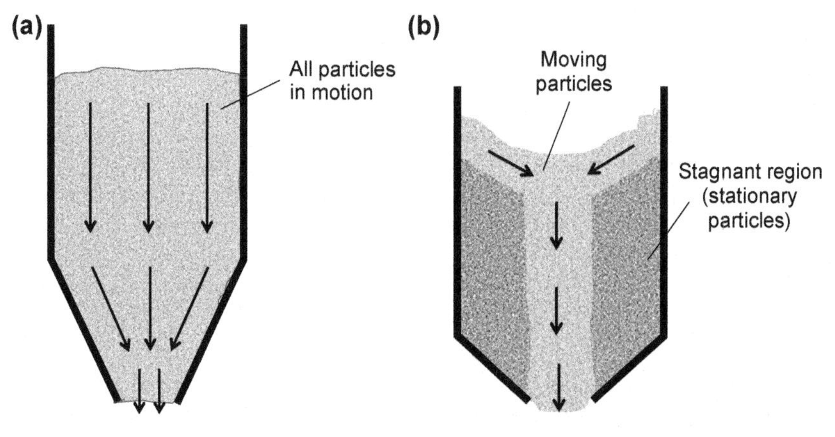

FIGURE 7.16

Illustration of different flow patterns during hopper flow: (a) mass flow and (b) funnel flow.

only be discharged by cascading into the central flowing region once a free surface is formed at the top of the material bed. This implies that in the funnel flow case the material fed into the vessel first may be discharged last, i.e., "first in last out." The material in the flowing region dilates (expands) slightly during discharge while densification takes place in the stagnant zone. In funnel flow, the material often discharges erratically in an uncontrollable manner, and flooding problems (i.e., rapid and sudden discharge) frequently occur, especially for fine and low-permeability powders. Because of the formation of stagnant zones, the vessel walls are to some degree protected from wear which would otherwise occur due to the moving bulk solids. Hence, for abrasive bulk solids, funnel flow may be the preferable flow pattern during discharge. Table 7.2 summarizes the differences between mass flow and funnel flow.

Generally, silos, bins, and hoppers should be designed to achieve mass flow. However, this is not always possible due to the complex frictional nature of bulk solids. Moreover, in industrial practice, the same vessel is often used to store and discharge various bulk solids with different physical properties. It is possible that different flow patterns are developed when the same vessel is used to handle different bulk solids. That is to say that funnel flow may dominate in a vessel initially designed for mass flow with a different material.

The stagnant zone in funnel flow may be small for bulk solids that are not very cohesive; hence the volume of material trapped in the stagnant zone can be small. The detrimental consequences can be easily compensated for by potential benefits, such as the reduced wall stresses and wear, less particle attrition, and smaller

Table 7.2 Comparison of Mass Flow and Funnel Flow of Bulk Solids (Seville et al., 1997)

	Mass Flow	**Funnel Flow**
Characteristics	No stagnant zones Flow through full cross-section of vessel First in, first out High particle velocities close to vessel walls	Formation of stagnant zones Flow occurs in the central region of the vessel First in, last out Very low particle velocities close to vessel walls
Advantages	Minimizing segregation and agglomeration of materials Complete discharge can be achieved	Low stresses on vessel walls during flow Reduced wall wear and particle attrition
Disadvantages	Large stresses on vessel walls during flow Attrition of particles Wear and erosion of vessel walls Small storage volume to vessel height ratio	High tendency to segregation and agglomeration during flow Discharge rates are less predictable as the flow region boundary changes with time

vessel height to hold the same volume. However, for cohesive bulk solids, funnel flow will lead to overconsolidation of material into rigid blocks in the stagnant zones, so-called "build-up". This will in turn reduce the central flow zone and can eventually result in termination of the discharge as a result of the blockage of the discharge orifice by the overconsolidated material.

7.4.2 MASS FLOW RATE

How fast a bulk solid can flow out of a storage vessel is of practical importance in many industrial processes and is normally characterized as a mass flow rate (kg/s).

When a bulk solid is discharged under gravity from a vessel, the top and bottom of the vessel (at the discharge point) are generally exposed to atmosphere. The resulting flow is referred to as gravity flow. The mass flow rate for gravity flow of bulk materials through a circular orifice can be described using the Beverloo equation (Seville et al., 1997):

$$M = C\rho_b g^{1/2}(D_O - kd)^{5/2} = C\rho_b g^{1/2} D_O^{5/2}\left(1 - k\frac{d}{D_O}\right)^{5/2} \tag{7.30}$$

where ρ_b is the bulk density of loosely packed bulk solids (kg/m^3), g is the acceleration due to gravity (m/s^2), D_O is the orifice diameter (m), d is the particle diameter (m), C is a dimensionless constant with a value typically in the range 0.57−0.60, and k is a proportional constant primarily related to particle shape: k is about 1.5 for spherical particles but with slightly larger values for angular particles (Seville et al., 1997). For angular particles, the value of k also depends on how the size of the particles is defined (see Chapter 3).

During discharge through an orifice, particles close to the edge of the orifice are inhibited from moving freely, and all flowing particles are channeled toward the center of the orifice., It is therefore considered that there is an "empty annulus," i.e., a region of retarded flow adjacent to the orifice edge (see Fig. 7.17). Consequently, the *effective* discharge area is considered to be smaller than the actual area of the orifice. The size of the empty annulus depends on particle shape and the ratio of particle size to the orifice size. The term $[1 - k(d/D_O)]$ in Eq. (7.30) can be regarded as a correction factor for the effective discharge area.

Using Eq. (7.30), if $M^{2/5}$ is plotted against D_O, as illustrated in Fig. 7.18, a straight line can be obtained. The intercept of this straight line with the horizontal axis gives the value of kd. Knowing the particle size d, the value of k can hence be determined.

It should be noted that Eq. (7.30) is applicable primarily for bulk solids of negligible cohesion and for discharge without an interstitial pressure gradient, i.e., it is applicable for discharges of particles with an average size larger than 0.5 mm, and for $D_O \geq 6d$. For smaller particles (<0.5 mm), cohesion due to van der Waals'

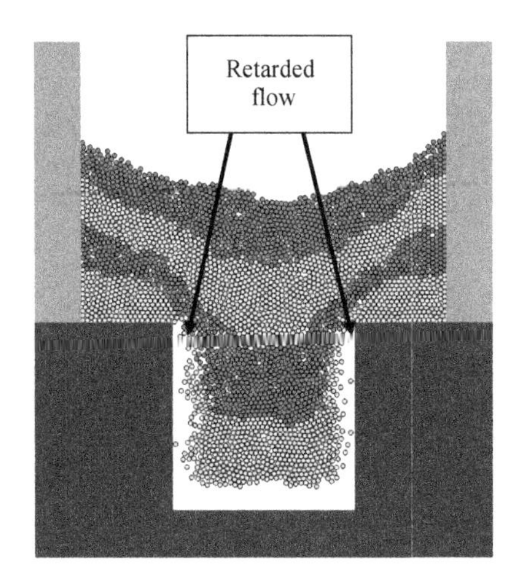

FIGURE 7.17

Formation of an empty annulus during discharge, obtained from discrete element method simulations (Guo et al., 2009).

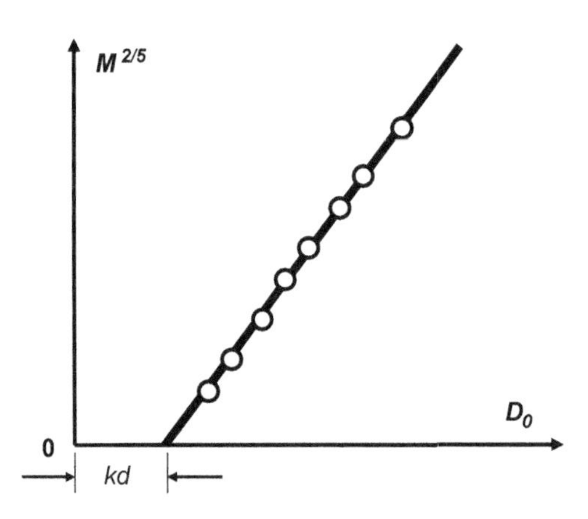

FIGURE 7.18

Graphical representation of the Beverloo equation.

forces, liquid bridges, etc. becomes more significant. The use of Eq. (7.30) for cohesive powders needs caution, as it may significantly overestimate the mass flow rate.

In order to obtain mechanistic understanding of the discharge of bulk solids from storage vessels, some analytical solutions for the mass flow rate are developed based on the concept of the "free-fall arch," as explained in Box. 7.3.

BOX 7.3 THE CONCEPT OF "FREE-FALL ARCH"

The "free-fall arch" theory assumes that, in the central flowing region, the particles at the bottom of the flowing stream detach from the powder mass and fall freely under gravity. This zone is the so called free-fall zone (Tüzün et al., 1982). Velocities of the free-falling particles are higher than those in the fast moving dilated region. The velocity difference implies that above the free-fall zone, an arch is developed, which resists the flow of material in the central moving region (Wu et al., 2003). This is the so-called free-fall arch, as illustrated in Fig. 7.19.

As a first approximation, the height of the free-fall arch is proportional to the diameter D_0 of the circular orifice. After particles detach from the arch, they are accelerated freely under gravity and reach a velocity of the order of $(gD_0)^{1/2}$ on passing through the orifice with a cross-sectional area, $\pi D_0^2/4$. The flow rate is therefore proportional to $g^{1/2}D_0^{5/2}$ (Guo et al., 2009), which is consistent with the prediction shown in Eq. (7.30).

The above analysis can also be used for discharge of bulk solids from storage vessels with a noncircular orifice. Let us consider a rectangular orifice of dimensions $b \times l$, where b is the width and l is the length of the orifice ($b < l$). The height of the free-fall arch is proportional to b (i.e., the smaller side), and the particle velocity passing through the orifice can be approximated as a function of $(gb)^{1/2}$. Since the orifice area is bl, the mass flow rate is proportional to $g^{1/2}lb^{3/2}$ if the empty annulus can be ignored. Now including the effect of the empty annulus, Seville et al. (1997) showed that the mass flow rate can be given as

$$M = 1.03\rho_b g^{1/2}(l - kd)(b - kd)^{3/2} \tag{7.31}$$

Brown and Richards (1965) analyzed the mass flow rate during discharge with a spherical free-fall arch. A mass flow rate equation for the discharge from a cylindrical bin with a circular orifice was obtained as follows (Seville et al., 1997)

$$M = \frac{\sqrt{2\pi}}{6}\rho_b g^{1/2} D_0^{5/2}\left(1 - \cos^{3/2}\beta\right)\Big/\sin^{5/2}\beta \tag{7.32}$$

where β is the half-angle of the flowing zone (see Fig. 7.20). Equation (7.32) indicates that the mass flow rate is proportional to $g^{1/2}D_0^{5/2}$, which is consistent with the form of the Beverloo equation (Eq. (7.30)).

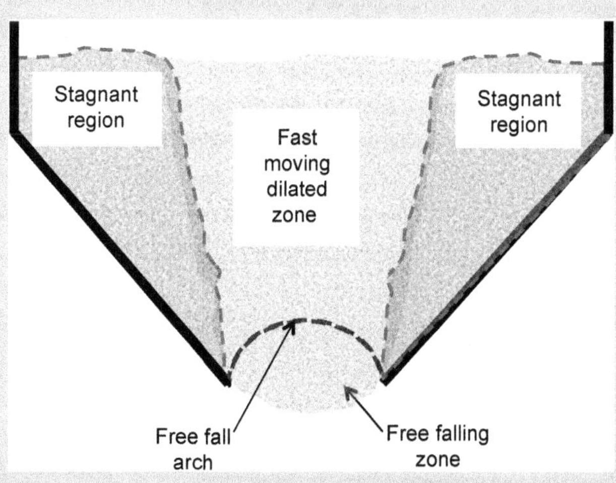

FIGURE 7.19

Illustration of the free-fall arch.

Continued

BOX 7.3 THE CONCEPT OF "FREE-FALL ARCH"—cont'd

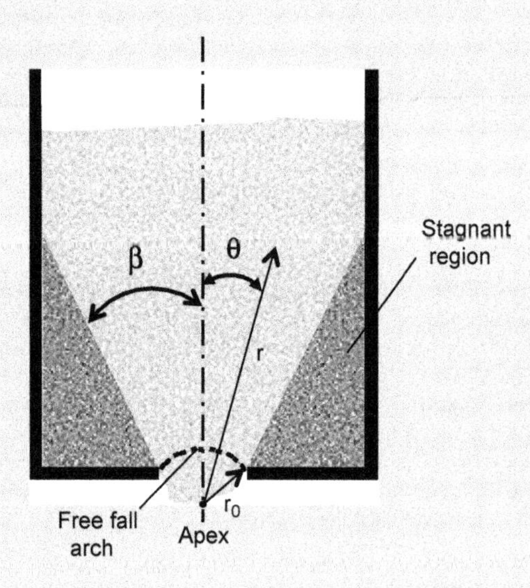

FIGURE 7.20

Schematic diagram of "free-fall arch" and "virtual apex" at the orifice of a cylindrical bin (Seville et al., 1997).

Using the same approach, the mass flow rate for discharge through a rectangular orifice ($b \times l$ with $b < l$) can be obtained. The radial velocity in plane strain can be given as (Seville et al., 1997; Guo et al., 2009)

$$v_r = \frac{g^{1/2} r_0^{3/2} \cos^{1/2}\theta}{r} \tag{7.33}$$

where r_0 is the radius of the free-fall arch (Fig. 7.20), r and θ are the radial and angular coordinates of a given point in the flowing zone. Thus, the mass flow rate can be obtained as

$$M = 2\rho_b \int_0^\beta v_r l\, r\, d\theta = 2\rho_b g^{1/2} l\, r_0^{3/2} \int_0^\beta \cos^{1/2}\theta\, d\theta \tag{7.34}$$

Since the width of the orifice $b = 2r_0 \sin\beta$, Eq. (7.34) can be rewritten as (Guo et al., 2009)

$$M = \frac{\sqrt{2}}{2} \rho_b g^{1/2} l\, b^{3/2} \frac{\int_0^\beta \cos^{1/2}\theta\, d\theta}{\sin^{3/2}\beta} \tag{7.35}$$

Davidson and Nedderman (1973) derived a similar equation for the mass flow rate for discharge from a chisel-shaped hopper with a rectangular orifice ($b \times l$ with $b < l$):

$$M = \frac{\rho_b g^{1/2} b^{3/2} l}{sin^{1/2}\alpha} \left[\frac{1 + K_p}{2(K_p - 2)} \right]^{1/2} \tag{7.36}$$

where α is the half-angle, K_p ($=1/K$) is a function of the angle of internal friction ϕ and has a value in the range of 2.5−7.5 for commonly encountered materials (Seville et al., 1997).

7.4.3 DISCHARGE WITH INTERSTITIAL PRESSURE GRADIENTS

The analysis presented in Section 7.4.2 focused on the mass flow rate for discharge from a storage vessel with negligible interstitial pressure gradient. However, in many applications, an interstitial pressure gradient is induced or present due to the intrinsic characteristics of the powder flow and the system setup. For example, interstitial pressure gradients can be induced by the expansion of the self-entrained gas in the voids of a powder mass as the powder mass flows and dilates (Nedderman, 1992; Seville et al., 1997; Crewdson et al., 1977; Wu et al., 2003). Pressure gradients can also be induced if the powder mass discharges into an airtight container (e.g., in die filling) or from a storage vessel with a closed top. For both cases, a negative pressure gradient is induced (i.e., the pressure below the orifice is higher than the interstitial pressure, so the pressure gradient is negative along the flowing solids streamline and opposes the solids flow) and acts on the particles passing through the orifice, which can significantly inhibit flow of the powder mass.

For discharge with an induced interstitial pressure gradient, Crewdson et al. (1977) proposed a modified Beverloo equation for the mass flow rate by considering that the flowing powder stream is subject to an extra body force induced by the pressure gradient at the orifice:

$$\widehat{M} = M\left(1 + \frac{1}{g\rho_b}\left.\frac{dp}{dr}\right|_{r_0}\right)^{1/2} \tag{7.37}$$

where M is the mass flow rate given by Eq. (7.30), $(dp/dr)|_{r_0}$ is the pressure gradient at the orifice (i.e., $r = r_0$) and can be approximated as (Seville et al., 1997):

$$\left.\frac{dp}{dr}\right|_{r_0} = \frac{\Delta P}{r_0} f[\mathrm{Re}] \tag{7.38}$$

where ΔP is the overall pressure difference, r_0 is the radius from the virtual apex to the orifice, $f[\mathrm{Re}] \to 1$ when the particle Reynolds number is low and $f[\mathrm{Re}] \to 3$ as $\mathrm{Re} \to \infty$. Equation (7.37) can hence be rewritten as

$$\widehat{M} = M\left(1 + \frac{C'\Delta P}{\rho_b g r_0}\right)^{1/2} \tag{7.39}$$

where C' is related to the Reynolds number and has a value of 1–3.

The magnitude of the pressure gradient at the orifice $(dp/dr)|_{r_0}$ induced by dilation during discharge of fine powders depends on particle size. Verghese and Nedderman (1995) showed that the magnitude of this negative pressure gradient increases as the particle size decreases; consequently the mass flow rate decreases. Based upon experimental measurements using sand particles of 150–2500 μm, the following empirical equation was obtained:

$$\widehat{M} = M\left(1 + \frac{\lambda}{d^2}\right)^{1/2} \tag{7.40}$$

where $\lambda = -1.46 \times 10^{-8}\ m^2$ for the particular experiments conducted and may depend upon the experimental setup and other physical properties of the materials.

Just as an unfavorable pressure gradient can induce counter-current gas flow which opposes discharge, an increase in the mass flow rate can be achieved by properly controlling the pressure gradient. A positive pressure gradient (i.e., favoring the solids flow) can be imposed by deliberately supplying air into the storage vessel, as indicated in Eqs (7.37) and (7.39). This is so-called pressurized flow.

Two kinds of pressurized flow can be realized: positive and negative (Jing and Li, 1999). Positive pressure flow takes place when the interstitial pressure is higher than atmospheric, as the powder in the hopper is aerated. For instance, Ferrari and Poletto (2002) observed a 10-fold increase in the mass flow rate of fine powders in the size range of 60–150 μm when they were fluidized (see Chapter 5).

Negative pressure flow occurs when the pressure below the orifice is higher than the interstitial pressure in the flowing powder stream. This can either prevent the formation of arches above the orifice or lead to the development of an arch further up in the hopper, creating a larger effective discharge area in steep angled hoppers where mass flow dominates (Wu et al., 2003). Hence a higher mass flow rate can be achieved compared to standard gravity flow (Jing and Li, 1999). However, a negative pressure always reduces the mass flow rate in shallow hoppers where funnel flow dominates, as predicted by Eqs (7.37) and (7.39). A negative pressure flow can also be developed when the top of the discharging vessel is sealed. As the powder flows out, gas must flow into the vessel to fill the space thereby created. As a result, a negative pressure gradient is induced at the orifice, which can significantly reduce the flow rate, as illustrated in Fig. 7.21.

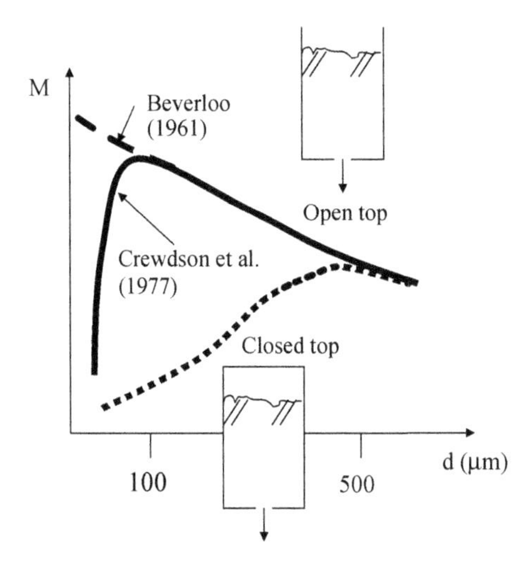

FIGURE 7.21

Effect of interstitial pressure and fine particle size on the flow rate (Seville et al., 1997).

Unfavorable pressure gradients also occur when discharging a bulk solid into a closed and confined space, for example, packing bulk materials into a container and depositing powders into a die before tableting (i.e., die filling). In these situations, as the discharge proceeds, the void volume in the receiving container or the die decreases, leading to a rapid build-up of air pressure that can be much higher than the interstitial pressure in the flowing bulk solids. This effect can significantly reduce the flow rate, especially for fine and light particles, as demonstrated by Guo et al. (2009), who performed a systematic numerical simulation of packing monosized particles into a die in air. For large and heavy particles, the effect of negative pressure is insignificant and the flow rate is essentially identical to that in vacuum. For fine and light particles, the effect becomes more significant as the particle size and density decrease (see Fig. 7.22). Guo et al. (2009) hence suggested that bulk solids can be classified into two groups: (1) air sensitive and (2) air inert. The demarcation between these two groups is defined using a dimensionless parameter:

$$\varsigma = \mathrm{Ar} \cdot \Phi_p \tag{7.41}$$

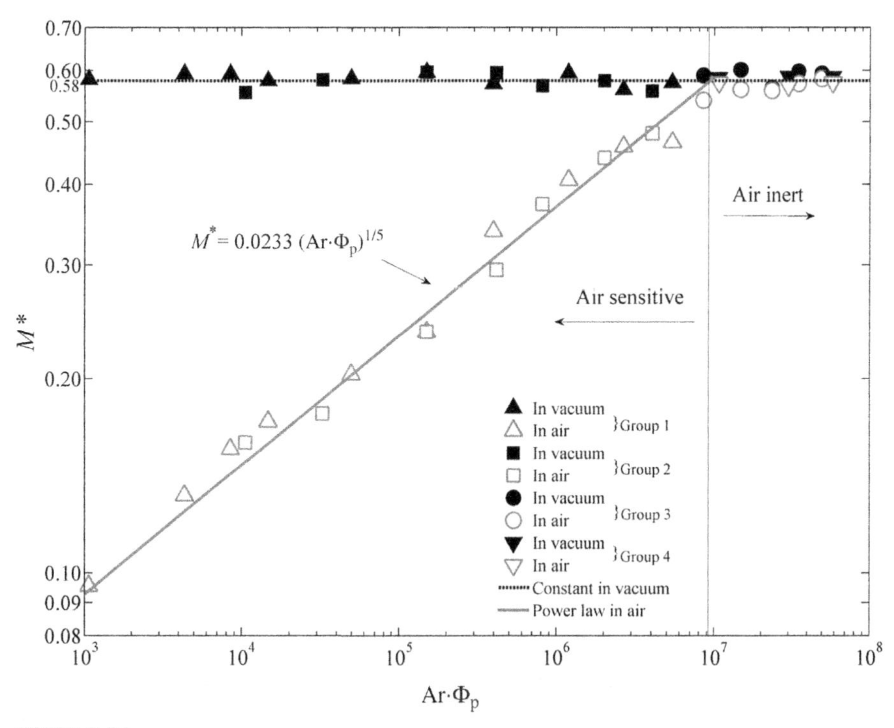

FIGURE 7.22

Dimensionless mass flow rate M^* versus $\mathrm{Ar} \cdot \Phi_p$ obtained from discrete element method analysis of flow of monosized particles into a die; different symbols represent materials with different particle density and size (Guo et al., 2009).

where Ar is the Archimedes number defined as

$$\mathrm{Ar} = \frac{\rho(\rho_p - \rho)gd^3}{\eta^2} \tag{7.42}$$

Φ_p is the ratio of particle density ρ_p to air density ρ:

$$\Phi_p = \rho_s/\rho_a \tag{7.43}$$

and η is the air viscosity. For the cases considered by Guo et al. (2009), the two distinct regimes (i.e., air sensitive and air inert) can also be classified with a critical value of $\varsigma = \mathrm{Ar}\cdot\Phi_p = 9.56 \times 10^6$. The flow rate in the air-inert regime can be predicted using the equations for gravity flow (i.e., Eq. (7.30)), while the flow rate in the air-sensitive regime can be given as

$$M^* = M_0(\mathrm{Ar}\cdot\Phi_p)^{1/5} \tag{7.44}$$

where M^* is the dimensionless mass flow rate and is equivalent to the Beverloo constant C in Eq. (7.30), i.e.,

$$M^* = \frac{M}{\rho_b g^{1/2} l b^{3/2}\left(1 - k\frac{d_p}{b}\right)^{3/2}} \tag{7.45}$$

for discharge through a rectangular orifice. M_0 is a proportional constant related to particle shape, size distribution, and cohesion and has a value of 0.0233 for the cases considered by Guo et al. (2009).

REFERENCES

Aydin, I., Briscoe, B.J., Sanlitürk, K.Y., 1996. The internal form of compacted ceramic components: a comparison of a finite element modelling with experiment. Powder Technology 89, 239–254.

Beverloo, W.A., Leniger, H.A., van de Velde, J., 1961. The flow of granular solids through orifices. Chemical Engineering Science 15, 260–269.

Brown, R.L., Richards, J.C., 1965. Kinematics of the flow of dry powders and bulk solids. Rheologica Acta 4, 153–165.

Coube, O., Riedel, H., 2000. Numerical simulation of metal powder die compaction with special consideration of cracking. Powder Metallurgy 43, 123–131.

Crewdson, B.J., Ormond, A.L., Nedderman, R.M., 1977. Air-impeded discharge of fine particles from a hopper. Powder Technology 16, 197–207.

Cunningham, J.C., Sinka, I.C., Zavaliangos, A., 2004. Analysis of tablet compaction. I. Characterization of mechanical behavior of powder and powder/tooling friction. Journal of Pharmaceutical Sciences 93 (8), 2022–2039.

Davidson, J.F., Nedderman, R.M., 1973. The hour-glass theory of hopper flow. Transactions of the Institution of Chemical Engineers 51, 29–35.

Ferrari, G., Poletto, M., 2002. The particle velocity field inside a two-dimensional aerated hopper. Powder Technology 123, 242–253.

Guo, Y., Kafui, K.D., Wu, C.Y., Thornton, C., Seville, J.P.K., 2009. A coupled DEM/CFD analysis of the effect of air on powder flow during die filling. AIChE Journal 55 (1), 49–62.

Han, L.H., Elliott, J.A., Bentham, A.C., Mills, A., Amidon, G.E., Hancock, B.C., May 15, 2008. A modified Drucker-Prager cap model for die compaction simulation of pharmaceutical powders. International Journal of Solids and Structures 45 (10), 3088–3106.

Jing, S., Li, H., 1999. Study on the flow of fine powders from hoppers connected to a moving-bed standpipe with negative pressure gradient. Powder Technology 101, 266–278.

Marston, J.O., Seville, J.P.K., Cheun, Y.-V., Ingram, A., Decent, S.P., Simmons, M.J.H., 2008. Effect of packing fraction on granular jetting from solid sphere entry into aerated and fluidized beds. Physics of Fluids 20, 023301.

Michrafy, A., Ringenbacher, D., Techoreloff, P., 2002. Modelling the compaction behaviour of powders: application to pharmaceutical powders. Powder Technology 127, 257–266.

Nedderman, R.M., 1992. Statics and Kinematics of Granular Materials. Cambridge University Press, New York.

Seville, J.P.K., Tüzün, U., Clift, R., 1997. Processing of Particulate Solids. Blackie Academic and Professional, London, pp. 330–348.

Sinka, I.C., Cunningham, J.C., Zavaliangos, A., 2003. The effect of wall friction in the compaction of pharmaceutical tablets with curved faces: a validation study of the Drucker-Prager cap model. Powder Technology 133, 33–43.

Train, D., 1957. Transactions of the Institution of Chemical Engineers 35, 258–266.

Tüzün, U., Houlsby, G.T., Nedderman, R.M., Savage, S.B., 1982. Flow of granular materials 2. Velocity distributions in slow flow. Chemical Engineering Science 37, 1691–1709.

Verghese, T.M., Nedderman, R.M., 1995. The discharge of fine sands from conical hoppers. Chemical Engineering Science 50, 3143–3153.

Wu, C.Y., Dihoru, L., Cocks, A.C.F., 2003. The flow of powder into simple and stepped dies. Powder Technology 134 (1–2), 24–39.

Wu, C.Y., Ruddy, O., Bentham, A.C., Hancock, B.C., Best, S.M., Elliott, J.A., 2005. Modelling the mechanical behaviour of pharmaceutical powders during compaction. Powder Technology 152 (1–3), 107–117.

CHAPTER

8 Particle–Particle Interaction

The mechanical behavior of bulk solids is determined by interactions between particles (i.e., particle–particle interaction) at the microscopic level. Therefore, understanding of particle–particle interactions is obviously important in any study of particle technology. In powder handling and processing, a particle can be in static contact with its neighboring particles (i.e., with negligible relative velocities), as in a packing, or particles can be moving at different velocities and colliding with each other as, for example, when blending in a mixer. When particles touch, a *contact* occurs; if the contact is at relative velocity it will be referred to here as an *impact*. In general, particles can make contact such that the direction of the contact force is along the contact normal (Fig. 8.1(a)), or the contact force direction can be oblique (Fig. 8.1(b)). For contacts in which the force direction is normal, the friction at the contact surface plays an insignificant role and can generally be ignored, while for oblique contacts, the interaction is significantly affected by the tangential force induced by friction.

Although real particles have irregular shapes and are rarely spherical, theoretical work on irregularly shaped particles is still scarce, so that only interactions between spherical particles are discussed here. Particle–particle interactions depend on the material properties of the particles, such as elasticity and plasticity, and on the interfacial properties, including friction, surface energy and the presence of liquids. Figure 8.2 shows the difference between the stress–strain relationships for elastic, and elastic-perfectly plastic materials. For elastic materials, the stress σ is proportional to the strain ε. For elastic-perfectly plastic materials, when the stress is smaller than the yield stress Y, only elastic deformation takes place in the materials. When the applied stress is beyond the yield stress, plastic deformation takes place and the strain increases even though the stress reaches the plateau; both of these cases are discussed in the present chapter. In addition, the effects of surface energy and the presence of liquids on particle–particle interactions are also discussed. We refer readers to Johnson (1985) for more detailed discussion on the interaction of visco-elastic and visco-plastic particles.

8.1 INTERACTION OF ELASTIC PARTICLES

8.1.1 NORMAL CONTACT OF ELASTIC SPHERES—HERTZ THEORY

The classical paper "on the contact of elastic solids" published by Heinrich Hertz in 1882 is the pioneering work on normal contact of elastic spheres, and his

Particle Technology and Engineering.

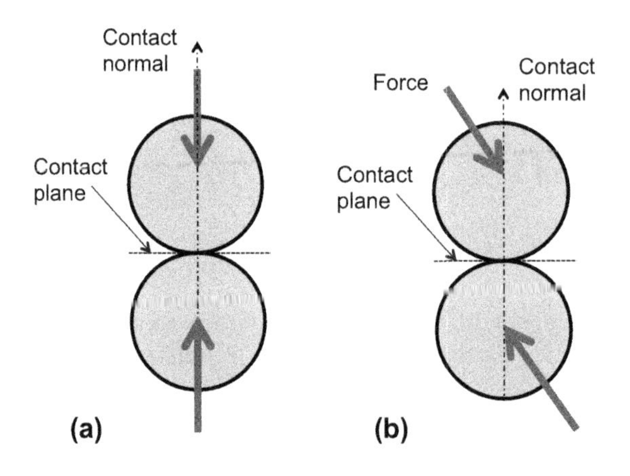

FIGURE 8.1

Illustration of (a) normal and (b) oblique interaction between two spheres.

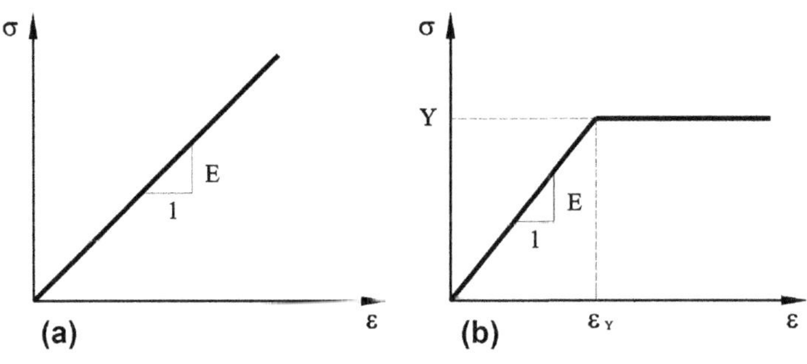

FIGURE 8.2

Stress–strain relationships for (a) elastic and (b) elastic-perfectly plastic materials, where E is the Young's modulus, Y is the yield stress and ε_y is the yield strain.

theoretical development of the equations of interaction for this case is usually referred to as the Hertz theory (Johnson, 1985). The theory describes a situation in which:

1. The contact area between the two spheres is circular.
2. The size of the contact area is very small compared to both the size of each contacting sphere and the radii of curvature of the contacting surfaces.
3. The contacting surfaces are frictionless, hence only a normal force is transmitted between the contacting spheres, i.e., neither tangential forces nor adhesive forces are considered.
4. The induced deformation (i.e. strain) is very small.
5. The contacting surfaces are continuous and can be described using quadratic formulae.

Hertz (1896) proposed that the contact pressure between two frictionless elastic solid spheres in contact can be given by

$$p_{(r)} = p_0 \left[1 - \left(\frac{r}{a}\right)^2\right]^{1/2} \tag{8.1}$$

where p_0 is the maximum pressure at the contact center; a is the contact radius; and r is the distance from the contact center. It is worth noting that the Hertz theory discussed here can also be applied to the contact between an elastic solid sphere and an elastic substrate or a plane. This pressure distribution over the contact area is presented in Fig. 8.3, in which the pressure distribution in a contact between an elastic sphere and an elastic substrate obtained from numerical analysis using the finite element method discussed in Section 10.1.2 is also superimposed.

The contact force F_n can then be calculated as

$$F_n = \int_0^a p_{(r)} 2\pi r \,\mathrm{d}r = \frac{2}{3}\pi a^2 p_0 \tag{8.2}$$

Hence

$$p_0 = \frac{3F_n}{2\pi a^2} \tag{8.3}$$

For an arbitrary point at a distance r from the contact center, the normal displacement w induced by the contact pressure is given as (Johnson, 1985)

$$w(r) = \frac{1 - \upsilon^2}{E} \frac{\pi p_0}{4a} \left(2a^2 - r^2\right) \tag{8.4}$$

where E and υ are the Young's modulus and Poisson's ratio of the material of the particle, respectively.

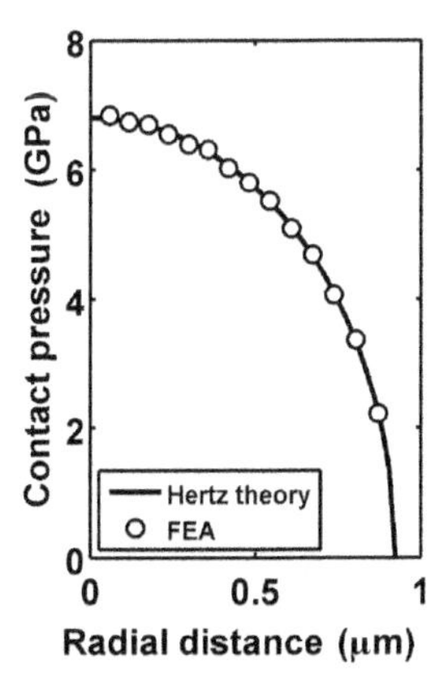

FIGURE 8.3

The pressure distribution for the contact between a 10-μm elastic sphere with an elastic substrate of the same material ($E = 208$ GPa, $\upsilon = 0.3$).

FEA data reproduced from Wu (2002).

For two contacting spheres the boundary condition for displacements within the contact area can be described as

$$w_1(r) + w_2(r) = \alpha - \frac{r^2}{2R^*} \tag{8.5}$$

where w_1 and w_2 are displacements of arbitrary points on the contact surfaces in the normal direction at a distance r from the contact center ($r \le a$), α is the relative approach (i.e., displacement) of the two sphere centers, and

$$\frac{1}{R^*} = \frac{1}{R_1} + \frac{1}{R_2} \tag{8.6}$$

where R^* is the effective radius, and R_1 and R_2 are the radii of the two spheres, respectively.

Using Eq. (8.4) and Eq. (8.5), we obtain

$$\frac{\pi p_0}{4aE^*}\left(2a^2 - r^2\right) = \alpha - \frac{r^2}{2R^*} \tag{8.7}$$

where E^* is the effective Young's modulus:

$$\frac{1}{E^*} = \frac{1 - \upsilon_1^2}{E_1} + \frac{1 - \upsilon_2^2}{E_2} \tag{8.8}$$

E_i and υ_i ($i = 1, 2$) denote Young's moduli and Poisson's ratios of the contacting spheres, respectively.

Equation (8.7) is applicable to any point within the contact area, i.e., independent of the value of r, so that

$$a = \frac{\pi p_0 R^*}{2E^*} \tag{8.9}$$

$$\alpha = \frac{\pi a p_0}{2E^*} \tag{8.10}$$

Equation (8.9) indicates that the maximum pressure at the contact center p_0 is proportional to the contact radius a, which has also been demonstrated by numerical analysis, as shown in Fig. 8.4.

The variation of the relative approach α with the contact radius a can be obtained using Eqs (8.9) and (8.10),

$$\alpha = \frac{a^2}{R^*} \tag{8.11}$$

From Eqs (8.2), (8.10) and (8.11), the contact force–displacement relationship can be obtained as follows

$$F_n = \frac{4E^*}{3R^*}a^3 = \frac{4}{3}E^* R^{*1/2} \alpha^{3/2} \tag{8.12}$$

It is clear that the contact force is proportional to $\alpha^{3/2}$ (see Fig. 8.5).

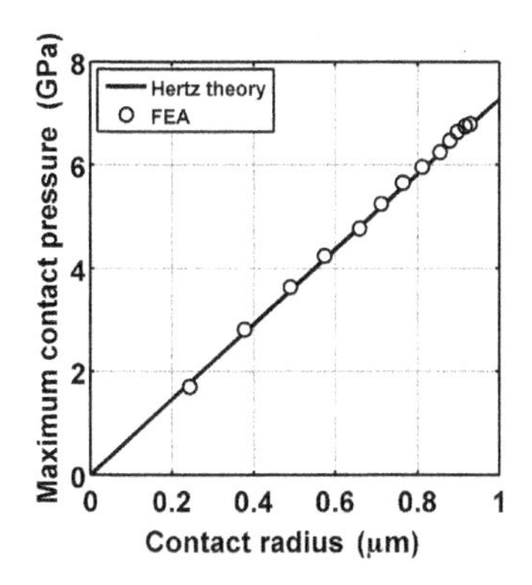

FIGURE 8.4

The variation of the maximum contact pressure p_0 with the contact radius a for elastic contact.

FEA data reproduced from Wu (2002).

8.1.2 NORMAL IMPACT OF ELASTIC SPHERES

For normal impact of frictionless elastic spheres, one question to be addressed is how long the two particles will be in contact during the impact, i.e., what is the duration of the impact? If the deformation induced is so small that it is restricted to the vicinity of the initial contact point and can be determined using the Hertz theory discussed in Section 8.1.1, and the effect of elastic wave motion in the contacting bodies is negligible, an approximate solution can be obtained. With these

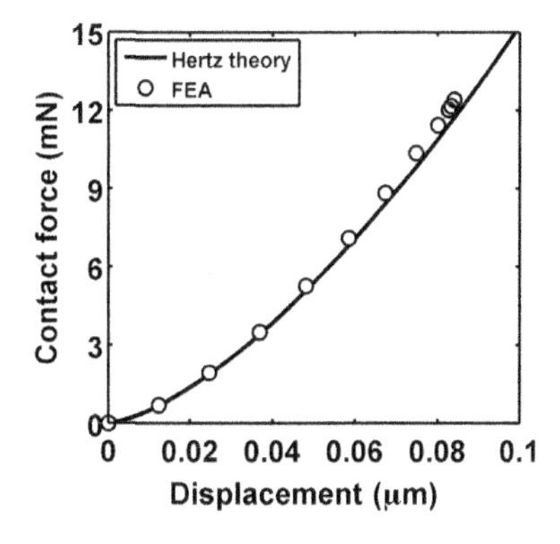

FIGURE 8.5

The force–displacement relationship for the impact of an elastic sphere with an elastic substrate at $V_{ni} = 5.0$ m/s.

FEA data reproduced from Wu (2002).

simplifications, the duration of impact is determined as (see Timoshenko and Goodier, 1951; Johnson, 1985)

$$t = 2.87\left(\frac{m^{*2}}{R^{*}E^{*2}v_{\text{ni}}}\right)^{1/5} \tag{8.13}$$

in which v_{ni} is the incident relative velocity of the two particles and

$$\frac{1}{m^{*}} = \frac{1}{m_1} + \frac{1}{m_2} \tag{8.14}$$

where m^* is the effective mass, and m_1 and m_2 denote the masses of the two contacting particles.

For the impact of an elastic sphere with an elastic planar surface, Eq. (8.13) reduces to

$$t = 5.09\left(\frac{\rho_{\text{s}}}{E^{*}}\right)^{2/5}\frac{R}{v_{\text{ni}}^{1/5}} \tag{8.15}$$

where ρ_{s} is the particle density (see Section 2.1.1). It is apparent that the duration of impact is proportional to the radius of the sphere and inversely proportional to $v_{\text{ni}}^{1/5}$, as shown in Fig. 8.6. This was confirmed by the experimental measurements of Andrews (1930) who investigated the impact of two equal sized metal spheres.

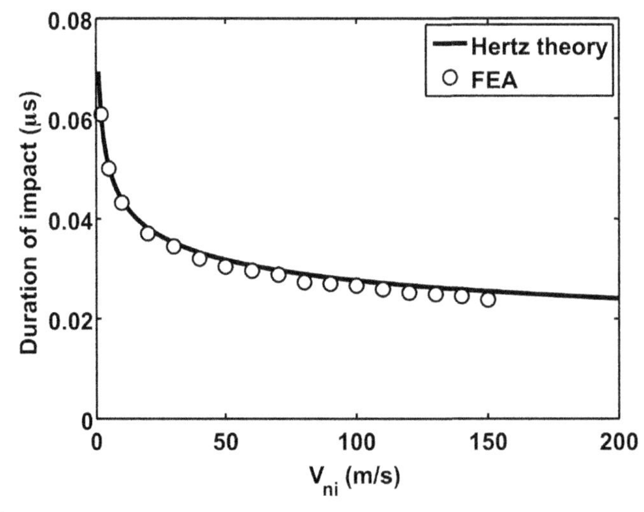

FIGURE 8.6

Duration of impact as a function of impact velocity.

FEA data reproduced from Wu (2002).

Generally, the overall duration of impact can be divided into two phases: an initial compression phase followed by a subsequent phase of *restitution*. When collision occurs between two elastic particles, the initial kinetic energy is transformed to the elastic strain energy stored in the contacting particles and the energy associated with the elastic waves which propagate during the compression phase. The relative velocity gradually reduces to zero at the instant when the maximum compression is reached. During restitution, the stored elastic strain energy is gradually recovered with the relaxation of deformation induced during compression, and the relative velocity correspondingly increases until the restitution is terminated when the contacting particles separate. The contacting particles then rebound with a rebound kinetic energy that is somewhat less than the initial kinetic energy, as a portion of the initial energy will be dissipated by the stress wave propagation.

The *coefficient of restitution*, *e*, which is used to characterize the change in kinetic energy during the impact, is among the important parameters needed to predict the rebound behavior of particles; *e* is defined as the ratio of the rebound velocity v_r to the incident velocity v_i:

$$e = \frac{v_r}{v_i} \tag{8.16}$$

Since for elastic impacts the energy dissipation due to elastic wave motion is generally negligible, the initial kinetic energy W_i is assumed to be completely converted to the elastic strain energy stored in the contacting particles during compression, i.e.,

$$W_i = \frac{1}{2} m^* v_{ni}^2 = \int_0^{\alpha^*} F_n d\alpha = \frac{8}{15} E^* R^{*1/2} \alpha^{*5/2} = \frac{8}{5} \frac{E^* a^{*5}}{R^{*2}} \tag{8.17}$$

where α^* and a^* are the maximum relative approach and maximum contact radius during the impact, respectively.

Due to the reversibility of elastic deformation, the elastic strain energy is recovered and becomes the rebound kinetic energy W_r of the contacting bodies during restitution. In the ideal case of a fully elastic sphere striking a fully elastic surface, the rebound velocity is almost identical to the initial impact velocity. Therefore, the coefficient of restitution is close to unity, i.e., $e \approx 1$, as also demonstrated experimentally by Andrews (1930).

The above analysis is based upon the assumption that the energy dissipated in stress wave motion is negligible. In order to satisfy this condition, Love (1952) suggested that the duration of the impact needs to be long enough to permit stress waves to traverse the impacting particles many times (i.e. the energy is not 'carried' away by the elastic waves). For the impact of a sphere of radius R, the time for a longitudinal wave to travel a distance of two sphere diameters (i.e., a complete lap: out and return) is

$$\overline{T} = 4R/c_0 \tag{8.18}$$

where c_0 is the velocity of the stress wave in one dimension and given as,

$$c_0 = \sqrt{E/\rho} \tag{8.19}$$

According to Love's criterion, the ratio of $\overline{T}$ to the duration of impact t should be much less than unity so the stress wave can travel several times across the sphere, i.e.,

$$\overline{T}/t \ll 1 \tag{8.20}$$

where t is given by Eq. (8.15). This implies that the initial impact velocity must be low enough to make Eq. (8.20) valid, as the duration of impact t is inversely proportional to $V_{\text{ni}}^{1/5}$, see Eqs (8.13) and (8.15).

8.1.3 FRICTIONAL CONTACTS BETWEEN ELASTIC PARTICLES WITH TANGENTIAL LOADING

We now consider the case of a monotonically increasing tangential force applied to two spheres when they are subject to a constant normal contact force and there is friction over the contact interface. It was shown by Mindlin (1949) that the result is a central region of the contact area in which there is no slip (the *stick* region), which is surrounded by an annular region in which there is slip. As the tangential force increases, the size of the stick region shrinks inward until sliding occurs everywhere and the particles start to move with increasing relative displacement in the tangential direction. Any point inside the contact area must be in one of the following two states for frictional contact with tangential loading:

1. *stick*, during which there is no relative motion between adjacent points in the two contacting bodies and the resulting tangential traction is less than μp (μ is the frictional coefficient and p is the normal contact pressure); and
2. *slip*, during which relative motion occurs and the resulting tangential traction is equal to μp and opposes the instantaneous direction of slip.

Mindlin (1949) also showed that when slip occurs over the contact surface, it starts at the circumference of the contact area and progresses radially inward so that the portion on which slip occurs is an annulus of outer radius a and inner radius c:

$$c = a\left(1 - \frac{F_{\text{t}}}{\mu F_{\text{n}}}\right)^{1/2} \tag{8.21}$$

where F_{n} and F_{t} are the normal and tangential forces, respectively. The distribution of tangential traction over the contact surface is given as

$$q_{(r)} = \left(\frac{3\mu F_{\text{n}}}{2\pi a^2}\right)\left(1 - \frac{r^2}{a^2}\right)^{1/2} \qquad c \le r \le a \tag{8.22a}$$

$$q_{(r)} = \left(\frac{3\mu F_{\text{n}}}{2\pi a^2}\right)\left[\left(1 - \frac{r^2}{a^2}\right)^{1/2} - \frac{c}{a}\left(1 - \frac{r^2}{c^2}\right)^{1/2}\right] \qquad 0 \le r \le c \tag{8.22b}$$

It is clear that the tangential traction in the slip region is given by $q'_{(r)} = \mu p_{(r)}$, where $p_{(r)}$ is defined in Eq. (8.1). The tangential traction distribution Eq. (8.22b) is obtained by superimposing on Eq. (8.22a) with a negative traction over the stick region:

$$q''_{(r)} = -\left(\frac{3\mu F_{\mathrm{n}}}{2\pi a^2}\right)\frac{c}{a}\left(1-\frac{r^2}{c^2}\right)^{1/2} \tag{8.22c}$$

as shown in Fig. 8.7.

The relative tangential displacement of the centers of the two spheres δ is given by

$$\delta = \frac{3\mu F_{\mathrm{n}}(2-\upsilon)}{8G^*a}\left[1-\left(1-\frac{F_{\mathrm{t}}}{\mu F_{\mathrm{n}}}\right)^{2/3}\right] \tag{8.23}$$

Introducing *compliance* as the inverse of stiffness with units of meter per Newton, the tangential compliance S can be calculated from Eq. (8.23) as

$$S = \frac{\mathrm{d}\delta}{\mathrm{d}F_{\mathrm{t}}} = \frac{2-\upsilon}{4G^*a}\left(1-\frac{F_{\mathrm{t}}}{\mu F_{\mathrm{n}}}\right)^{1/3} \tag{8.24}$$

where G^* is the effective shear modulus of the contacting bodies and given as

$$\frac{1}{G^*} = \frac{2-\upsilon_1}{G_1} + \frac{2-\upsilon_2}{G_2} \tag{8.25}$$

8.1.4 OBLIQUE IMPACT OF ELASTIC SPHERES

Mindlin's analyses reported above were extended to other loading conditions by Mindlin and Deresiewicz (1953). Their analysis showed that for the contact of two spherical particles the traction distribution at the interface at any instant is determined not only by the value of the normal and tangential forces but also the previous loading history. Hence, for oblique impact with friction, an accurate prediction of the rebound behavior must take the contact deformation into account.

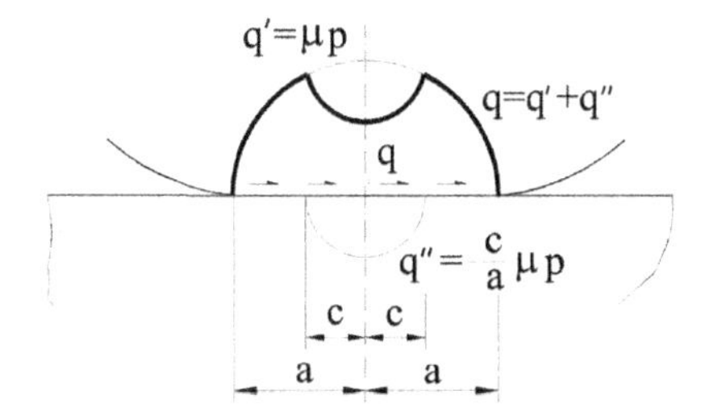

FIGURE 8.7

Tangential traction distribution (Mindlin, 1949).

The contact radius, contact pressure, and tangential traction should be calculated in an incremental manner, with the normal interaction being analysed using the Hertz theory (see Section 8.1.1). This approach was employed by Maw et al. (1976) who analyzed the oblique impact of two elastic spheres. Here this analysis is referred to as the elastic model for oblique impact.

Consider an oblique impact of an elastic sphere with an elastic surface with motion in the y-z plane (see Fig. 8.8), in which the sphere collides with the surface with an initial velocity v_i and angular velocity ω_i and impact angle θ_i. After the impact, the sphere bounces back with a rebound velocity v_r and rebound angular velocity ω_r. Note that v_i and v_r are the incident and rebound velocities at the sphere center. The corresponding velocities at the contact patch are denoted by $\hat{v}_i$ and $\hat{v}_r$.

During the impact the maximum contact radius a^* is determined using Eq. (8.17)

$$a^* = \left(\frac{15}{16}\frac{m^* R^{*2} v_{ni}^2}{E^*}\right)^{1/5} \tag{8.26}$$

For calculation purposes, the maximum contact area is discretized into n annuli. At any given time t, the total tangential traction q at a radius r can be equated to the following summation:

$$q_{(r)} = \sum_{i=j}^{n} q_{ti}\left[1 - \frac{n^2 r^2}{a^{*2} i^2}\right]^{1/5} \tag{8.27}$$

where j is the smallest integer larger than nr/a^* and the traction coefficients q_{ti} are the parameters to be determined. A solution for each time step is obtained once the n traction coefficients q_{ti} are found by solving a system of n equations for q_{ti}.

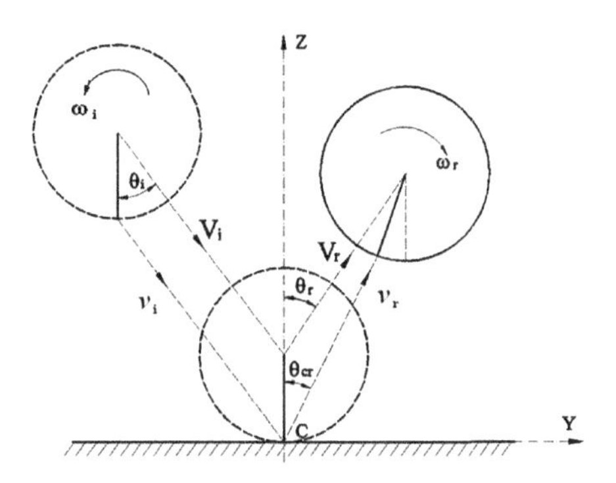

FIGURE 8.8

Illustration of the oblique impact of a sphere with a surface.

As the impact proceeds, each contact annulus is treated as sticking, sliding, or not being in contact. In the sticking region, the tangential displacement is prescribed as the sum of the displacement of the sphere during the time increment and that before the time increment. Certain equations in terms of q_{ti} can then be established. For the sliding region, tangential traction is given by multiplying the coefficient of friction by the normal pressure distribution. $q_{ti} = 0$ is given for the region outside the contact area i.e., not in contact. Therefore, one equation can be developed for each of the n annuli, and a system of equations is established for determining the n traction coefficients q_{ti}. Using this approach, the mixed boundary value problem involving displacements and tractions can be solved from these simultaneous equations.

Once every annulus has been described as either sticking, or sliding, or not in contact, the total traction is assessed for consistency over each annulus. The sticking annuli are checked to ensure that the tangential traction does not exceed the product of the friction coefficient and the normal pressure. For the sliding region, the direction of sliding is checked to ensure that it is as expected. If either of these conditions is not satisfied, the status of the annulus is changed, for which the traction is defined. The process is repeated until all the conditions are satisfied. Then the tangential force F_t at time t is calculated by integrating the traction distribution over the contact area:

$$F_t = \frac{2\pi a^2}{3} \sum_{i=1}^{n} \frac{i^2 q_{ti}}{n^2} \tag{8.28}$$

It is then assumed that the tangential force is constant during that time step and it is used to calculate the tangential relative velocity of the bodies at the contact patch.

Using the above approach, the rebound behavior of the elastic sphere can be determined. The response of oblique elastic impacts can be characterized using two nondimensional parameters. One of these parameters is related to the material and geometrical properties of the contacting bodies as follows

$$\chi = \frac{(1-\upsilon)\left(1+1/K^2\right)}{2-\upsilon} \tag{8.29}$$

where

$$K^2 = \frac{I}{mR^2} \tag{8.30}$$

and I is the mass moment of inertia.

A further parameter is the dimensionless impact angle related to the impact velocity,

$$\psi_i = \frac{\kappa v_{ti}}{\mu v_{ni}} \tag{8.31}$$

where v_{ni} and v_{ti} are the normal and tangential components of the initial velocity, respectively and κ is the ratio of the tangential to normal stiffness,

$$\kappa = \frac{(1-\upsilon_1)/G_1 + (1-\upsilon_2)/G_2}{\left(1-\frac{\upsilon_1}{2}\right)/G_1 + \left(1-\frac{\upsilon_2}{2}\right)/G_2} \tag{8.32}$$

where G_1 and G_2 are the shear moduli, υ_1 and υ_2 are the Poisson's ratios of the two contacting particles.

Figure 8.9 shows the variation of the tangential force throughout the impact of identical spheres under different incident conditions. For impact angles that are smaller than the angle of friction ($\psi_i \leq 1$) there is no sliding at the start of the impact. For larger impact angles $[1 < \psi_i < (4\chi - 1)]$ the impact starts and finishes with sliding; in between no sliding is observed. If the impact angle is sufficiently high $[\psi_i \geq (4\chi - 1)]$, sliding occurs throughout the impact. For the cases in which sliding does not take place throughout, the tangential force undergoes a reversal during the impact, whereas the normal force only completes a half cycle. Figure 8.10 shows the variation of the reflection angle of the contact patch with the dimensionless impact angle. The reflection angle of the contact patch is also given in a corresponding nondimensional form

$$\psi_r = \frac{\kappa \hat{v}_{tr}}{\mu v_{ni}} \tag{8.33}$$

where v_{tr} is the tangential component of the rebound velocity at the contact patch. The reflection angle at the contact patch is found to be mainly negative except for $\psi_i \geq 4\chi$. This phenomenon has been substantiated by various experiments: Maw et al., 1981; Johnson, 1983; Gorham and Kharaz, 2000. For impacts at large impact angles, sliding occurs throughout the impact.

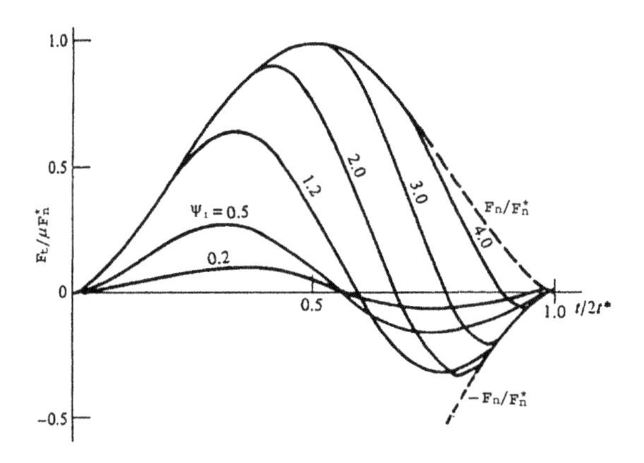

FIGURE 8.9

Tangential force evolution during the impact of solid spheres Maw et al. (1976, 1981).

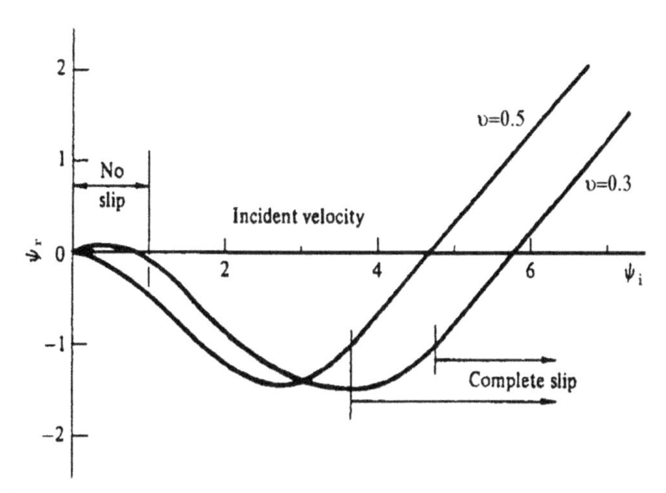

FIGURE 8.10

The reflection angle of the contact patch for the oblique of solid spheres Maw et al. (1976, 1981).

8.1.5 RIGID BODY DYNAMICS FOR OBLIQUE IMPACT

Rigid body dynamics is based on the principle of impulse-momentum conservation, with the tangential effects governed by Coulomb friction. Using this approach, a single-point impulse is applied to each contacting body at the contact point during the impact. The impulse is generally decomposed into normal and tangential components, P_n and P_t (see Fig. 8.1). The normal and tangential rebound velocities of the particles at their centers of mass can be calculated from the following impulse-momentum equations:

$$P_n = m_1\left(v_{nr}^1 - v_{ni}^1\right) = -m_2\left(v_{nr}^2 - v_{ni}^2\right) \tag{8.34a}$$

$$P_t = m_1\left(v_{tr}^1 - v_{ti}^1\right) = -m_2\left(v_{tr}^2 - v_{ti}^2\right) \tag{8.34b}$$

where m_i is the mass of the particle i ($i = 1, 2$), subscripts 1 and 2 refer to contacting particles 1 and 2, respectively. Subscripts n and t refer to the components in the normal and tangential directions, and i and r denote the initial and rebound velocities of the sphere centers, respectively. The angular momentum of each particle is conserved about the contact point, i.e.,

$$I_1\left(\omega_r^1 - \omega_i^1\right) = P_t R_1 \tag{8.35a}$$

$$I_2\left(\omega_r^2 - \omega_i^2\right) = P_t R_2 \tag{8.35b}$$

where ω is the angular velocity, I_i is the mass moment of inertia of the body i ($i = 1$, 2), R_1 and R_2 are the radii of the spheres 1 and 2, respectively.

It can be seen from Eqs (8.34) and (8.35) that there are six unknowns, but only four equations. In order to solve for the rebound velocities, two more equations are required. One of these two required equations can be obtained by relating the normal components of the rebound velocities with those of the initial velocities, for which the normal coefficient of restitution e_{n} is defined as

$$e_{\mathrm{n}} = -\frac{v_{\mathrm{nr}}}{v_{\mathrm{ni}}} \tag{8.36}$$

It should be noted that a negative sign is introduced on the right-hand side of Eq. (8.36) since the normal component of the velocity changes its direction during the impact (i.e., v_{nr} and v_{ni} are in opposite directions) and the normal coefficient of restitution usually has a positive value. In rigid body dynamics, it is often assumed that the coefficient of restitution e_{n} is constant.

The other required equation can generally be obtained by considering the correlation between normal and tangential response. Conventional rigid body dynamics assumes that (Whittaker, 1904)

$$\hat{v}_{\mathrm{tr}} = 0 \quad \text{for} \quad P_{\mathrm{t}} \leq \mu P_{\mathrm{n}} \tag{8.37a}$$

while if $\hat{v}_{\mathrm{tr}} > 0$

$$P_{\mathrm{t}} = \mu P_{\mathrm{n}} \tag{8.37b}$$

where the lower case v denotes the velocity of the contact patch. However, it has been shown by Maw et al. (1976, 1981) and Johnson (1983, 1985) that Eq. (8.37) is not valid for impact at small impact angles, as the tangential velocity reverses its direction so the rebound velocity of the contact patch can be negative.

Brach (1988) introduced an impulse ratio to correlate tangential and normal interactions, together with the normal coefficient of restitution e_{n}. The impulse ratio is defined as the ratio of tangential impulse to normal impulse

$$f = \frac{P_{\mathrm{t}}}{P_{\mathrm{n}}} = \frac{\int F_{\mathrm{t}} \mathrm{d}t}{\int F_{\mathrm{n}} \mathrm{d}t} \tag{8.38}$$

where F_{n} and F_{t} are the normal and tangential contact forces, respectively.

Brach (1988) also pointed out that it is necessary to make a distinction between the impulse ratio f and the friction coefficient μ for a proper analysis of oblique impact. If f is known *a priori*, the system of equations can be solved explicitly to obtain the rebound parameters.

Since rigid body dynamics is primarily governed by the impulse-momentum law, it does not consider the influence of the contact deformation and the tangential compliance of the bodies. It therefore cannot be used to analyse the stresses and forces induced during the impact. If oblique impact of spheres is to be analyzed accurately, contact deformation must be taken into account.

8.1.6 KINEMATIC MODEL FOR OBLIQUE IMPACT

A kinematic model for rebound behavior during oblique impacts of spheres has been developed by Wu et al. (2009), based on rigid body dynamics. Similarly to Eq. (8.36), a tangential coefficient of restitution e_t is introduced:

$$e_t = v_{tr}/v_{ti} \tag{8.39}$$

where v_{ti} and v_{tr} are the tangential components of the impact and rebound velocities, respectively. The tangential coefficient of restitution can have either a negative or positive value because the sphere can bounce backward under certain impact conditions, especially with initial spin. The restitution coefficients e_n and e_t defined in Eqs (8.36) and (8.39) represent the partial recovery of translational kinetic energy in the normal and tangential directions, respectively.

Similarly, a total coefficient of restitution e is also introduced to represent the recovery of total translational kinetic energy (see Fig. 8.8),

$$e = \frac{v_r}{v_i} = \sqrt{\frac{v_{nr}^2}{v_{ni}^2/\cos^2\theta_i} + \frac{v_{tr}^2}{v_{ti}^2/\sin^2\theta_i}} = \sqrt{e_n^2 \cos^2\theta_i + \sin^2\theta_i} \tag{8.40}$$

Substituting Eq. (8.34) into Eq. (8.38), and using Eqs (8.36) and (8.39) leads to

$$e_t = 1 - f(1 + e_n)/\tan\theta_i \tag{8.41}$$

Substituting Eq. (8.34) into Eq. (8.35) yields,

$$\omega_r = \omega_i - mR(v_{ti} - v_{tr})/I \tag{8.42}$$

For a solid sphere, $I = 2mR^2/5$. Hence

$$\omega_r = \omega_i - \frac{5(v_{ti} - v_{tr})}{2R} = \omega_i - \frac{5v_{ti}(1 - e_t)}{2R} \tag{8.43}$$

Substituting Eq. (8.41) into Eq. (8.43) leads to

$$\omega_r = \omega_i - \frac{5f(1 + e_n)v_{ni}}{2R} \tag{8.44}$$

v_{tr} can be expressed as

$$\hat{v}_{tr} = v_{tr} + R\omega_r \tag{8.45}$$

Substituting Eq. (8.44) into Eq. (8.45) yields

$$\hat{v}_{tr} = v_{tr} + R\omega_i - \frac{5}{2}f(1 + e_n)v_{ni} \tag{8.46}$$

Combining Eqs (8.39), (8.41), and (8.46) gives

$$\hat{v}_{tr} = \hat{v}_{ti} - \frac{7}{2}f(1 + e_n)v_{ni} \tag{8.47a}$$

or

$$\hat{v}_{\mathrm{tr}} = \hat{v}_{\mathrm{ti}} - \frac{7}{2} f(1 - e_{\mathrm{t}}) v_{\mathrm{ti}} \tag{8.47b}$$

Equation (8.47b) can be rewritten as

$$e_{\mathrm{t}} = \frac{5}{7} + \frac{2\hat{v}_{\mathrm{tr}}}{7v_{\mathrm{ti}}} - \frac{2R\omega_{\mathrm{i}}}{7v_{\mathrm{ti}}} \tag{8.48}$$

The rebound angle θ_{r} can be determined from (see Fig. 8.8)

$$\tan\theta_{\mathrm{r}} = \frac{v_{\mathrm{tr}}}{v_{\mathrm{nr}}} = -\frac{e_{\mathrm{t}}}{e_{\mathrm{n}}} \tan\theta_{\mathrm{i}} \tag{8.49}$$

It can be seen from Eqs (8.41), (8.44), and (8.47) that the rebound kinematics are determined by the impact angle θ_{i}, the initial impact speed v_{i}, the initial rotational speed ω_{i}, the normal coefficient of restitution e_{n} and the impulse ratio f. For oblique impact of elastic spheres, the normal coefficient of restitution e_{n} is essentially unity as the energy dissipation by elastic wave propagation is generally negligible. A close examination of these equations reveals that the rebound parameters are not all independent but are related to each other. For instance, Eq. (8.44) can be rewritten as

$$f = -\frac{2R(\omega_{\mathrm{r}} - \omega_{\mathrm{i}})}{5(1 + e_{\mathrm{n}}) v_{\mathrm{ni}}} \tag{8.50}$$

Substituting Eq. (8.50) into Eqs (8.41) and (8.47a) leads to

$$e_{\mathrm{t}} = 1 + \frac{2R(\omega_{\mathrm{r}} - \omega_{\mathrm{i}})}{5v_{\mathrm{ni}} \tan\theta_{\mathrm{i}}} = 1 + \frac{2}{5} \frac{R(\omega_{\mathrm{r}} - \omega_{\mathrm{i}})}{\mu v_{\mathrm{ni}}} \frac{\mu}{\tan\theta_{\mathrm{i}}} \tag{8.51}$$

$$\hat{v}_{\mathrm{tr}} = \hat{v}_{\mathrm{ti}} + \frac{7}{5} R(\omega_{\mathrm{r}} - \omega_{\mathrm{i}}) \tag{8.52}$$

Using Eq. (8.52), the following equation can be obtained

$$\frac{\hat{v}_{\mathrm{tr}}}{\mu v_{\mathrm{ni}}} = \frac{\hat{v}_{\mathrm{ti}}}{\mu v_{\mathrm{ni}}} + \frac{7}{5} \frac{R(\omega_{\mathrm{r}} - \omega_{\mathrm{i}})}{\mu v_{\mathrm{ni}}} = \frac{\tan\theta_{\mathrm{i}}}{\mu} + \frac{7}{5} \frac{R\omega_{\mathrm{r}}}{\mu v_{\mathrm{ni}}} - \frac{2}{5} \frac{R\omega_{\mathrm{i}}}{\mu v_{\mathrm{ni}}} \tag{8.53}$$

It is clear from Eqs (8.50)–(8.53) that both the tangential coefficient of restitution e_{t} and the tangential rebound velocity at the contact patch $\hat{v}_{\mathrm{tr}}$ depend on the rebound rotational angular velocity ω_{r}.

Introducing dimensionless angular velocities Ω_{i} and Ω_{r}, a dimensionless rebound tangential velocity at the contact patch Ψ_{r}, and a dimensionless impact angle Θ:

$$\Omega_{\mathrm{r}} = \frac{2R}{5(1 + e_{\mathrm{n}}) \mu v_{\mathrm{ni}}} \cdot \omega_{\mathrm{r}} \tag{8.54a}$$

$$\Omega_{\mathrm{i}} = \frac{2R}{5(1 + e_{\mathrm{n}}) \mu v_{\mathrm{ni}}} \cdot \omega_{\mathrm{i}} \tag{8.54b}$$

$$\Psi_r = \frac{2}{(1+e_n)\mu v_{ni}} \cdot \hat{v}_{tr} \tag{8.54c}$$

$$\Theta = \frac{2}{(1+e_n)\mu} \cdot \tan\theta_i \tag{8.54d}$$

Equations (8.44), (8.51), and (8.53) become

$$\Omega_r = \Omega_i - \frac{f}{\mu} \tag{8.55}$$

$$e_t = 1 + \frac{2(\Omega_r - \Omega_i)}{\Theta} = 1 - \frac{2f}{\Theta\mu} \tag{8.56}$$

$$\Psi_r = \Theta + 7\Omega_r - 2\Omega_i = \Theta + 5\Omega_i - \frac{7f}{\mu} \tag{8.57}$$

Equations (8.55)–(8.57) indicate that e_t, Ψ_r, and Ω_r are functions of Ω_i, Θ, and f/μ, respectively. Wu et al. (2009) performed finite element analysis (FEA) of oblique impacts of an elastic sphere with an elastic substrate. Based on the FEA data, the impulse ratios at various impact angles were calculated using Eq. (8.38), and the normal and tangential impulses were determined by integrating the normal and tangential contact forces over time. The typical variation of the impulse ratio with the normalized impact angle is presented in Fig. 8.11. It is shown that f/μ increases when Θ is small, while at large values of Θ, f/μ is essentially equal to unity, indicating that sliding takes place throughout the impact, i.e., persistent sliding occurs. There is a critical normalized impact angle Θ_c above which persistent sliding impact occurs. Using the ratio of the tangential to normal stiffness κ defined in Eq. (8.32), Θ_c is given as

$$\Theta_c = \frac{7\kappa - 1}{\kappa} \tag{8.58}$$

which is consistent with the criterion of Maw et al. (1976, 1981). When $\Theta \geq \Theta_c$

$$f/\mu = 1 \tag{8.59}$$

For $\Theta < \Theta_c$, as sliding does not occur throughout the impact, the behavior can be referred to as a nonpersistent sliding impact. For nonpersistent sliding impacts, f/μ is determined by the dimensionless impact angle Θ. The correlation between f/μ and Θ can be approximated using the following expression (Wu et al., 2009):

$$f/\mu = c_1 + c_2 \cdot \tanh(c_3 + c_4\Theta) \tag{8.60}$$

where c_1, c_2, c_3, and c_4 are parameters related to the properties of the colliding particles. Table 8.1 gives the values of these parameters obtained from multivariate fitting to the FEA data.

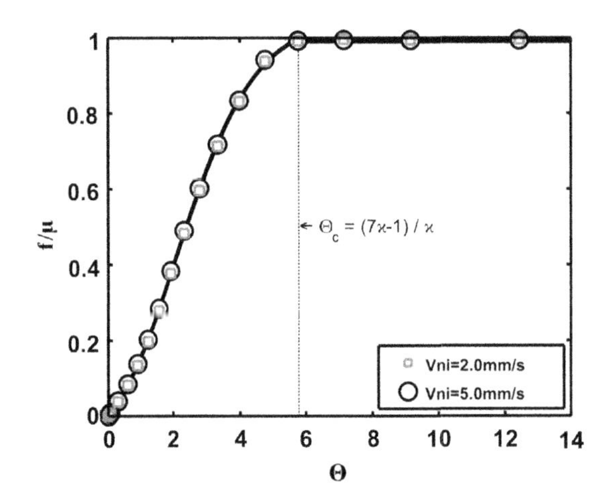

FIGURE 8.11

The variation of f/μ with the dimensionless impact angle for the impacts of an elastic sphere with an elastic substrate at different initial impact velocities (Wu et al., 2009).

Table 8.1 Parameters for the Impact of an Elastic Sphere with an Elastic Substrate

c_1	c_2	c_3	c_4	Θ_c^*
0.446	−0.602	0.952	−0.434	5.658

Equating (8.59) and (8.60) gives the critical impact angle Θ_c^*, which is listed in the last column of Table 8.1. The value of Θ_c^* is comparable to Θ_c determined using Eq. (8.58).

Substituting Eqs (8.59) and (8.60) into Eqs (8.55)–(8.57) gives the complete rebound kinematics as follows:

$$\Omega_r = \begin{cases} -[c_1 + c_2 \tanh(c_3 + c_4\Theta)] & (\Theta < \Theta_c) \\ -1 & (\Theta \geq \Theta_c) \end{cases} \tag{8.61}$$

$$e_t = \begin{cases} 1 - \dfrac{2}{\Theta}[c_1 + c_2 \tanh(c_3 + c_4\Theta)] & (\Theta < \Theta_c) \\ 1 - \dfrac{2}{\Theta} & (\Theta \geq \Theta_c) \end{cases} \tag{8.62}$$

$$\Psi_r = \begin{cases} \Theta - 7[c_1 + c_2 \tanh(c_3 + c_4\Theta)] & (\Theta < \Theta_c) \\ \Theta - 7 & (\Theta \geq \Theta_c) \end{cases} \tag{8.63}$$

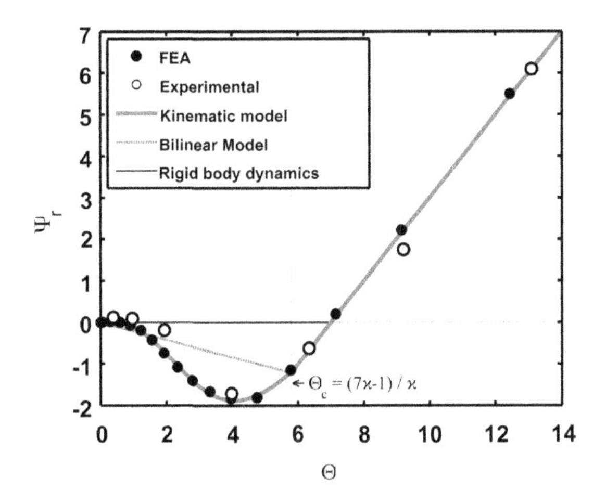

FIGURE 8.12

The variation of the dimensionless tangential rebound velocity at the contact patch with the dimensionless impact angle for elastic impacts (Wu et al., 2009).

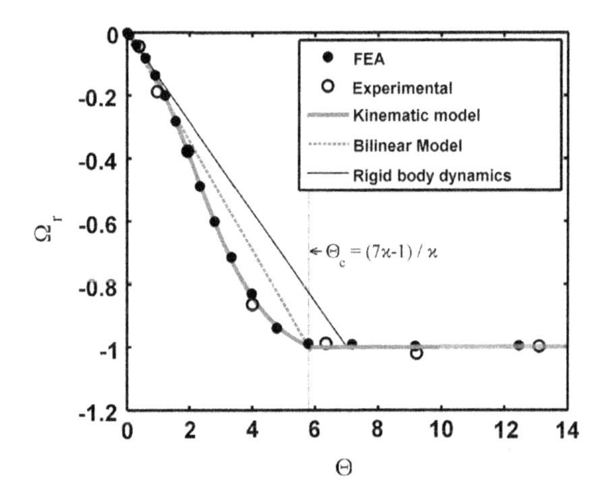

FIGURE 8.13

The variation of the dimensionless rebound angular velocity with the dimensionless impact angle for elastic impacts (Wu et al., 2009).

The rebound kinematics predicted using Eqs (8.61)–(8.63) are shown in Figs 8.12–8.14, in which FEA results of Wu et al. (2009), experimental data reported in the literature (Kharaz et al., 2001; Maw et al., 1976; Thornton et al., 2001) and the predictions of the rigid body dynamics (Box 8.1) and bilinear models (Box 8.2) are also superimposed. It is clear that the predictions agree well with the numerical and experimental results, indicating that Eqs (8.61)–(8.63) can accurately predict the rebound behavior of elastic spheres during impact with an elastic substrate.

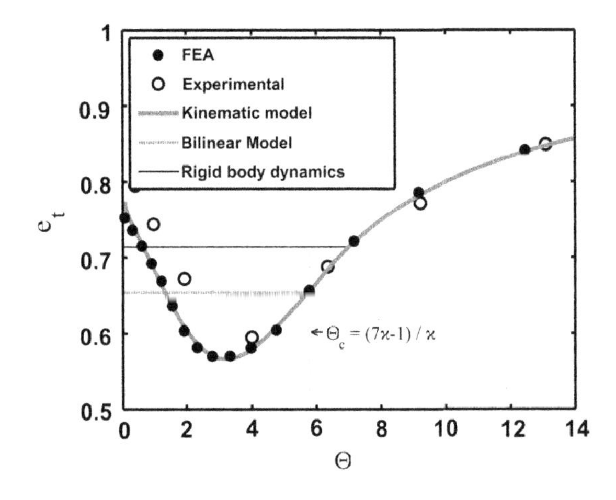

FIGURE 8.14

The variation of e_t with the dimensionless impact angle for elastic impacts (Wu et al., 2009).

BOX 8.1 RIGID BODY DYNAMICS: A REVISIT

It is worth reexamining the rigid body dynamics discussed in Section 8.1.5, in which it was assumed (Brach, 1988) that the rebound tangential velocity at the contact patch is either zero, i.e.,

$$\hat{v}_{\mathrm{tr}} = 0 \quad \text{for } f < \mu \tag{8.64a}$$

or

$$\hat{v}_{\mathrm{tr}} \geq 0 \quad \text{if } f = \mu \tag{8.64b}$$

This implies that

$$f/\mu = \begin{cases} \dfrac{\Theta}{7} & (\Theta < \Theta_c) \\ 1 & (\Theta \geq \Theta_c) \end{cases} \tag{8.65}$$

with $\Theta_c = 7$.

Substituting Eq. (8.65) into Eqs (8.56) and (8.57) gives

$$\Omega_r = \begin{cases} -\dfrac{\Theta}{7} & (\Theta < \Theta_c) \\ -1 & (\Theta \geq \Theta_c) \end{cases} \tag{8.66}$$

$$e_t = \begin{cases} \dfrac{5}{7} & (\Theta < \Theta_c) \\ 1 - \dfrac{2}{\Theta} & (\Theta \geq \Theta_c) \end{cases} \tag{8.67}$$

The predictions from these equations are superimposed in Figs 8.12–8.14. It is clear that using rigid body dynamics the critical impact angle Θ_c is overestimated. For persistent sliding impacts, rigid body dynamics can predict the rebound behavior satisfactorily. However, for nonpersistent sliding impacts, rigid body dynamics fails to predict the rebound behavior accurately as it gives a constant tangential coefficient of restitution e_t (Fig. 8.14), and overestimates the tangential rebound velocity at the contact patch (Fig. 8.12) and the rebound rotational angular speeds (Fig. 8.13).

BOX 8.2 BILINEAR MODEL FOR OBLIQUE IMPACT

Walton and Braun (1986) and Walton (1992) introduced a rotational coefficient of restitution e_r, defined as the ratio of the tangential rebound velocity to the initial tangential impact velocity at the contact patch,

$$e_r = -\frac{\hat{v}_{tr}}{\hat{v}_{ti}} \tag{8.68}$$

It was shown that for persistent sliding impacts (i.e., sliding throughout the impact), e_r is proportional to tan θ, as indicated by Maw et al. (1976, 1981). For nonpersistent sliding impacts, a constant limiting value of the rotational coefficient of restitution e_{r0} is assumed.

$$e_r = \begin{cases} -1 + \mu\left(1 + \frac{1}{K}\right)(1 + e_n)\frac{\hat{v}_{ni}}{\hat{v}_{ti}} & \text{For persistent sliding impacts} \\ e_{r0} & \text{For nonpersistent sliding impacts} \end{cases} \tag{8.69}$$

where K is defined in Eq (8.30). Thus, the variation of the dimensionless tangential rebound velocity with the dimensionless impact angle can be represented using a bilinear model, i.e.,

$$\Psi_r = -\beta\Theta \quad (\Theta < \Theta_c) \tag{8.70}$$

and

$$\Psi_r = \Theta - 7 \quad (\Theta \geq \Theta_c) \tag{8.71}$$

where β is a constant. Using Eqs (8.58), (8.70), and (8.71), we have

$$\beta = \frac{1}{7\kappa - 1} \tag{8.72}$$

Substituting Eqs (8.70) and (8.71) into Eq. (8.57) gives

$$f/\mu = \begin{cases} \frac{(1+\beta)}{7}\Theta & (\Theta < \Theta_c) \\ 1 & (\Theta \geq \Theta_c) \end{cases} \tag{8.73}$$

Substituting Eq (8.73) into Eqs (8.55) and (8.56) yields

$$\Omega_r = \begin{cases} -\frac{(1+\beta)}{7}\Theta & (\Theta < \Theta_c) \\ -1 & (\Theta \geq \Theta_c) \end{cases} \tag{8.74}$$

$$e_t = \begin{cases} 1 - \frac{2(1+\beta)}{7}\Theta & (\Theta < \Theta_c) \\ 1 - \frac{2}{\Theta} & (\Theta \geq \Theta_c) \end{cases} \tag{8.75}$$

Labous et al. (1997) measured the impact parameters using a high-speed video experiment. Following the bilinear model discussed above, three impact parameters were determined as $e_{r0} = 0.5 \pm 0.1$, $e_n = 0.97 \pm 0.07$, and $\mu = 0.175 \pm 0.1$. Using these parameters, the predictions of the bilinear model were superimposed in Figs 8.12–8.14. It can be seen that, similarly to rigid body dynamics, the bilinear model accurately predicts the critical impact angle and the rebound behavior for persistent sliding impacts, but it cannot predict the complex rebound behavior of nonpersistent sliding impacts.

8.2 INTERACTION OF ELASTIC-PLASTIC PARTICLES

8.2.1 YIELD

In many industrial processes, particles are subject to high forces or impact at high speeds, as a consequence of which plastic deformation can take place. To define the onset of plastic deformation, the theory of plasticity introduces two yield criteria:

1. The von Mises yield criterion, in which plastic deformation is initiated when the distortion energy exceeds the distortion energy for yield in simple tension, i.e., when the second invariant J_2 of the stress deviator tensor (s_{ij}) reaches a critical value:

$$J_2 = \frac{1}{2} s_{ij} s_{ij} = \frac{1}{6}\left[(\sigma_1 - \sigma_2)^2 + (\sigma_2 - \sigma_3)^2 + (\sigma_3 - \sigma_1)^2\right] = k^2 = \frac{Y^2}{3} \tag{8.76}$$

   where σ_1, σ_2, and σ_3 are the principal stresses in the state of complex stress, and k and Y are the yield stress in simple shear and in simple tension, respectively.

2. The Tresca criterion, which states that plastic deformation occurs when the maximum shear stress reaches the same value as the maximum shear stress for yield under simple tension,

$$\max\{\, |\sigma_1 - \sigma_2|, \quad |\sigma_2 - \sigma_3|, \quad |\sigma_3 - \sigma_1| \,\} = 2k = Y \tag{8.77}$$

For axisymmetric contacts between two spheres with Poisson's ratio $\upsilon = 0.3$, both the von Mises and Tresca criteria predict that plastic deformation starts when (Johnson, 1985)

$$p_{y0} = 1.60Y \tag{8.78}$$

where p_{y0} is the maximum contact pressure at the initiation of plastic deformation. In terms of the mean contact pressure p_{ym}, plastic deformation starts to occur when (Tabor, 1951)

$$p_{ym} = 1.10Y \tag{8.79}$$

8.2.2 CONTACT PRESSURE

The solution for a Hertz contact remains valid until plastic deformation occurs in one or both particles, depending on their yield strengths. Yield will first develop inside the particle at some distance from the center of the contact surface and a small amount of plastic flow occurs within the larger elastic region around it (Tabor, 1951). With further increase of the load and contact pressure, the plastic zone gradually reaches the contact surface, and eventually the whole of the material around the

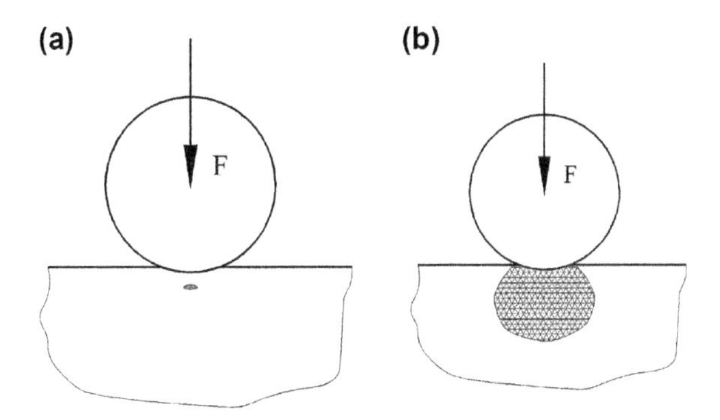

FIGURE 8.15

The plastic deformation process during the compression of a spherical particle into a plastic solid: (a) the onset of plastic deformation which starts beneath the surface; (b) the plastic zone expands to the contact surface and fully plastic deformation occurs.

contact area undergoes plastic deformation (see Fig. 8.15). This implies that during the overall process the deformation changes from purely elastic to elastic-plastic and then to fully plastic. Thus, the contact deformation process can be divided into three phases:

1. Elastic: the deformations of both contacting particles are elastic and the Hertz theory (Section 8.1.1) is still applicable;
2. Elastic-plastic: this is a transitional phase from purely elastic to fully plastic. It begins once plastic deformation occurs in the particle and terminates in the fully plastic phase;
3. Fully plastic: this becomes dominant once the plastic zone reaches the surface and the material surrounding the contact area undergoes plastic deformation.

Note that for contacts between two elastic-plastic particles, plastic deformation will normally be initiated first in the particle with the lower yield strength, but as the contact deformation proceeds, the maximum contact pressure increases and, as soon as it exceeds $1.60Y$ in the other particle, this begins to deform plastically as well. As a result, both particles will be permanently deformed.

For contacts between elastic bodies, or elastic-plastic bodies within the elastic region, the contact pressure can be determined using Hertz theory as discussed in Section 8.1.1. A solution for a fully plastic contact was obtained using the slip line method by Tabor (1951), who suggested that the pressure over the contact area is not uniform but is somewhat higher in the center than at the edge. The mean contact pressure p_m for a fully plastic contact is

$$p_m = 2.8Y \sim 3.0Y \tag{8.80}$$

An expanding cavity model was proposed to describe the contact pressure in the transitional elastic-plastic phase, i.e., from elastic to fully plastic (see Johnson, 1985; Hill, 1950). The expanding cavity model assumes that, if the contact area has a diameter $d = 2a$, the contact surface is encased in a hemispherical core of radius a. Within the core, the material is under hydrostatic pressure of the same value as the mean contact pressure p_m. Plastic deformation will not take place in the material under hydrostatic pressure. Outside the core, the material behaves like an elastic-plastic body with a spherical cavity subject to a pressure p_m. From the boundary of the core, plastic deformation spreads into the surrounding material, and the plastic strains gradually diminish until they match the elastic strains at some radius c ($c > a$), at which the plastic-elastic boundary lies. Johnson (1985) showed that the pressure in the hydrostatic core is a function of the critical parameter $(E^*/Y)(a/R^*)$ for a spherical particle of radius R:

$$\frac{p_m}{Y} = \frac{2}{3}\left\{1 + \ln\left[\frac{1}{3}\frac{E^*}{Y}\frac{a}{R^*}\right]\right\} \tag{8.81}$$

Figure 8.16 illustrates how the mean contact pressure increases from $p_m \sim 1.1Y$ (the onset of plastic deformation) to $\sim 3.0Y$ (fully plastic deformation) as the size of the contact a/R^* increases, based on the prediction of Eq. (8.81). Fully plastic deformation occurs at a value of $E^*a/YR^* \approx 40$, which is about 16 times greater than the value for the occurrence of yield.

The typical pressure distribution involving plastic deformation obtained from FEA (Wu et al., 2005) is shown in Fig. 8.17. It can be seen from Fig. 8.17(a) that the pressure distribution can be predicted using the Hertz theory when the maximum pressure p_0 is lower than *1.6Y*. When p_0 is greater than *1.6Y*, the pressure profile

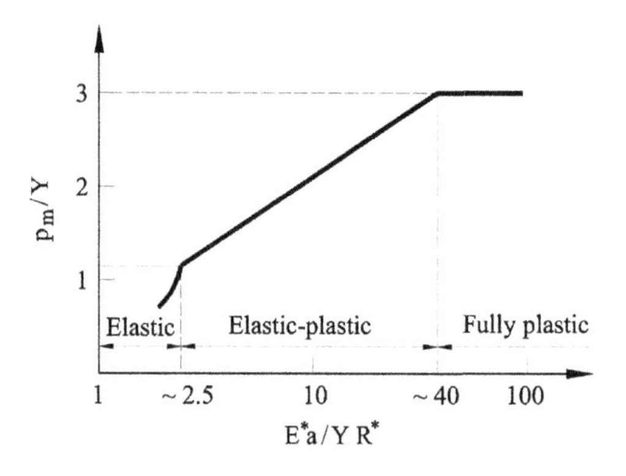

FIGURE 8.16

The variation of p_m/Y with E^*a/YR^* for the contact of elastic-plastic particles (Johnson, 1985).

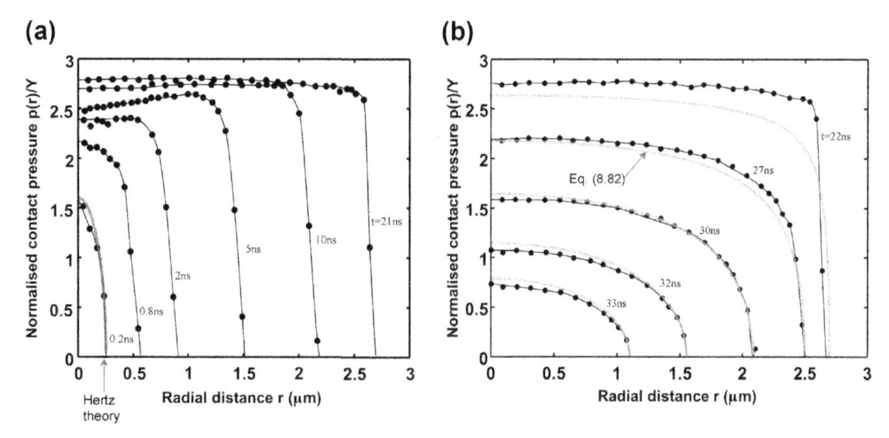

FIGURE 8.17

Evolution of contact pressure distributions during the impact of an elastic sphere with an elastic-perfectly plastic substrate ($V_i = 30$ m/s) at various time instants during (a) compression and (b) restitution, obtained from FEA. The solid line in (a) is obtained from the Hertz theory, while those in (b) are predictions of Eq. (8.82) (Wu et al., 2005).

becomes significantly flattened, which is caused by plastic deformation. The maximum pressure remains approximately constant at 2.7~3.0 times the yield stress Y once the flattened pressure profile is fully established.

Mesarovic and Johnson (2000) derived an analytical expression for the contact pressure distribution during unloading:

$$p_{(r)} = \frac{2p^*}{\pi} \sin^{-1}\left[\frac{a^2 - r^2}{a^{*2} - r^2}\right]^{1/2} \tag{8.82}$$

where p^* and a^* are the pressure and contact radius at the end point of loading or the starting point of unloading. Figure 8.17b shows the evolution of the contact pressure distribution during unloading, in which the predictions of Eq. (8.82) using $p^* = 2.8Y$ and $a^* = 2.68$ μm, obtained from the best fit to the FEA data, are also superimposed. It is clear that Eq. (8.82) can satisfactorily predict the pressure distribution during unloading of elastic-plastic spheres.

8.2.3 CONTACT FORCE–DISPLACEMENT RELATIONSHIP

The contact force–displacement relationship is crucial for understanding the interaction of elastic-plastic particles. The analytical derivation of the contact force–displacement relationship should be based on the determination of both the contact pressure distribution and the relationship between the relative approach and the contact radius. For fully plastic contact, it is generally accepted that the mean contact pressure is constant (2.8–3.0Y), and the relative approach is related to the contact radius by $\alpha = a^2/2R^*$ provided that plastic deformation is small (see Johnson, 1985; Tabor, 1951). The force–displacement relationship can hence be determined.

For the transition regime from purely elastic to fully plastic contact (i.e., the elastic-plastic regime), an accurate determination of the contact pressure distribution and the relationship between relative approach and contact radius becomes more complicated. Ritter (1963) assumed that, once plastic deformation is initiated, the pressure becomes constant and the area that is loaded to that constant pressure increases upon further increase in the relative approach. A flattened contact pressure distribution obtained by truncating the Hertzian pressure profile was then proposed by Thornton (1997) and Thornton and Ning (1998), who also assumed that the Hertzian substitution $a^2 = R^*\alpha$ is still valid for the elastic-plastic regime.

Figure 8.18 shows the typical time evolutions of the contact force and corresponding force–displacement relationship during impacts at various impact velocities. For impact at higher impact velocities, greater maximum contact forces and shorter durations of impact are obtained (see Fig. 8.18(a)). It is also interesting to note that the duration of compression slightly reduces with increasing impact velocity owing to the decrease in the duration for elastic compression at higher impact velocity. The restitution phase is essentially elastic and its duration decreases with increasing impact velocity as indicated by Eq. (8.15).

In Fig. 8.18(b), the corresponding contact force F is plotted against displacement α. The Hertz theory prediction is also superimposed in this figure. It can be seen that, in the elastic impact region, the loading curves obtained from the theoretical and

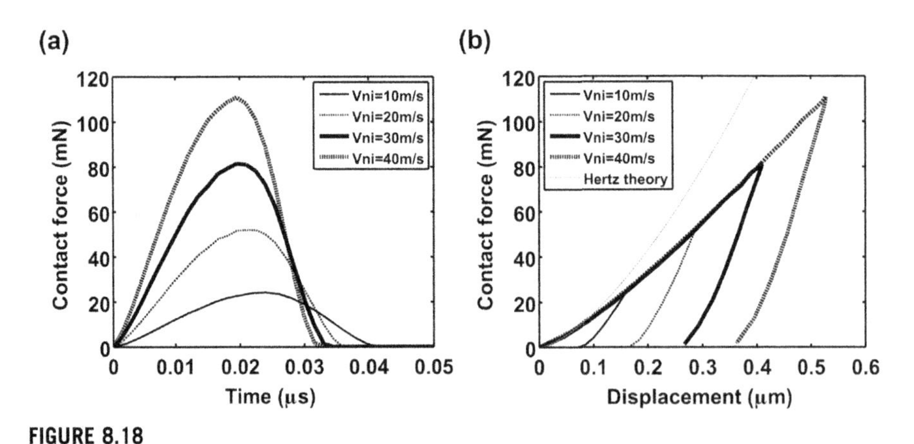

FIGURE 8.18

(a) Time histories of contact force and (b) Contact force–displacement relationships obtained from FEA, for the impact of an elastic sphere with an elastic-perfectly plastic substrate at various impact velocities.

numerical methods are almost identical. However, when plastic deformation occurs, the contact force–displacement curve diverges from the Hertz theory. The loading curves exhibit an upward curvature followed by a nearly constant slope at large displacements, which is consistent with the experimental observations of Goldsmith and Lyman (1960). A stiffer initial slope of the unloading curves is observed for impact at a higher velocity.

8.2.4 COEFFICIENT OF RESTITUTION

When collision occurs between two elastic-plastic particles, if the initial kinetic energy is so high that plastic deformation initiates, the initial kinetic energy converts to:

1. energy associated with elastic deformation of the material (i.e. elastic strain energy);
2. energy associated with plastic deformation (plastic strain energy);
3. energy associated with elastic wave propagation.

Since plastic deformation is irreversible, only the elastic strain energy can be recovered as rebound kinetic energy. Some initial kinetic energy is thus dissipated by plastic deformation and stress wave propagation. Below a certain impact velocity, the energy dissipation due to plastic deformation and stress wave propagation is negligible. Hence the rebound velocity is almost the same as the initial impact velocity and the coefficient of restitution is close to unity. As the impact velocity is increased, more energy is dissipated due to both plastic deformation and elastic wave propagation, and the coefficient of restitution decreases.

Hutchings (1979) showed that only a few percent of the initial kinetic energy is normally dissipated by stress waves during plastic impacts. Where plastic deformation occurs, the energy associated with it is normally much larger than that associated with stress wave propagation. For example, for the impact of a hard steel sphere with a mild steel block at a velocity of about 70 m/s, the measured coefficient of restitution is about 0.4, in which only about 3% of the kinetic energy is dissipated by stress wave propagation.

Several theoretical models have been developed to predict the coefficient of restitution for the impact of elastic-plastic spheres. Thornton (1997) developed a theoretical model for the normal impact of two elastic-perfectly plastic spheres, in which a Hertzian pressure distribution with a cut-off was introduced. In addition, a constant cut-off pressure p_y was assumed during the loading. An analytical solution for the coefficient of restitution was obtained as follows,

$$e_n = \left(\frac{6\sqrt{3}}{5}\right)^{1/2}\left[1-\frac{1}{6}\left(\frac{v_y}{v_{ni}}\right)^2\right]^{1/2}\left[\frac{v_y/v_{ni}}{v_y/v_{ni}+2\sqrt{6/5-(1/5)(v_y/v_{ni})^2}}\right]^{1/4} \tag{8.83}$$

where v_{ni} is the initial velocity and v_{y} is the impact velocity below which the contact deformation is elastic and is defined by

$$v_{\mathrm{y}} = \left(\frac{\pi}{2E^*}\right)^2 \left(\frac{8\pi R^{*3} p_{\mathrm{y}}^5}{15m^*}\right)^{1/2} \tag{8.84}$$

It should be noted that p_{y} and v_{y} do not represent the maximum contact pressure and the impact velocity at the instant when plastic deformation is initiated. Instead, they correspond to the instant when significant flattening of the contact pressure distribution commences (See Fig. 8.17(a)), and $p_{\mathrm{y}} \approx 2.8Y$ (Thornton et al., 2001).

For impact dominated by fully plastic deformation, Johnson (1985) proposed a simplified model by introducing the following simplifications: (1) the relative approach is related to the contact radius by:

$$\alpha = \frac{a^2}{2R^*} \tag{8.85}$$

(2) the mean contact pressure p_{m} is constant and equal to 3.0Y. An expression for the coefficient of restitution e_{n} was obtained as

$$e_{\mathrm{n}}^2 = \frac{3\sqrt{2}\pi^{5/4}}{5} \left(\frac{p_{\mathrm{m}}}{E^*}\right) \left(\frac{\frac{1}{2}m^* v_{\mathrm{ni}}^2}{p_m R^{*3}}\right)^{-1/4} \tag{8.86}$$

For the impact of a sphere with a surface, Eq. (8.86) reduces to

$$e_{\mathrm{n}} = 1.718 \left(\frac{p_{\mathrm{m}}^5}{E^{*4}\rho}\right)^{1/8} v_{\mathrm{ni}}^{-1/4} \tag{8.87}$$

Equation (8.87) indicates that e_{n} is proportional to $v_{\mathrm{ni}}^{-1/4}$.

Wu et al. (2003) systematically analyzed the impact of elastic-plastic particles using the finite element method. The coefficients of restitution obtained from numerical modeling of elastic-plastic impacts at various conditions are plotted against the normalized impact velocity $v_{\mathrm{ni}}/v_{\mathrm{y}}$ in Fig. 8.19. This shows that the data for various impact cases follow a single master curve, implying that the coefficient of restitution depends primarily upon $v_{\mathrm{ni}}/v_{\mathrm{y}}$ for impacts of elastic-perfectly plastic particles. A better fit to the data at high impact velocities (i.e., $v_{\mathrm{ni}}/v_{\mathrm{y}} \geq 100$) gives:

$$e_{\mathrm{n}} = \bar{c} \left(\frac{v_{\mathrm{ni}}}{v_{\mathrm{y}}}\right)^{-1/4} \quad \text{where } \bar{c} = 2.08 \tag{8.88}$$

as shown by the dashed line in Fig. 8.19. This is consistent with the prediction of Johnson (1985), see also Eq. (8.87).

Many experimental data for the collision of elastic-plastic spheres have also been reported in the literature (see, e.g., Tabor, 1948; Bridges et al., 1984;

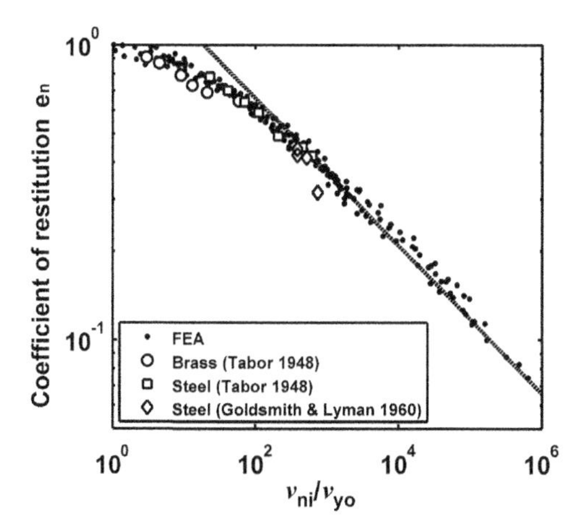

FIGURE 8.19

Coefficient of restitution as a function of v_{ni}/v_{yo} for elastic-plastic impacts. FEA, finite element analysis.

Labous et al., 1997). The experimental results for coefficient of restitution were reported by Tabor (1948) for cast steel and drawn brass balls with impact velocities up to 4.5 m/s, which are also plotted in Fig. 8.19. It is clear that the coefficient of restitution is not constant in general. Bridges et al. (1984) reported some experimental results on the coefficient of restitution for the collision of ice particles at impact velocities relevant to the formation of Saturn's rings, in order to understand ice-particle collision processes. They obtained the following empirical expression for the coefficient of restitution:

$$e_n = 0.32 v_{ni}^{-0.23} \tag{8.89}$$

for the range of impact velocity 0.015 cm/s $< v_{ni} <$ 5.1 cm/s. The impact of two nylon spheres was experimentally investigated by Labous et al. (1997) using a high-speed video system. The size and velocity dependences of the coefficient of restitution were investigated. It was suggested that there were two basic energy dissipation regimes: (i) plastic deformation at high impact velocities and (ii) visco-elastic dissipation at low impact velocities.

8.2.5 THE EFFECT OF TANGENTIAL LOADING

For contacts of elastic-plastic particles with tangential loading, the effect of friction is to superimpose a stress induced by the tangential traction. This tangential traction alters the stresses in the contacting bodies and hence changes the loading condition at which plastic deformation is initiated. The stresses due to the combined effect of normal pressure and tangential traction in sliding contacts were analyzed by Johnson and Jefferis (1963), Hamilton and Goodman (1966), and Bryant and Keer (1982).

The analysis showed that, due to the effect of friction, the point of maximum shear stress moves closer to the contacting surfaces. Both Tresca and von Mises yield criteria were adopted to determine the contact pressure for the onset of yield in sliding contacts by Johnson and Jefferis (1963). Their results showed that for low values of the coefficient of friction ($\mu \leq 0.25$ by the Tresca criterion and $\mu \leq 0.3$ by the von Mises criterion), the yield point is still reached at a position beneath the contact surface (see Fig. 8.15). For large values of μ, plastic deformation first occurs at the interface between the contacting bodies, rather than in the subsurface.

Due to the complexity of elastic-plastic contact behavior with tangential loading and oblique impact, theoretical analysis becomes very difficult and there is consequently little literature on this topic. Nevertheless, the kinematic model discussed in Section 8.1.6 is generic so it can be applied to oblique impacts with elastic-plastic particles, for which the normal coefficient of restitution e_n can be defined as discussed in Section 8.2.4.

8.3 INTERACTION OF PARTICLES WITH ADHESION

8.3.1 INTERACTION BETWEEN ADHESIVE RIGID SPHERES—HAMAKER THEORY

When two particles approach each other, interparticle attraction can result due to the van der Waals potential between the molecules of the two bodies. As a result, particles can adhere to each other, especially for fine particles, i.e., less than 100 μm. As discussed in detail in Seville et al. (1997), the van der Waals attractive potential for two molecules in vacuum can be given as

$$W(r) = -\frac{C}{r^6} \tag{8.90}$$

Using pair-wise addition, Hamaker (1937) analyzed the adhesive force between two rigid spherical particles at a separation distance s, for which the interaction energy is calculated by

$$W = -\int_{V_1}\int_{V_2} \frac{Cq^2}{r^6} \mathrm{d}V_1 \mathrm{d}V_2 \tag{8.91}$$

where V_1 and V_2 are the volumes of the two particles and q is the number density of molecules in the solids (i.e., the number of molecules in a unit volume). Integration of Eq. (8.91) gives

$$W(S) = -\frac{AR^*}{6s} \tag{8.92}$$

where A ($=\pi^2 Cq^2$) is the Hamaker constant, and R^* is the effective radius defined in Eq. (8.6). Table 8.2 lists the typical values of Hamaker constant for the interaction of various materials in vacuum and in water.

Table 8.2 Hamaker Constants in Vacuum and Water of Various Inorganic Materials Interacting Against Four Materials at Room Temperature (Bergström, 1997)

	Hamaker constant (10^{-20} J)							
	In Vacuum				In Water			
Materials	**Silica**	**Silica Nitride**	**Alumina**	**Mica**	**Silica**	**Silica Nitride**	**Alumina**	**Mica**
$BaTiO_3$	10.10	16.50	15.20	21.40	0.62	4.84	3.55	1.98
BeO	9.67	15.40	14.80	11.90	0.95	3.87	3.50	2.06
Diamond	13.70	22.00	21.10	17.00	1.71	7.94	7.05	4.03
$CaCO_3$	8.07	12.90	12.30	9.94	0.69	2.53	2.17	1.35
CaF_2	6.70	10.60	10.30	8.26	0.45	1.17	1.10	0.73
CdS	8.03	13.10	12.00	9.86	0.72	3.12	2.15	1.43
KCl	5.94	9.53	9.00	7.31	0.37	0.73	0.51	0.46
$MgAl_2O_4$	9.05	14.50	13.80	11.20	0.85	3.39	2.97	1.79
MgF_2	6.15	9.74	9.42	7.57	0.36	0.66	0.69	0.50
MgO	8.84	14.20	13.50	10.90	0.81	3.26	2.79	1.69
Mica	8.01	12.80	12.20	9.86	0.69	2.45	2.15	1.34
NaCl	6.45	10.30	9.77	7.93	0.44	1.17	0.88	0.66
PbS	5.37	8.88	7.90	6.57	−0.08	0.64	−0.20	−0.03
6H–SiC	12.60	20.30	19.20	15.50	1.52	7.22	6.05	3.54
β-Si_3N_4	10.80	17.30	16.50	13.30	1.17	5.13	4.43	2.61
SiO_2	7.59	12.10	11.60	9.35	0.63	2.07	1.83	1.16
$SrTiO_3$	9.44	15.40	14.20	11.60	0.57	4.02	2.98	1.69
TiO_2	9.46	15.40	14.20	11.60	0.69	4.26	3.11	1.83
Y_2O_3	9.24	14.90	14.00	11.40	0.89	3.80	3.11	1.89
ZnO	7.38	12.00	11.10	9.06	0.58	2.30	1.58	1.08
ZnS	9.69	15.70	14.60	11.90	1.02	4.56	3.55	2.19
3Y–ZrO_2	11.40	18.40	17.40	14.10	1.25	5.89	4.95	2.89

Hence the adhesive force between two particles can be obtained as

$$F(s) = \frac{\mathrm{d}W(s)}{\mathrm{d}s} = \frac{AR^*}{6s^2} \tag{8.93}$$

Equation (8.93) indicates that the adhesive force increases as the minimum separation distance decreases. It also implies that the adhesive force becomes infinite as the minimum separation distance approaches zero. Thus, Eq. (8.93) can only be applied up to a limit value of the minimum separation distance, i.e., a cut-off separation. Israelachvili (1991) suggested that a cut-off separation of 1.65×10^{-10} m gives a good approximation to the surface energies measured experimentally. This can be explained as follows: (1) the effect of surface roughness on the interaction will become significant when the separation between particles reduces below a certain value; (2) At small separations, the overlapped electron clouds associated with molecules leads to a very strong repulsion (see Fig. 8.20). The repulsive forces increase significantly as the separation distance decreases, so that the attractive force given in Eq. (8.93) can be balanced.

8.3.2 INTERACTION BETWEEN ADHESIVE ELASTIC PARTICLES

In the Hamaker analysis discussed above, the particles are taken as rigid, i.e., the deformation induced by the attractive and repulsive forces is ignored. However, in reality, particles will deform as they are subjected to external forces, as shown in the earlier parts of this chapter, and the forces themselves will be modified by the resulting deformation. For this reason it is necessary to consider the deformation of particles in a full treatment of adhesive interaction. Therefore, theoretical models for normal and oblique interactions of elastic particles developed by Johnson et al. (1971) and Savkoor and Briggs (1977), respectively, are introduced in this section.

8.3.2.1 Normal Contact—JKR Theory

The JKR theory (Johnson et al., 1971) was developed as an extension of the Hertz theory for the contact of elastic spheres by considering surface adhesion at the

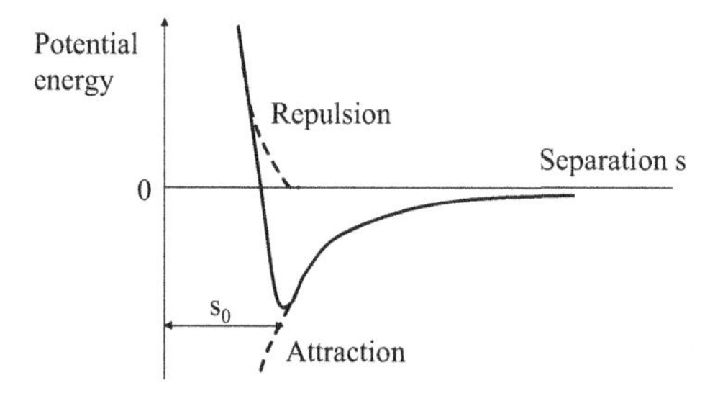

FIGURE 8.20

The variation of potential energy with the separation between two particles attracted by van der Waals forces (s_0 denotes the equilibrium separation) (Seville et al., 1997).

interface. For the contact of two elastic spheres under a normal force F_n, if there is no surface adhesion, the contact radius a can be obtained from Eq. (8.12), i.e.,

$$a^3 = \frac{3F_nR^*}{4E^*} \tag{8.94}$$

If the two spheres are adhesive, an additional adhesive force will arise. As a result, the equilibrium contact area will increase. By introducing an equivalent Hertzian force $\hat{F}_n$ (i.e. an augmented normal contact force that would produce the same contact area as the equilibrium contact area using the Hertz theory), Eq. (8.94) can be rewritten as

$$a^3 = \frac{3\hat{F}_nR^*}{4E^*} \tag{8.95}$$

Johnson et al. (1971) showed that

$$\hat{F}_n = F_n + 6\pi\gamma R^* \pm \sqrt{12\pi\gamma R^* F_n + (6\pi\gamma R^*)^2} \tag{8.96}$$

where γ is the energy per unit area of the surfaces in contact.

Substituting Eq. (8.96) into Eq. (8.95) and rearranging, an expression for the applied force F_n can be obtained as

$$F_n = \frac{4E^*a^3}{3R^*} - 4\sqrt{\pi\gamma E^* a^3} \tag{8.97}$$

The contact pressure distribution for the contact of two adhesive spheres can be given by the following equation (Johnson, 1976)

$$p_{(r)} = \frac{3\hat{F}_n}{2\pi a^2}\left[1 - \left(\frac{r}{a}\right)^2\right]^{1/2} - \frac{\hat{F}_n - F_n}{2\pi a^2}\left[1 - \left(\frac{r}{a}\right)^2\right]^{-1/2} \tag{8.98}$$

As shown by Johnson (1985), the first term of Eq. (8.98) results in a normal displacement over the contact area as

$$\mu_i' = \frac{3(1 - v_i^2)\hat{F}_n}{8E_ia^3}(2a^2 - r^2) \tag{8.99}$$

and the second term of Eq. (8.98) leads to the following uniform displacement over the contact area

$$\mu_i'' = -\frac{(1 - v_i^2)(\hat{F}_n - F_n)}{2E_ia} \tag{8.100}$$

where $i = 1$ or 2, denoting the contacting particles.

The relative approach of the two spheres can then be determined as

$$\alpha = \mu_1' + \mu_2' + \mu_1'' + \mu_2'' + \frac{r^2}{2R^*} = \frac{3\hat{F}_n(2a^2 - r^2)}{8E^*a^3} - \frac{\hat{F}_n - F_n}{2E^*a} + \frac{r^2}{2R^*} \tag{8.101}$$

Substituting Eqs (8.96) and (8.97) into Eq. (8.101), the relative approach α can be expressed as a function of the contact radius a, i.e.,

$$\alpha = \frac{a^2}{R^*} - \sqrt{\frac{2\pi\gamma a}{E^*}} \tag{8.102}$$

When the applied force F_n is zero, solving Eq. (8.97) gives

$$a_0^3 = \frac{9\pi\gamma (R^*)^2}{E^*} \tag{8.103}$$

For $\gamma = 0$, i.e., no adhesion between the two particles, a_0 becomes zero, as predicted by the Hertz theory. However, for adhesive contacts, i.e., $\gamma \neq 0$, Eq. (8.103) gives a finite contact area (i.e., $a_0 > 0$). The contact area can be *reduced* when the applied load F_n becomes negative (i.e., a pulling force). In this case, to ensure that a real solution can be obtained from Eq. (8.96), the following condition has to be met:

$$(F_n + 6\pi\gamma R^*)^2 - F_n^2 \geq 0 \tag{8.104}$$

from which we have

$$F_n \geq -3\pi\gamma R^* \tag{8.105}$$

Johnson et al. (1971) showed that the two spheres will separate if the applied pulling force reaches a critical value of F_c,

$$F_c = -3\pi\gamma R^* \tag{8.106}$$

where F_c is referred to as the pull-off force. When $F_n = -F_c$, the contact radius a_c becomes

$$a_c^3 = \frac{3F_c R^*}{4E^*} = \frac{9\pi\gamma R^{*2}}{4E^*} \tag{8.107}$$

and the corresponding relative approach is given as

$$\alpha_c = \left[-\left(\frac{4}{3}\right)^{1/3} + \left(\frac{9}{16}\right)^{1/3}\right]\left(\frac{F_c^2}{E^{*2}R^*}\right)^{1/3} = -0.275\left(\frac{F_c^2}{E^{*2}R^*}\right)^{1/3} \tag{8.108}$$

For the contact of two adhesive elastic spheres, Johnson (1976) provided the following expression relating the contact force F_n to the relative approach α (Thornton and Ning, 1998):

$$\frac{\alpha}{\alpha_s} = \frac{3\left(\frac{F_n}{F_c}\right) + 2 \pm 2\left(1 + \frac{F_n}{F_c}\right)^{1/2}}{3^{2/3}\left[\left(\frac{F_n}{F_c}\right) + 2 \pm 2\left(1 + \frac{F_n}{F_c}\right)^{1/2}\right]^{1/3}} \tag{8.109}$$

where α_s is the relative approach when separation occurs.

$$\alpha_s = \left(\frac{3F_c^2}{16E^{*2}R^*}\right)^{1/3} = \frac{3}{2}\left(\frac{\pi^2\gamma^2R^*}{2E^{*2}}\right)^{1/3} \tag{8.110}$$

The force–displacement relationship given in Eq. (8.109) is graphically illustrated in Fig. 8.21. It is worth mentioning that an alternative model for normal interaction of adhesive elastic spheres was developed by Derjaguin et al. (1975), which is known as the DMT theory. The difference between the JKR theory and the DMT theory is discussed in Box 8.3.

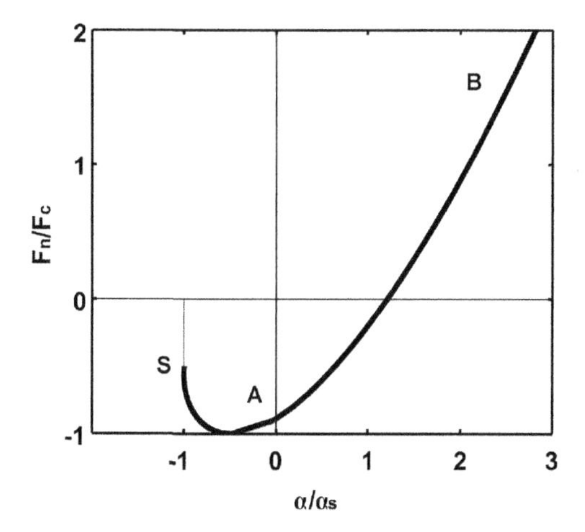

FIGURE 8.21

Force–displacement relationship for the contact of two adhesive spheres (Thornton and Ning, 1998).

BOX 8.3 JKR THEORY VERSUS DMT THEORY (LIAN, 1994)

In the DMT theory, it is assumed that the attractive force does not affect the deformation of the contacting surfaces. In addition, the DMT theory considers the adhesive force and Hertz theory separately, so that the contact area is determined by the Hertz theory and the adhesive force is only considered outside the contact area where the separation distance between surfaces is small. This is a different approach from the JKR theory which takes the adhesive force as acting within the contact area and contributing to the deformation of the contacting surfaces. So a larger contact area than that predicted by the Hertz theory is obtained (see Eqs (8.95) and (8.96)). In the DMT theory, no analytical expression for the adhesive force is available so that this needs to be obtained numerically. Nevertheless, the pull-off force is obtained as (Derjaguin et al., 1975)

$$F_c = 4\pi\gamma R^* \tag{8.111}$$

which is similar to the JKR prediction but with a higher constant. There has been intense debate about the relative merits of the JKR and DMT theories. Tabor (1977) performed experiments with mica surfaces glued onto glass cylinders and examined the shape of the deformed surface and the way the surfaces separate. The experimental observations were more consistent with the analysis of the JKR

Continued

BOX 8.3 JKR THEORY VERSUS DMT THEORY (LIAN, 1994)—cont'd

theory than the DMT theory. Muller et al. (1980, 1983) revisited the DMT model and relaxed the assumption that the adhesive force does not change the contact deformation. This further analysis revealed that the pull-off force is a continuous function of a parameter ζ, with the DMT and JKR solutions as special cases. The parameter ζ is defined as

$$\zeta = \frac{32}{3\pi}\left[\frac{\Gamma^2 R^*}{\pi s_0^3 E^{*2}}\right] \tag{8.112}$$

where s_0 is the equilibrium separation, Γ is the interfacial surface energy and given as

$$\Gamma = \frac{A}{16\pi s_0^2} \tag{8.113}$$

where A is the Hamaker constant. Muller et al. (1980) showed that the DMT theory holds for $\zeta \ll 1$, i.e., for hard solids of small radius and low surface energy, while the JKR theory is valid for $\zeta \gg 1$, i.e., for soft solid spheres of large radius and high surface energy. This was further confirmed experimentally by Cappella and Dietler (1999), who examined the interaction between the tip of an atomic force microscope probe and a sample, and found that the JKR theory can accurately predict the pull-off force for highly adhesive systems with a low stiffness and large tip radii, while the DMT model is valid for systems with low adhesion and small tip radii.

8.3.2.2 Normal Impact

According to the JKR theory (see Fig. 8.21), when two adhesive elastic spheres collide, as a result of the van der Waals forces a finite normal contact force will be induced at $\alpha = 0$ (i.e., point A in Fig. 8.21). During the loading/compression phase, the velocity of the spheres decreases and the initial kinetic energy is converted into elastic energy stored in the deforming spheres (energy dissipation due to stress wave propagation is very small and can be ignored). When the velocity reduces to zero, the contact force reaches the maximum (point B in Fig. 8.21) and the compression stage terminates. During the restitution stage, the stored elastic energy is converted back to kinetic energy, enabling the spheres to move apart. When the relative approach returns to $\alpha = 0$, all stored kinetic energy is recovered. However, the spheres remain sticking to each other. Further work needs to be done to separate them. As shown in Fig. 8.21, separation occurs at point S and hence the work required to break the contact W_s can be determined (Thornton and Ning, 1998) as

$$W_s = \int_0^{\alpha_s} F_n \mathrm{d}\alpha \tag{8.114}$$

Substituting Eqs (8.97) and (8.102) into Eq. (8.114) gives,

$$W_s = \int_{a_0}^{a_s} \left(\frac{4E^* a^3}{3R^*} - 4\sqrt{\pi\gamma E^* a^3}\right)\left(\frac{2a}{R^*} - \sqrt{\frac{\pi\gamma}{E^* a}}\right) \mathrm{d}a \tag{8.115}$$

In Eq. (8.115), when $\alpha = 0$

$$a = a_0 = \left(\frac{4\pi\gamma R^{*2}}{E^*}\right)^{1/3} \tag{8.116}$$

When $\alpha = \alpha_{s,}$

$$a = a_s = \left[\frac{\pi\gamma R^{*2}}{4E^*}\right]^{1/3} \tag{8.117}$$

Equation (8.115) can be integrated to obtain (Lian, 1994; Thornton and Ning, 1998)

$$W_s = 14.18\left(\frac{\gamma^5 R^{*4}}{E^{*2}}\right)^{1/3} = 0.936 F_c \alpha_s \tag{8.118}$$

During the impact, the only energy dissipated is the work done in separating the two spheres W_s. Therefore, applying an energy balance,

$$\frac{1}{2}m^* v_{ni}^2 - \frac{1}{2}m^* v_{nr}^2 = W_s \tag{8.119}$$

Equation (8.119) shows that when the two spheres adhere (i.e., are sticking), $v_{nr} = 0$, and the impact velocity $v_{ni} = v_s$, below which sticking takes place. Therefore, using Eqs (8.118) and (8.119), the sticking criterion can be obtained as

$$v_s = \left(\frac{14.18}{m^*}\right)^{1/2}\left(\frac{\gamma^5 R^{*4}}{E^{*2}}\right)^{1/6} \tag{8.120}$$

If $v_{ni} > v_s$, the spheres will rebound and Eq. (8.119) can be rewritten as

$$1 - \left(\frac{v_{nr}}{v_{ni}}\right)^2 = \left(\frac{v_s}{v_{ni}}\right)^2 \tag{8.121}$$

Introducing the coefficient of restitution defined in Eq. (8.36), Eq. (8.121) can be rewritten as

$$e_n = \left[1 - \left(\frac{v_s}{v_{ni}}\right)^2\right]^{1/2} \tag{8.122}$$

The variation of e_n with v_s/v_{ni} is presented in Fig. 8.22. It can be seen that the coefficient of restitution increases as the velocity ratio increases, and will approach unity if the velocity is sufficiently high.

8.3.2.3 Oblique Contact

The oblique interaction of adhesive elastic spheres was first analyzed by Savkoor and Briggs (1977), in which the JKR theory was extended to consider the tangential loading. They argued that, with the presence of adhesion, the tangential traction

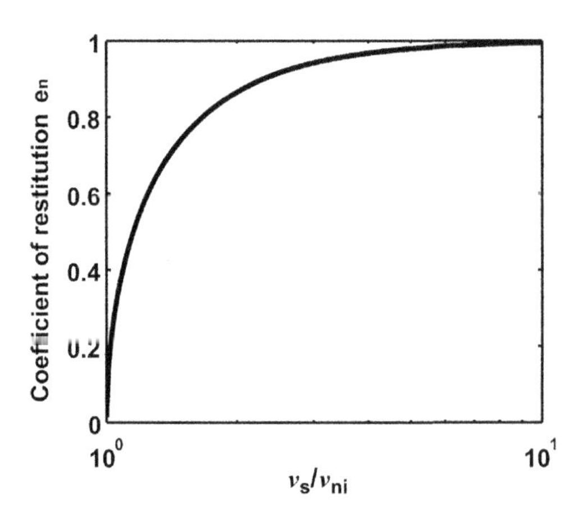

FIGURE 8.22

Coefficient of restitution for the impact of two adhesive spheres.

distribution in the contact area could be approximated with a "no-slip" solution of Mindlin (1949), i.e.,

$$q_{(r)} = \frac{F_t}{2\pi a}\left(a^2 - r^2\right)^{-1/2} \quad r \le a \tag{8.123}$$

in which the tangential force F_t is given by

$$F_t = 8G^* a\delta \tag{8.124}$$

It is further assumed that under the tangential force F_t the mechanical potential energy is $-F_t\delta$ and the elastic energy is $F_t\delta/2$. Thus, the tangential force induces a reduction of the potential energy by

$$U_T = -\frac{1}{2}F_t\delta = -\frac{F_t^2}{16G^* a} \tag{8.125}$$

The total potential energy of the system then becomes

$$\begin{aligned} U &= U_e + U_s + U_m + U_T = -\frac{1}{2}F_t\delta \\ &= \frac{1}{15}\left[\frac{9}{16E^{*2}R^*\hat{F}_n}\right]^{1/3}\left(\hat{F}_n^2 - 5\hat{F}_nF_n - 5F_n^2 - 30\pi\gamma R^*\hat{F}_n - \frac{5E^*}{4G^*}F_t^2\right) \end{aligned} \tag{8.126}$$

The system reaches equilibrium when $\partial U/\partial \hat{F}_n = 0$, from which the equivalent Hertzian force can be obtained with a stable solution as

$$\hat{F}_n = F_n + 6\pi\gamma R^* + \sqrt{(F_n + 6\pi\gamma R^*)^2 - F_n^2 - \frac{E^*}{4G^*}F_t^2} \tag{8.127}$$

Using Eq. (8.106), Eq. (8.127) can be rewritten as

$$\hat{F}_n = F_n + 2F_c + \sqrt{4F_nF_c + 4F_c^2 - \frac{E^*}{4G^*}F_t^2} \tag{8.128}$$

Substituting Eq. (8.128) into Eq. (8.95), the contact radius can then be determined as follows

$$a^3 = \frac{3\hat{F}_nR^*}{4E^*} = \frac{3R^*}{4E^*}\left(F_n + 2F_c + \sqrt{4F_nF_c + 4F_c^2 - \frac{E^*}{4G^*}F_t^2}\right) \tag{8.129}$$

It can be seen from Eq. (8.129) that the contact area decreases as the tangential force F_t increases. This is the so-called peeling mechanism proposed by Savkoor and Briggs (1977). They also suggested that the peeling process terminates when the tangential force F_t reaches

$$\tilde{F}_t = 4\sqrt{\frac{G^*}{E^*}}\left(F_nF_c + F_c^2\right) \tag{8.130}$$

If F_t exceeds $\tilde{F}_t$, there is no real solution to Eq. (8.129). The corresponding contact radius at the critical tangential force is

$$\tilde{a}^3 = \frac{3R^*}{4E^*}(F_n + 2F_c) \tag{8.131}$$

Equation (8.131) gives the smallest contact radius for oblique interaction of adhesive elastic spheres.

8.4 INTERACTION OF PARTICLES WITH LIQUID BRIDGES

Many industrial applications or natural systems involve mixtures of particles and liquids, as discussed in Chapter 6. Depending on the concentration and chemical compositions of the constituents, the liquid may be present as a mobile phase or an immobile adsorbed layer (Lian, 1994). The mobile liquid may be present in various states ranging from *pendular*, *funicular* or *capillary* liquid bridges to liquid drops, as the degree of saturation of the system increases (see Fig. 8.23). If only a small amount of liquid is present, a film of mobile liquid can be formed on the surface of a particle, leading to the formation of individual liquid bridges at the contact points between particles. This is referred to as the pendular state (Fig. 8.23(a)). As the amount of liquid increases, some voids in the particle system will be filled with the mobile liquid, in addition to the pendular liquid bridges; this is the funicular state (Fig. 8.23(b)). As the amount of liquid increases further, all voids in the particle system are completely filled with the liquid and a capillary state is reached (Fig. 8.23(c)). Further increase in the liquid leads to the formation of liquid droplets in which the particles are dispersed (Fig. 8.23(d)).

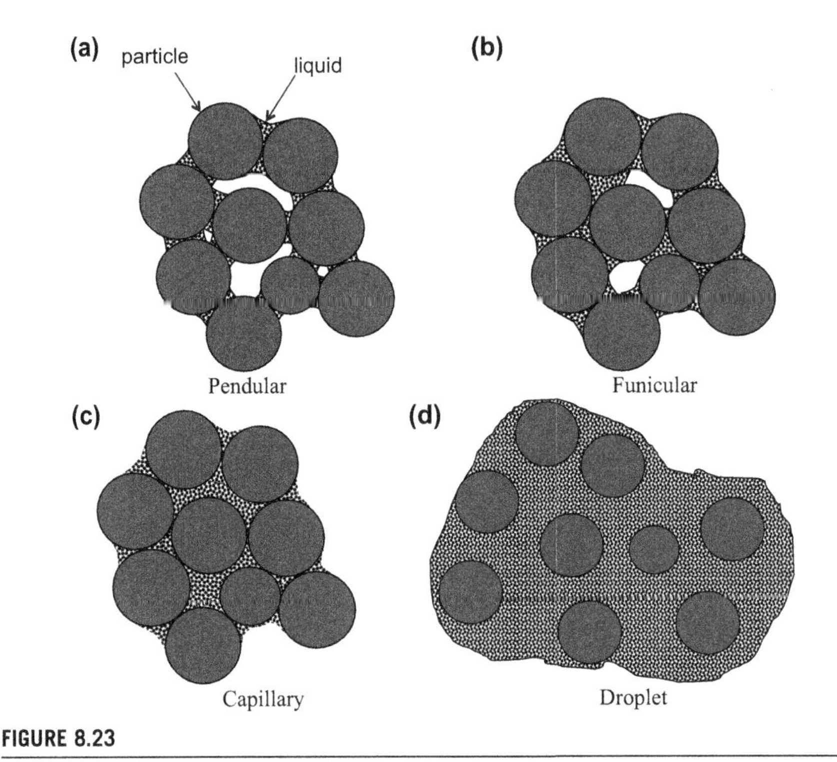

FIGURE 8.23

Distribution of liquids in particle systems at various degrees of saturation.

8.4.1 CAPILLARY FORCE

In this section, only the particle–particle interaction due to the presence of pendular liquid bridges will be discussed. Consider first a pendular liquid bridge between two identical spheres of radius R (Fig. 8.24). The separation distance between the two spheres is $2s$, and the liquid bridge has a half-angle β and contact angle θ at the particle surface. It is assumed that the buoyancy force and any distortion of the shape of the liquid bridge due to gravity can be ignored. The total capillary force of the pendular liquid bridge F_L is the sum of two components: (1) a surface tension force F_1, due to the surface tension forces acting at the circular gas-liquid interface, and (2) a pressure deficit force F_2, which arises from the fact that the pressure within the liquid bridge is generally less than that outside. The two are additive, i.e.,

$$F_L = F_1 + F_2 \tag{8.132}$$

As a first approximation, the liquid bridge can be approximated as toroidal with two principal radii of curvature: r_1 (the radius of the bridge surface in the plane of the figure) and r_2 (the radius of the neck of the bridge in the plane of symmetry, i.e., a

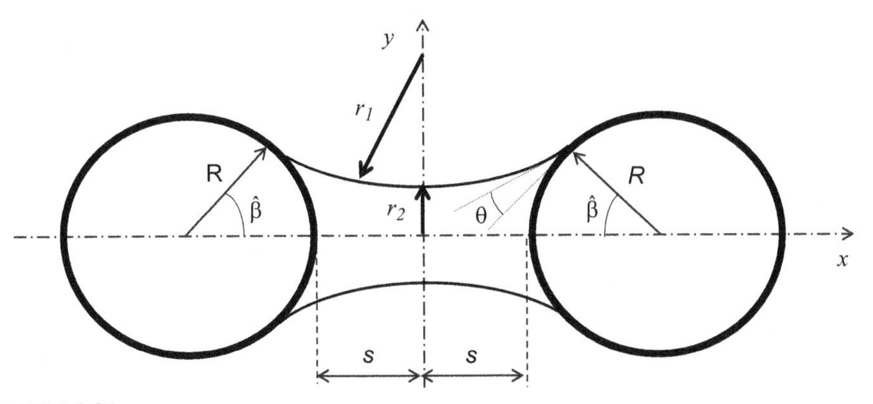

FIGURE 8.24

Illustration of a pendular liquid bridge between two equal sized spheres.

"cut" through the neck).[1] Fisher (1926) determined the axial surface tension force F_1 and the pressure deficit force F_2 acting at the plane of symmetry for contact (i.e., $s = 0$) as follows:

$$F_1 = 2\pi r_2 \gamma_L \tag{8.133}$$

$$F_2 = \pi r_2^2 \Delta P \tag{8.134}$$

where γ_L is the surface tension of the liquid-gas interface, hereafter termed the liquid surface tension, ΔP is the reduced hydrostatic pressure (i.e., the difference between the pressures inside and outside the bridge). Hence the total capillary force is

$$F_L = F_1 + F_2 = 2\pi r_2 \gamma_L + \pi r_2^2 \Delta P \tag{8.135}$$

For perfect wetting (i.e., the contact angle θ is zero), the reduced hydrostatic pressure within the liquid bridge ΔP can be determined according to the Laplace equation as (Seville et al., 1997)

$$\Delta P = \gamma_L \left[\frac{1}{r_1} - \frac{1}{r_2} \right] \tag{8.136}$$

Substituting Eq. (8.136) into Eq. (8.135) gives,

$$F_L = \pi r_2 \gamma_L \left(\frac{r_1 + r_2}{r_1} \right) \tag{8.137}$$

[1]Closer consideration of the geometry shows that the bridge cannot be truly toroidal since according to the Laplace equation (Eq 8.136) $1/r_1$ - $1/r_2$ = a constant, since the pressure is the same everywhere in the bridge. It is apparent that r_1 depends on where on the bridge it is evaluated, being smallest at the neck, so r_2 cannot be a constant and the shape of the bridge is not an arc in the plane of the figure. Nevertheless, the toroidal approximation is satisfactory for many applications especially if the liquid loading is small.

By trigonometry,

$$r_1 = R\left(\sec\hat{\beta} - 1\right) \tag{8.138}$$

$$r_2 = R\left(1 + \tan\hat{\beta} - \sec\hat{\beta}\right) \tag{8.139}$$

Substituting Eqs (8.138) and (8.139) into Eq. (8.137) yields

$$F_\mathrm{L} = \frac{2\pi R\gamma_\mathrm{L}}{1 + \tan\left(\hat{\beta}/2\right)} \tag{8.140}$$

Equation (8.140) indicates that the capillary force in a pendular liquid bridge increases as the liquid surface tension γ_L increases. It also increases as the half-angle $\hat{\beta}$ decreases, i.e., as the size of the liquid bridge decreases. Equation (8.140) implies that a maximum value of the capillary force F_L is obtained as $\hat{\beta}$ approaches zero, i.e., when the liquid content is zero. This is counter-intuitive, as one would expect wet powders to be more cohesive than dry ones. For example, one can build a strong sandcastle with wet beach sand but not with dry sand. The prediction of maximum bridge strength at zero liquid content is the so-called Fisher's paradox. The problem arises from the assumption that the minimum separation distance s is zero (Pietsch, 1968). In reality, due to the surface roughness, there is always an effective finite separation (i.e., $s > 0$, as illustrated in Fig. 8.25), which should be considered in the theory. With a finite separation distance $s > 0$, Eqs (8.138) and (8.139) can be rewritten as

$$r_1 = R\left[(1 + s')\sec\hat{\beta} - 1\right] \tag{8.141}$$

$$r_2 = R\left[1 + (1 + s')\tan\hat{\beta} - (1 + s')\sec\hat{\beta}\right] \tag{8.142}$$

where $s' = s/R$. The variations of F_L with $\hat{\beta}$ at various separation distances are shown in Fig. 8.26. It can be seen that even a very small separation distance can make a dramatic difference to the prediction of the capillary force. With a finite separation distance, the capillary force reduces to zero as β approaches zero, and there is a maximum capillary force at intermediate half-angles.

For more general cases (e.g., when the contact angle is not zero, and/or the shape of the liquid bridge is more complex), the reduced hydrostatic

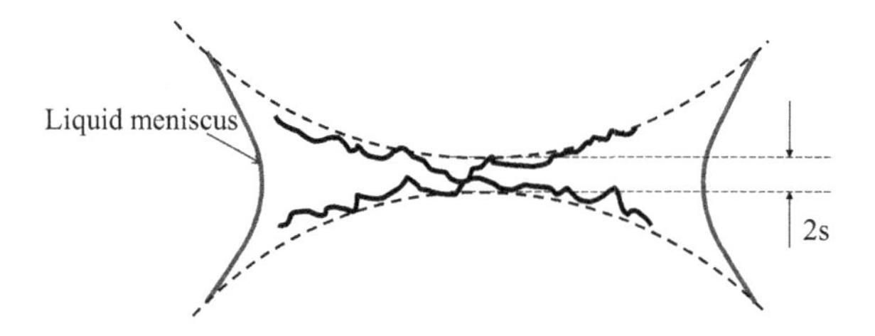

FIGURE 8.25

Illustration of a liquid bridge between two equal sized spheres with rough surfaces.

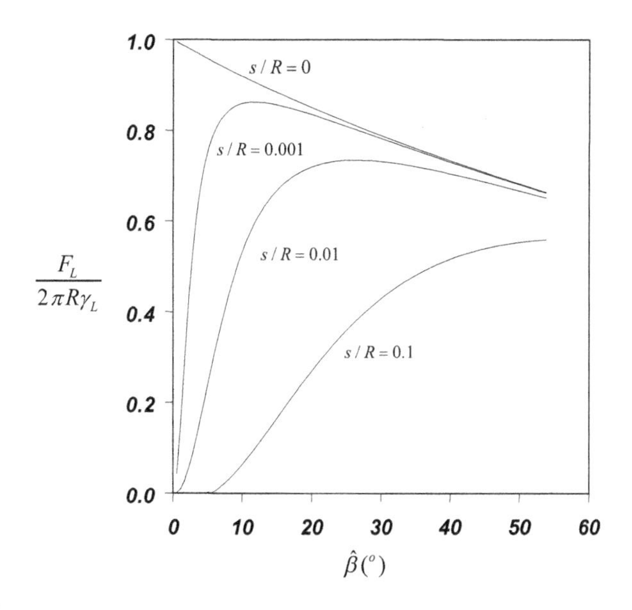

FIGURE 8.26

The variation of normalized capillary force $F_L/2\pi R\gamma_L$ with the bridge half-angle $\hat{\beta}$ at various dimensionless separation s/R.

pressure ΔP can be determined from the Laplace-Young equation expressed as (Lian et al., 1993)

$$\Delta P = \frac{\gamma_L}{R}\left[\frac{\ddot{X}}{\left(1+\dot{X}^2\right)^{3/2}} - \frac{1}{X\left(1+\dot{X}^2\right)^{1/2}}\right] \tag{8.143}$$

where X is the shape function of the liquid bridge denoting the dimensionless coordinates with respect to the sphere radius R and the dot notation denotes differentiation with respect to x. Equation (8.143) can be solved numerically, as illustrated by Lian et al. (1993) who also obtained an expression for the capillary force as

$$F_L = 2\pi R\gamma_L X_0(1 + H^* X_0) \tag{8.144}$$

where H^* is the dimensionless mean curvature and given as

$$H^* = \frac{\Delta PR}{2\gamma_L} \tag{8.145}$$

and

$$X_0 = C \quad (H^* = 0) \tag{8.146}$$

$$X_0 = \frac{-1 + \sqrt{1 + 4H^* C}}{2H^*} \quad (H^* \neq 0) \tag{8.147}$$

where

$$C = \sin\hat{\beta}\sin\left(\hat{\beta}+\theta\right) + H^* \sin^2\hat{\beta} \tag{8.148}$$

If the shape of the pendular liquid bridge can be determined or approximated analytically, i.e., the function X is known, Eqs (8.143)–(8.148) can be used to calculate the capillary force. Using this approach, Lian et al. (1993) also analyzed the capillary force between two equal spheres with an initial separation $2s$ and a contact angle θ. Based on the toroidal approximation, the capillary force F_L is given as

$$F_L = 2\pi\gamma_L r_2(1+Hr_2) \tag{8.149}$$

where H is the mean curvature of the liquid bridge evaluated at the neck and

$$H = \frac{\Delta P}{2\gamma_L} = \frac{r_2 - r_1}{2r_1 r_2} \tag{8.150}$$

r_1 and r_2 can be determined from the geometry of the liquid bridge:

$$r_1 = \frac{s + R\left(1-\cos\hat{\beta}\right)}{\cos\left(\hat{\beta}+\theta\right)} \tag{8.151}$$

$$r_2 = R sin\,\beta - r_1[1 - sin(\beta+\theta)] \tag{8.152}$$

Based on numerical integration of the Laplace–Young equation, Willett et al. (2000) obtained a closed form approximation for the capillary force between two equal spheres:

$$F_L = \frac{2\pi R\gamma_L\cos\theta}{1 + 2.1\hat{s} + 10\hat{s}^2} \tag{8.153}$$

where $\hat{s}$ is the dimensionless half-separation distance and given as

$$\hat{s} = \frac{s}{\sqrt{V_L/R}} \tag{8.154}$$

and V_L is the volume of the liquid bridge.

For the contact of spheres of unequal sizes, the application of the toroidal approximation becomes more complicated. Israelachvili (1991) suggested that the Derjaguin approximation could be used to simplify the calculation of capillary forces. In the Derjaguin approximation, a harmonic mean sphere radius R_{12} is introduced and defined by

$$\frac{2}{R_{12}} = \frac{1}{R_1} + \frac{1}{R_2} \tag{8.155}$$

With the substitution of R by R_{12}, the expressions derived for equal spheres can be used to estimate the capillary forces between spheres of different radii R_1 and R_2. It has been demonstrated by Willett et al. (2000) that this approximation is reasonable for nonzero separation distance within the limit of rupture distance (i.e., $0 < s < s_c$), where s_c is the rupture distance (see Section 8.4.2).

8.4.2 RUPTURE DISTANCE

For pendular liquid bridges, it is reasonable to assume (in the absence of evaporation or drying) that the volume of the liquid bridge V_L is constant, and the liquid surface tension γ_L and the contact angle θ are intrinsic properties of the liquid and solid, thus independent of the separation distance. As the separation distance increases, the radius of the neck of the liquid bridge (i.e., r_2) decreases. Eventually the bridge will rupture (i.e., break). The stability of liquid bridges and their rupture distance is of importance in practical applications, such as agglomeration (Section 6.3). Lian et al. (1993) showed that the rupture distance of a liquid bridge between two equal sized spheres is proportional to the cube root of the liquid volume, i.e.,

$$s_c = \frac{1}{2}(1 + 0.5\theta)V_L^{1/3} \tag{8.156}$$

where s_c is half of the rupture distance and the contact angle θ is in radians.

Extending the analysis, Willett et al. (2000) proposed a similar expression for the rupture distance of liquid bridges between two equal sized spheres:

$$s_c = \frac{1}{2}(1 + 0.5\theta)\left(V_L^{1/3} + \frac{V_L^{2/3}}{10}\right) \tag{8.157}$$

The predictions from Eqs (8.156) and (8.157) are compared in Fig. 8.27, which shows that a slightly greater rupture distance is obtained from Eq. (8.157) than that from Eq. (8.156).

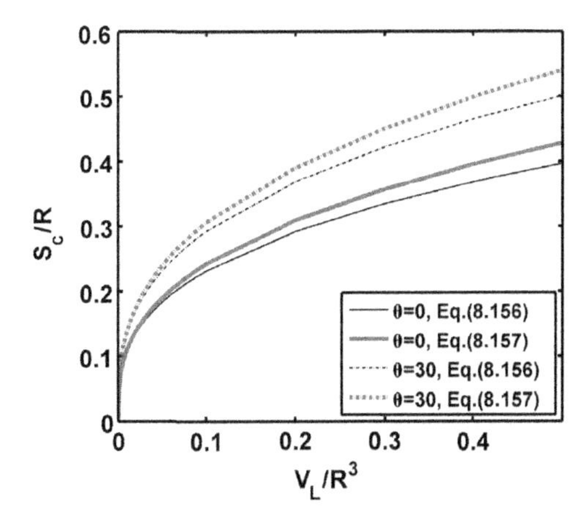

FIGURE 8.27

The variation of s_c/R with V_L/R^3 for liquid bridges with different contact angles. Predicted from Eqs (8.156) and (8.157).

For pendular liquid bridges between two spheres of different sizes R_1 and R_2 ($R_1 > R_2$), the rupture distance can be approximated by (Willett et al., 2000)

$$s_c = \frac{1}{2}\left[1 + \frac{\theta}{4}\left(\frac{R_2}{R_1} + 1\right)\right]\left[V_L^{1/3} + \left(\frac{R_2}{2R_1} - \frac{2}{5}\right)V_L^{2/3}\right] \quad (8.158)$$

8.4.3 DYNAMIC FORCES

When two spheres of radius R and separation $2s$ are separated at a rate of $2v$ (where $v = ds/dt$), the dynamic liquid bridge force F_d, also known as the viscous force, can be determined using the Reynolds lubrication equation (Seville et al., 1997):

$$F_d = \frac{3\pi\mu_L R^2 v}{2s} \quad (8.159)$$

where μ_L is the viscosity of the liquid. The Reynolds lubrication equation for a sphere of radius R at a separation distance s from a flat surface and moving at a velocity v is given as

$$F_d = \frac{6\pi\mu_L R^2 v}{s} \quad (8.160)$$

8.4.4 IMPACT WITH THE PRESENCE OF A LIQUID LAYER

The impact of particles onto a surface covered by a liquid layer occurs in many industrial processes, such as agglomeration, fluidized bed coating, filtration, sedimentation, and wet grinding, and in nature, e.g., pollen capture in plants. The involvement of a liquid layer during the impact leads to a much more complicated impact behavior than the impact of dry particles as discussed in previous sections. In particular, it will affect the rebound behavior of the particle, which is more likely to stick to a wet surface.

Consider the simple case of a spherical particle of radius R impacting on a flat surface covered with a liquid layer at velocity v, as illustrated in Fig. 8.28. The thickness of the liquid layer is h. At time t, the separation between the sphere and the surface is s, and the equations of motion of the particle are given as

$$\frac{ds}{dt} = -v(t) \quad (8.161)$$

and

$$m\frac{dv}{dt} = -F(t) \quad (8.162)$$

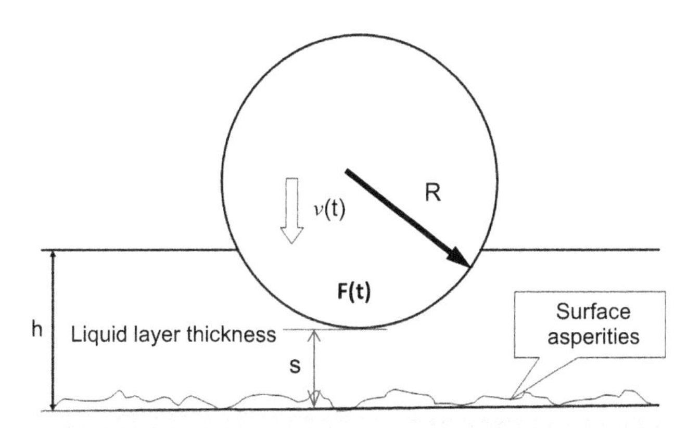

FIGURE 8.28

Illustration of a sphere colliding with a flat surface covered with a liquid layer of thickness *h*. (Barnocky and Davis, 1988).

The forces acting on the particle generally include the hydrodynamic force due to direct interaction with the liquid, the gravitational force, and the van der Waals force. The hydrodynamic force generally dominates over the others while the particle is partially immersed in the liquid layer (Seville et al., 1997). Kantak and Davis (2006) showed that the hydrodynamic force could be determined as

$$F(t) = \frac{6\pi\mu_L R^2 v}{s(t)}\left[1 - \frac{s(t)}{h}\right]^2 \tag{8.163}$$

They also pointed out that the second term in the bracket of Eq. (8.163) can be neglected once the particle penetrates a significant distance into the liquid layer (i.e., $s \ll h$). In this case Eq. (8.163) is reduced to the Reynolds lubrication equation, i.e., Eq. (8.160).

Substituting Eq. (8.163) into Eq. (8.162), and integrating with the initial condition $v = v_0$ at $s = h$ gives (Kantak and Davis, 2006)

$$\frac{v}{v_0} = 1 - \frac{1}{St}\left[\ln\left(\frac{h}{s}\right) + 2\frac{s}{h} - \frac{1}{2}\left(\frac{s}{h}\right)^2 - \frac{3}{2}\right] \tag{8.164}$$

where *St* is the Stokes number representing the relative importance of the inertial and viscous forces (see also Box 4.3) and defined as

$$St = \frac{mv_0}{6\pi\mu_L R^2} \tag{8.165}$$

Equation (8.164) indicates, as expected, that the particle slows down as it penetrates the liquid layer, due to viscous resistance. There are now three possibilities: (1)

the particle may be slowed down so much by the viscous drag of the liquid that it loses essentially all its momentum before contacting the surface beneath and is "captured", (2) the particle may reach the surface, rebound but have insufficient energy to escape the liquid layer, and is again captured, or (3) the particle may retain enough energy after rebound that it can escape the liquid layer and emerge into the gas.

The hydrodynamic force can be calculated using the Reynolds lubrication equation, i.e., Eq. (8.160), if the following conditions are met (Barnocky and Davis, 1988):

1. The pressure in the liquid layer is not high enough to cause deformation of the surfaces.
2. The Reynolds number based on the gap dimension is small, i.e.,

$$Re = \frac{\rho v}{\mu_L} << 1 \tag{8.166}$$

3. The separation distance is very small compared to the radius of the particle, i.e.,

$$s \ll R \tag{8.167}$$

4. The nose of the particle must be wetted over an area of radius $(Rs)^{1/2}$, which is true when $s = \hat{s} = 2h/3$.

Based upon the above conditions, Eqs (8.162) and (8.163) can be used to obtain

$$m\frac{dv}{ds} = \frac{F}{v} = \frac{6\pi\mu_L R^2}{s} \tag{8.168}$$

Integrating Eq. (8.168) with the initial condition $v = \hat{V}$ at $s = \hat{s}$ yields

$$\frac{v}{\hat{V}} = 1 - \frac{1}{St}\ln\left(\frac{\hat{s}}{s}\right) \tag{8.169}$$

in which $\hat{V}$ can be approximated with v_0 (i.e., neglecting the slowing of the particle up to $s = \hat{s}$).

Equation (8.169) implies that there is a critical Stokes number below which capture occurs, i.e., $v = 0$ at $St = St_c$, and

$$St_c = \ln\left(\frac{\hat{s}}{s_m}\right) \tag{8.170}$$

where $s_m = \left(\frac{4\mu_L v_0 R^{3/2}}{\pi E^*}\right)^{2/5}$ is the minimum separation.

In practice, particle surfaces are rough to some extent. It is very likely that asperities on the rough surfaces are greater than s_m. For this case, the surfaces contact on the tips of the asperities (Kantak and Davis, 2004), and the corresponding critical Stokes number St_c, below which particles will be captured on the rough surfaces is

$$St_c = \ln\left(\frac{\hat{s}}{s_a}\right) \tag{8.171}$$

where s_a is the asperity height. Moreover, the resistance of the liquid layer after rebound from the surface must be taken into account, as the pressures generated in the liquid are not low enough to induce cavitation (Seville et al., 1997). Using Eq. (8.169), the velocity at the point of first contact with the asperities can be given as

$$v_{io} = \hat{V}\left[1 - \frac{1}{St}\ln\left(\frac{\hat{s}}{s_a}\right)\right] \tag{8.172}$$

Assuming the coefficient of restitution is e_n, the rebound velocity is hence

$$v_{ro} = -e_n v_{io} = -e_n \hat{V}\left[1 - \frac{1}{St}\ln\left(\frac{\hat{s}}{s_a}\right)\right] \tag{8.173}$$

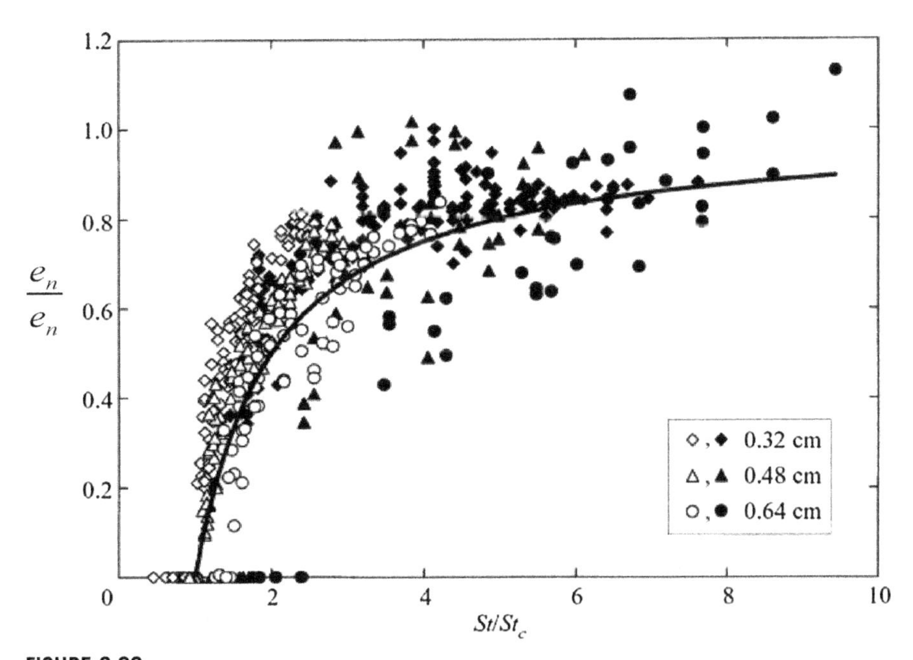

FIGURE 8.29

The ratio of the coefficients of restitution for wet and dry collisions $\tilde{e}_n/e_n$ versus the ratio St/St_c the solid line is the prediction of Eq. (8.174) (Davis et al., 2002).

Integrating Eq. (8.168) with $v = v_{ro}$ at $s = s_a$ and $-v$=0 at $s = \hat{s}$, we have

$$St_c = \left(1 + \frac{1}{e_n}\right) \ln\left(\frac{\hat{s}}{s_a}\right) \tag{8.174}$$

If $e_n \approx 1$, Eq. (8.174) becomes $St_c \approx 2\ln(\hat{s}/s_a)$. The experimental results of Barnocky and Davis, 1988, See Fig. 8.29) demonstrate that Eq. (8.174) can predict the critical Stokes number. The above analysis can be used to understand the "capture" of particles in processes such as agglomeration and filtration

Davis et al. (2002) developed an elasto-hydrodynamic theory for the collision of elastic spheres with wet surfaces, which is coupled with the Hertz theory to calculate the elastic deformation. The coefficient of restitution for the wet impact $\tilde{e}_n$ is given as

$$\begin{cases} \tilde{e}_n = e_n\left(1 - \dfrac{St_c}{St}\right) & St > St_c \\ \tilde{e}_n = 0 & St \le St_c \end{cases} \tag{8.175}$$

where e_n is the corresponding coefficient of restitution for dry collisions (i.e., without the liquid layer). Figure 8.29 shows the comparison of the prediction of Eq. (8.176) with the experimental measurements, demonstrating that Eq. (8.176) gives a good prediction of the coefficient of restitution for wet impacts.

REFERENCES

Andrews, J.P., 1930. Experiments on impact. Proceedings of the Physical Society 43, 8–17.

Barnocky, G., Davis, R.H., 1988. Elastohydrodynamic collision and rebound of spheres: experimental verification. Physics of Fluids 31, 1324–1329.

Bergström, L., 1997. Hamaker constants of inorganic materials. Advances in Colloid and Interface Science 70, 125–169.

Bitter, J.G.A., 1963. A study of erosion phenomena, part I. Wear 6, 5–21.

Brach, R.M., 1988. Impact dynamics with applications to solid particle erosion. International Journal of Impact Engineering 7, 37–53.

Bridges, F.G., Hatzes, A., Lin, D.N.C., 1984. Structure, stability and evolution of Saturn's rings. Nature 309, 333–335.

Bryant, M.D., Keer, L.M., 1982. Rough contact between elastically and geometrically identical curved bodies. Transactions of the American Society of Mechanical Engineers, and Journal of Applied Mechanics 49, 345–352.

Cappella, B., Dietler, G., 1999. Force-distance curves by atomic force microscopy. Surface Science Reports 34 (1–3), 1–104.

Davis, R.H., Rager, D.A., Good, B.T., 2002. Elastohydrodynamic rebound of spheres from coated surfaces. Journal of Fluid Mechanics 468, 107–119.

Derjaguin, B.V., Muller, V.M., Toporov, YuP., 1975. Effect of contact deformations on the adhesion of particles. Journal of Colloid and Interface Science 53, 314.

Fisher, R.A., 1926. On the capillary forces in an ideal soil, correction of formulae given by W.B. Haines. Journal of Agricultural Science 16, 492–505.

Goldsmith, W., Lyman, P.T., 1960. The penetration of hard-steel spheres into plane metal surfaces. Transactions of the American Society of Mechanical Engineers, and Journal of Applied Mechanics 27, 717–725.

Gorham, D.A., Kharaz, A.H., 2000. Measurement of particle rebound characteristics. Powder Technology 112, 193–202.

Hamaker, H.C., 1937. The London – van der Waals attraction between spherical particles. Physica 4 (10), 1058–1072.

Hamilton, G.M., Goodman, L.E., 1966. The stress field created by a circular sliding contact. Transactions of the American Society of Mechanical Engineers, and Journal of Applied Mechanics 33, 371–376.

Hertz, H., 1896. In: Jones, Schott (Eds.), Miscellaneous Papers. Macmillan and Co, London.

Hill, R., 1950. The Mathematical Theory of Plasticity. Oxford Univ. Press, London.

Hutchings, I.M., 1979. Energy absorbed by elastic waves during plastic impact. Journal of Physics D: Applied Physics 12, 1819–1824.

Israelachvili, J.N., 1991. Intermolecular and Surface Forces. Elsevier, London.

Johnson, K.L., 1976. Adhesion at the contact of solids. In: Koiter (Ed.), Theoretical and Applied Mechanics, Proc. 4th IUTAM Congress. North Holland, Amsterdam, p. 133.

Johnson, K.L., 1983. The bounce of 'superball'. International Journal of Mechanical Engineering Education 111, 57–63.

Johnson, K.L., 1985. Contact Mechanics. Cambridge University Press.

Johnson, K.L., Jefferis, J.A., 1963. Plastic flow and residual stresses in rolling and sliding contact. In: Proc. I. Mech. E. Symp. On Rolling Contact Fatigue. Institute of Mechanical Engineers, London, pp. 54–65.

Johnson, K.L., Kendall, K., Roberts, A.D., 1971. Surface energy and the contact of elastic solids. Proceedings of the Royal Society of London A324, 30.

Kantak, A.A., Davis, R.H., 2004. Oblique collisions and rebound of spheres from a wetted surface. Journal of Fluid Mechanics 509, 63–81.

Kantak, A.A., Davis, R.H., 2006. Elastohydrodynamic theory for wet oblique collisions. Powder Technology 168 (1), 42–52.

Kharaz, A.H., Gorham, D.A., Salman, A.D., 2001. An experimental study of the elastic rebound of spheres. Powder Technology 120, 281–291.

Labous, L., Rosato, A.D., Dave, R.N., 1997. Measurements of collisional properties of spheres using high-speed video analysis. Physical Review. E 56, 5717–5725.

Lian, G., 1994. Computer Simulation of Moist Agglomerate Collisions (Ph.D. thesis). University of Aston.

Lian, G., Adams, M.J., Thornton, C., 1996. Elastohydrodynamic collisions of solid spheres. Journal of Fluid Mechanics 311, 141–152.

Lian, G., Thornton, C., Adams, M.J., 1993. A theoretical study of the liquid bridge forces between two rigid spherical bodies. Journal of Colloid and Interface Science 161, 138–147.

Love, A.E.H., 1952. A Treatise on the Mathematical Theory of Elasticity, fourth ed. Cambridge University Press, Cambridge.

Maw, N., Barber, J.R., Fawcett, J.N., 1976. The oblique impact of elastic spheres. Wear 38, 101–114.

Maw, N., Barber, J.R., Fawcett, J.N., 1981. The role of elastic tangential compliance in oblique impact. Transactions of the American Society of Mechanical Engineers, and Journal of Lubrication Technology 103, 74–80.

Mesarovic, S.D., Johnson, K.L., 2000. Adhesive contact of elastic-plastic spheres. Journal of the Mechanics and Physics of Solids 48, 2009–2033.

Mindlin, R.D., 1949. Compliance of elastic bodies in contact. Transactions of the American Society of Mechanical Engineers, and Journal of Applied Mechanics 16, 259–268.

Mindlin, R.D., Deresiewicz, H., 1953. Elastic spheres in contact under varying oblique force. Transactions of the American Society of Mechanical Engineers, and Journal of Applied Mechanics 20, 327–344.

Muller, V.M., Yuschenko, V.S., Derjaguin, B.V., 1980. On the influence of molecular forces on the deformation of an elastic sphere and its sticking to a rigid plane. Journal of Colloid and Interface Science 77, 91.

Muller, V.M., Yushchenko, V.S., Derjaguin, B.V., 1983. General theoretical consideration of the influence of surface forces on contact deformation. Journal of Colloid and Interface Science 92, 92.

Pietsch, W.B., 1968. Tensile strength of granular materials. Nature (London) 217, 736.

Savkoor, A.R., Briggs, G.A.D., 1977. Effect of tangential force on the contact of elastic solids in adhesion. Proceedings of the Royal Society of London A 356, 103–114.

Seville, J.P.K., Tüzün, U., Clift, R., 1997. Processing of Particulate Solids. Blackie Academic & Professional, London.

Tabor, D., 1948. A simple theory of static and dynamic hardness. Proceedings of the Royal Society of London A192, 247–274.

Tabor, D., 1951. Hardness of Metals. Oxford University Press, Oxford.

Tabor, D., 1977. Surface forces and surface interactions. Journal of Colloid and Interface Science 58, 2.

Thornton, C., 1997. Coefficient of restitution for collinear collisions of elastic-perfectly plastic spheres. Transactions of the American Society of Mechanical Engineers, and Journal of Applied Mechanics 64, 383–386.

Thornton, C., Ning, Z., 1998. A theoretical model for the stick/bounce behaviour of adhesive, elastic-plastic spheres. Powder Technology 99, 154–162.

Thornton, C., Ning, Z., Wu, C.-Y., Nasrullah, M., Li, L.-Y., 2001. Contact mechanics and coefficients of restitution. In: Poschel, T., et al. (Eds.), Granular Gases. Springer, pp. 56–66.

Timoshenko, S., Goodier, J.N., 1951. Theory of Elasicity, third ed. McGraw-Hill, New York.

Walton, O.R., 1992. Numerical simulation of inelastic, frictional particle–particle interactions. In: Roco, M.C. (Ed.), Particulate Two-phase Flow, Ch. 25. Butterworth-Heinemann, Boston.

Walton, O.R., Braun, R.L., 1986. Viscosity, granular-temperature, and stress calculations for shearing assemblies of inelastic, frictional disks. Journal of Rheology 30, 949–980.

Whittaker, E.T., 1904. A Treatise on the Analytical Dynamics of Particles and Rigid Bodies. Cambridge University Press, London.

Willett, C.D., Adams, M.J., Johnson, S.A., Seville, J.P.K., 2000. Capillary bridges between two spherical bodies. Langmuir 16 (24), 9396–9405.

Wu, C.Y., 2002. Finite Element Analysis of Particle Impact Problems (Ph.D. thesis). University of Aston.

Wu, C.Y., Li, L.Y., Thornton, C., 2003. Rebound behaviour of spheres for plastic impacts. International Journal of Impact Engineering 28 (9), 929–946.

Wu, C.Y., Li, L.Y., Thornton, C., 2005. Energy dissipation during normal impact of elastic-plastic spheres. International Journal of Impact Engineering 32 (1–4), 593–604.

Wu, C.Y., Thornton, C., Li, L.Y., 2009. A semi-analytical model for oblique impacts of elastoplastic spheres. Proceedings of the Royal Society of London A 465, 937–960.

CHAPTER

Discrete Element Methods

9

The discrete element method (DEM), also know as the discrete particle method, is a numerical method for modeling micromechanics and dynamics of particle systems (i.e., assemblies of individual particles). It originated with the seminal paper of Cundall and Strack on "a discrete numerical model for granular assemblies" in 1979. Since then, it has been significantly advanced by many researchers. Although DEM was initially developed for analyzing soil mechanics and geotechnical problems, it has been adapted and used extensively to study the mechanics of bulk materials and to model powder handling and processing.

In DEM, a bulk material is treated as a system consisting of many individual solid particles that interact with each other according to certain interaction laws, depending on their material and interfacial properties. The particles may have complex shapes, and significant advances have been made in modeling irregular-shaped particles using DEM in recent years. More usually, though, particles have been modeled as disks in 2D or spheres in 3D in most DEM codes. The reason that spheres are often used in DEM modeling is clearly because rigorous interaction models based upon contact mechanics are well established for spheres, as discussed in Chapter 8. Moreover, for spherical particles, contact detection and modeling can be easily implemented and efficiently executed so that a large number of particles (say >1,000,000 particles) can be modeled in a reasonable timescale.

9.1 HARD-SPHERE AND SOFT-SPHERE DEMs

The core of DEM is to model the interactions between particles, which are generally dynamic processes, and an explicit time-dependent finite difference technique (see Section 9.2.1) is normally used to analyze their progressive motion. According to the choice of particle interaction model used, DEM can be classified into two branches: (1) hard-sphere DEM and (2) soft-sphere DEM.

In hard-sphere DEM, it is assumed that all particle interactions are *binary* and *instantaneous* and there is no enduring contact (i.e., contacts are collisional and the contact duration is assumed to be very small). The kinematics (e.g., velocities and positions) of particles after each collision are determined from their incident conditions and the collisional models governing the instantaneous interactions between particles, such as instantaneous momentum exchange. In other words, the particle rebound velocities are determined from the incident velocities, coefficients

Particle Technology and Engineering.

of restitution, and friction. The contact force between particles is neither needed nor calculated. Hard-sphere DEM can be used to simulate dilute, rapid granular flow dominated by particle collisions, such as that observed in *granular gases* (Cundall and Hart, 1992).

In soft-sphere DEM, it is assumed that particles can deform microscopically at the contact point as a result of friction and stresses. It is further assumed that the extent of contact deformation (i.e., penetration or overlap) is very small compared to the particle dimensions (Cundall and Hart, 1992), so that macroscopic deformation of the particle is negligible. A contact stiffness is then used to describe the relationship between the contact deformation and the magnitude of the contact force. The soft-sphere method is therefore capable of modeling multiple particle contacts simultaneously, in particular for quasi-static systems (Zhu et al., 2007). It is also suitable for modeling mechanical and dynamic behavior of particulate materials that is governed by particle–particle interactions, especially for dense particle systems and applications in which contact forces and stresses are of great concern. The discussion in this book is therefore constrained to cover only soft-sphere DEM, which is referred to simply as DEM hereafter.

9.2 THE PRINCIPLE OF DEM

DEM generally involves cyclic calculations with a very small time increment (i.e., time step Δt). At each time step (t_i), the evolution of the system is advanced to obtain new particle positions and velocities based upon the updated forces acting on the particles. A typical DEM algorithm is illustrated in Fig. 9.1. Each calculation cycle consists of three key steps:

1. Updating particle positions (see Section 9.2.1);
2. Contact detection (Section 9.2.2);
3. Contact modeling, i.e., updating the contact forces (Section 9.2.3).

These steps are discussed in detail in the following subsections.

9.2.1 UPDATING PARTICLE POSITIONS

The motion of each particle (say particle i) is governed by Newton's second law of motion. Based upon the resultant force $\mathbf{F}_i$ acting on the particle i with mass m_i and moment of inertia I_i, its translational and rotational motions are determined as follows:

$$m_i \frac{d\mathbf{v}_i}{dt} = \mathbf{F}_i \tag{9.1}$$

$$I_i \frac{d\boldsymbol{\omega}_i}{dt} = \mathbf{T}_i \tag{9.2}$$

in which, $\mathbf{v}_i$ and $\boldsymbol{\omega}_i$ are the translational and angular velocities, respectively, and $\mathbf{T}_i$ is the total torque. The resultant force $\mathbf{F}_i$ is the net force resulting from the sum of the

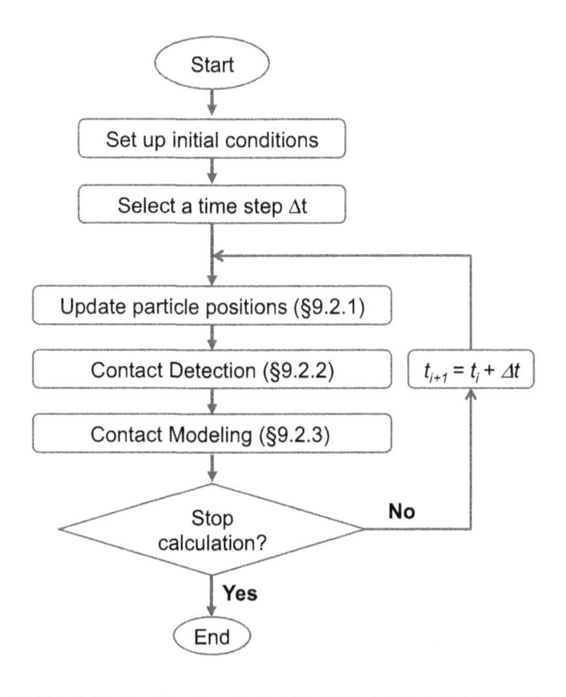

FIGURE 9.1

A typical DEM algorithm.

various individual forces acting on the particles, including gravitational forces, mechanical contact forces, electrostatic forces, fluid–particle interaction forces and cohesive forces.

Numerical integration of Eqs (9.1) and (9.2) is usually achieved using a central finite difference scheme, in which a fixed time step Δt is used and new velocities and positions of each particle are calculated as follows:

$$\mathbf{v}_i^{t+\Delta t/2} = \mathbf{v}_i^{t-\Delta t/2} + \frac{d\mathbf{v}_i}{dt}\Delta t \tag{9.3a}$$

$$\boldsymbol{\omega}_i^{t+\Delta t/2} = \boldsymbol{\omega}_i^{t-\Delta t/2} + \frac{d\boldsymbol{\omega}_i}{dt}\Delta t \tag{9.3b}$$

and

$$\mathbf{x}_i^{t+\Delta t} = \mathbf{x}_i^{t} + \mathbf{v}_i^{t+\Delta t/2}\Delta t \tag{9.4a}$$

$$\theta_i^{t+\Delta t} = \theta_i^{t} + \theta_i^{t+\Delta t/2}\Delta t \tag{9.4b}$$

where x_i and θ_i are the coordinates and rotational displacements of particle *i*, respectively.

9.2.2 CONTACT DETECTION

Once the particle positions are updated, it is necessary to detect whether new contacts between particles have been established and if any existing contact has been lost. Since DEM simulations routinely involve thousands or even millions of particles and the contact detection needs to be performed for each particle in the system in each calculation cycle, an efficient contact-searching algorithm is critical to reducing the computing time for contact detection and to improving the overall computing efficiency. For this purpose, two-phase contact searching schemes are commonly employed, which include a presorting phase and a subsequent precise contact detection phase.

The purpose of the presorting phase is to establish the neighborhood for each particle because contacts are only possible between a particle and those in the local vicinity (i.e., in its neighborhood). In other words, presorting aims to identify any potential contacting particle. The size of the neighborhood depends on the nature of the contact: for mechanical interaction that occurs when particles physically touch each other, the neighborhood can be confined to the immediate vicinity, while for interaction induced by long-range forces, such as electrostatic interaction, the neighborhood has to be large enough to ensure that all potential interactions are accounted for (Pei et al., 2015a). The presorting can be performed using two approaches: (1) a cell-based method (Fig. 9.2(a)), also known as the grid method or the boxing method, and (2) a Verlet list method (Fig. 9.2(b)).

In the cell-based method, the computation domain is divided into a number of regular cells or boxes. All particles are mapped into the cells according to their positions, and each cell has a link list of associated particles. A particle is mapped into a cell if there is any overlap between them. For example, in Fig. 9.2(a), particle 1 is only mapped into cell C1; particle 2 is mapped into two cells: B3 and C3, while particle 3 is mapped into four cells: D3, D4, E3, and E4. With the cell-based method, precise contact detection only needs to be carried out between the particle and those

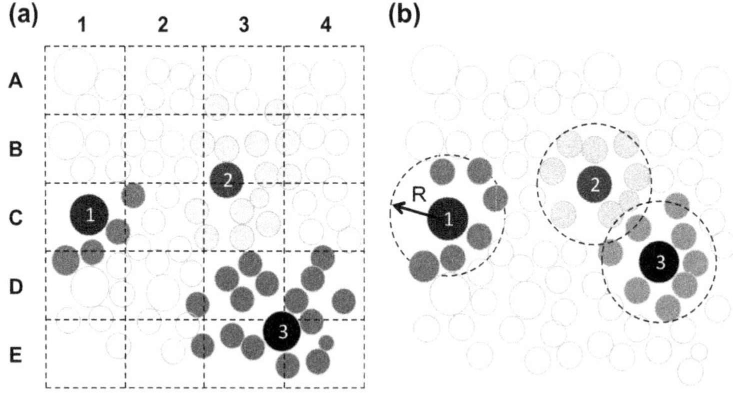

FIGURE 9.2

Illustration of (a) the cell-based method and (b) the Verlet list method for presorting.

sharing the same cell(s) with it, i.e., those mapped into the same cell(s). For the example illustrated in Fig. 9.2(a), precise contact detection will be performed to check whether particle 1 is making contact with the particles mapped into cell C1 (i.e., 4 particles will be checked); particle 2 with those particles mapped into cells B3 and C3, and so on. It is clear that, if the cell is too large, many potential neighboring particles will be identified, which could lead to a long computing time. If the cell is too small, many particles will be mapped into multiple cells so that the particles in all these cells must be scanned, and consequently the computing time could also be very long. Hence there is an optimal cell size to achieve efficient contact detection, which can be estimated using a heuristic method or through sensitivity studies. As an approximate indication, a cell size of 3–5 particle diameters can be used to achieve a good contact searching efficiency.

In the Verlet list method, named after its originator, French physicist Loup Verlet, each particle has a list that identifies all its neighboring particles. Although no box or cell is introduced, a cut-off radius R is used to construct and maintain the list. Each particle has an associated catchment space (area in 2D; volume in 3D) of radius R, and particles are said to be its neighbors and should be added into its Verlet list if the centers of these particles are located inside the catchment space, as illustrated in Fig. 9.2(b). In contrast to the cell-based method, precise contact detection is only needed between the particles and those in its Verlet list. The cut-off distance (radius of the catchment space) needs to be carefully chosen: it should not be too large otherwise more particles are included and more time is needed for precise contact searching; it should not be so small as to exclude any potential neighboring particles.

The purpose of the fine detection phase is to identify all contacts of a particle with its neighboring particles. For spherical particles, there is a mechanical contact if the distance between the centers of two particles with radius R_1 and R_2 is equal to or less than the sum of the two radii (see Fig. 9.3(a)), i.e.,

$$L \leq R_1 + R_2 \tag{9.5}$$

Contact detection for nonspherical particles (e.g., ellipsoids and polygons) is more complicated due to the more complex contact geometry. For elliptical particles

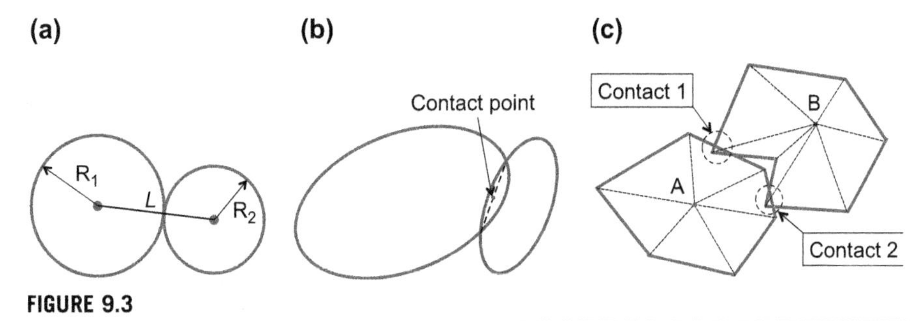

FIGURE 9.3

Contact detection between (a) spherical particles, (b) elliptical particles, and (c) polygonal particles.

in 2D, Ting et al. (1993) proposed a contact detection method in which the contact point between two elliptical particles is regarded as the midpoint of the line connecting the intersection points of the two ellipses (Fig. 9.3(b)); the intersection points can be mathematically determined using the equations defining the shapes. Similar approaches can be used for contact detection between other shapes that can be defined mathematically (Lin and Ng, 1995; Ouadfel and Rothenburg, 1999). For polygonal particles, contact detection becomes more complicated, because many types of contact may exist, including face–face, face–corner, corner–corner, and multiple contacts (see Fig. 9.3(c)). Wu and Cocks (2006) treated polygonal particles as a collection of triangles and a contact search scheme based upon each vertex was proposed. This method is able to detect each contacting vertex and multiple contacts between two particles, but it is very computationally intensive. Boon et al. (2013) proposed a contact search method for arbitrary 3D convex-shaped particles, in which the particles were constructed using an assembly of planes defined by mathematical functions, and the contact detection was performed by solving constrained minimization problems. They also showed that the computational time for contact detection for nonspherical particles is several orders of magnitude greater than that for spherical particles.

9.2.3 CONTACT MODELING

Contact modeling needs to be performed once a contact between two particles is identified, so that the contact forces acting on each particle in each time increment can be updated. The mechanical interaction between two particles can be modeled using either the contact laws based upon theoretical contact mechanics (such as the Hertz theory, and the JKR model described in Chapter 8), or simplified phenomenological models, such as linear spring or spring-dashpot models. Strictly speaking, the models based upon theoretical contact mechanics can only be used to model contacts between spherical particles. For particles of irregular shape, the interaction is very complicated and no general analytical model is available. Consequently, irregular particles are often simply modeled as spheres with full contact interactions or phenomenological models are used in estimating the contact forces between those particles.

The contact laws based upon theoretical contact mechanics described in Chapter 8 can be used to model the mechanical interaction between spherical particles in DEM. For elastic spherical particles, Thornton and his coworkers (Thornton and Ning, 1997; Thornton and Yin, 1991) implemented the Hertz theory (Section 8.1.1) to model the normal force–displacement relationship and the theory of Mindlin and Deresiewicz (Section 8.1.3) for the tangential force–displacement relationship. This is often called the Hertz–Mindlin–Deresiewicz model (or the HMD model, Thornton et al., 2011), and is briefly described in Box 9.1.

For the interaction between spheres with adhesion, the JKR model (Johnson et al., 1971) discussed in Section 8.3.2.1 and the theory of Thornton (1991) can be used to describe the normal and tangential force–displacement relationships in DEM, as introduced in Box 9.2.

BOX 9.1 THE HERTZ–MINDLIN–DERESIEWICZ MODEL

Using Eq. (8.12), the normal force–displacement (F_n–δ_n) relationship between two spheres is given as

$$F_n = \frac{4}{3} E^* (R^*)^{1/2} \delta_n^{3/2} \tag{9.6}$$

where E^* and R^* are defined in Eqs (8.6) and (8.8), respectively.

The normal stiffness k_n can be determined by differentiating Eq. (9.6) with respect to δ_n and is given as

$$k_n = \frac{dF_n}{d\delta_n} = 2E^* \sqrt{R^* \delta_n} \tag{9.7}$$

indicating that *the normal stiffness* k_n *is not a constant* but varies with δ_n.

As introduced in Section 8.1.3, the interaction between frictional elastic spheres in contact under varying oblique forces was analyzed by Mindlin and Deresiewicz (1953), and solutions were obtained in the form of instantaneous compliances. As a consequence, the solutions depend not only on the current state but also the previous loading history and cannot be integrated *a priori*. However, through examining several loading sequences involving variations of both normal and tangential forces, some general procedural rules were identified.

The recommended procedure uses an incremental approach to update the normal force and contact radius using Eqs (9.6) and (8.12). The tangential incremental force ΔF_t is then calculated using the tangential incremental relative surface displacement $\Delta\delta_t$, and the new values of F_n and a. Thornton and his coworkers (Thornton and Yin, 1991; Thornton and Randall, 1988; Thornton et al., 2011) reanalyzed the loading scenarios considered by Mindlin and Deresiewicz (1953), and showed that the tangential incremental displacement can be expressed in a general form as

$$\Delta\delta_t = \frac{1}{8G^* a}\left(\pm \mu \Delta F_n + \frac{\Delta F_t \cdot \mu \Delta F_n}{\theta} \right) \tag{9.8}$$

where G^* is given by Eq. (8.25). The tangential stiffness can be obtained by rearranging Eq. (9.8):

$$k_t = \frac{\Delta F_t}{\Delta \delta_t} = 8G^* a\theta \pm \mu(1-\theta)\frac{\Delta F_n}{\Delta \delta_t} \tag{9.9}$$

The negative sign in Eq. (9.9) is only necessary during unloading. If $\Delta F_n > 0$ and $|\Delta\delta_t| < \frac{\mu \Delta F_n}{8G^* a}$, θ should be set to one in Eq. (9.9); otherwise, θ is given as

$$\theta^3 = 1 - \frac{(F_t + \mu\Delta F_n)}{\mu F_n} \quad \Delta\delta_t > 0(\text{loading}) \tag{9.10a}$$

$$\theta^3 = 1 - \frac{(F_t^* - F_t + 2\mu\Delta F_n)}{2\mu F_n} \quad \Delta\delta_t < 0(\text{unloading}) \tag{9.10b}$$

$$\theta^3 = 1 - \frac{(F_t - F_t^{**} + 2\mu\Delta F_n)}{2\mu F_n} \quad \Delta\delta_t > 0(\text{reloading}) \tag{9.10c}$$

where F_t^* and F_t^{**} specify the forces for the transitions from loading to unloading and unloading to reloading, respectively, and need to be continuously updated to account for the effect of varying normal force:

$$F_t^* = F_t^* + \mu\Delta F_n \tag{9.11a}$$

$$F_t^{**} = F_t^{**} - \mu\Delta F_n \tag{9.11b}$$

BOX 9.2 CONTACT MODELS FOR ADHESIVE SPHERES

In the JKR model (Johnson et al., 1971), the radius of the contact area a is obtained using Eq. (8.95); the applied normal force and the relative normal displacement are functions of the contact radius according to Eqs (8.97) and (8.102), i.e.,

$$F_n = \frac{4E^* a^3}{3R^*} - 4\left(\pi\gamma E^* a^3\right)^{1/2} \tag{9.12}$$

$$\delta_n = \frac{a^2}{R^*} - 2\left(\frac{\pi\gamma a}{E^*}\right)^{1/2} \tag{9.13}$$

The normal contact stiffness k_n can be obtained by differentiating Eqs (9.12) and (9.13) with respect to a and using $k_n = \frac{dF_n}{d\delta_n}$. Hence

$$k_n = 2E^* a\left(\frac{3\sqrt{\hat{F}_n} - 3\sqrt{F_c}}{3\sqrt{\hat{F}_n} - \sqrt{F_c}}\right) \tag{9.14}$$

where $\hat{F}_n$ and F_c are defined by Eqs (8.96) and (8.106), respectively.

In order to model the tangential interaction with adhesion, Thornton (1991) developed a model that integrates the theories of Savkoor and Briggs (1977) and Mindlin and Deresiewicz (1953). In this model, the contact radius reduces with increasing tangential force F_t, i.e., the contact area shrinks as the tangential force increases. This can be visualized as the two contacting particles peeling gradually away from each other with increasing tangential force. During the peeling process, the contact radius is given as

$$a^3 = \frac{3R^*}{4E^*}\left(F_n + 2F_c \pm \sqrt{4F_n F_c + 4F_c^2 - \frac{F_t^2 E^*}{4G^*}}\right) \tag{9.15}$$

and the tangential stiffness k_t can be obtained using the no-slip solution of Mindlin (1949) and is given as

$$k_t = 8G^* a \tag{9.16}$$

The peeling process continues until the following condition is satisfied

$$F_t^{peel} = 4\sqrt{\frac{G^*}{E^*}\left(F_n F_c + F_c^2\right)} \tag{9.17}$$

and then the contact radius becomes

$$a^3 = \frac{3R^*}{4E^*}(F_n + 2F_c) \tag{9.18}$$

Thornton (1991) pointed out that peeling must occur before sliding, and there should be a smooth transition to sliding. If at the end of the peeling process the tangential force F_t^{peel} given by (9.17) is less than the sliding force, a subsequent slip annulus is assumed to spread radially inwards and the micro-slip solution of Mindlin and Deresiewicz (1953) should then be applied until sliding occurs. The equations for this case are Eqs (9.9) to (9.12) with substitution of $F_n + 2F_c$ for F_n. If the tangential force at the end of the peeling process is greater than the sliding force, the tangential force is immediately set to the sliding force.

Two sliding criteria are proposed to define the tangential force for sliding, F_t^{slide}:

$$F_t^{slide} = \mu\hat{F}_n\left(1 - \frac{\hat{F}_n - F_n}{3\hat{F}_n}\right) \quad F_n < -0.3F_c \tag{9.19a}$$

$$F_t^{slide} = \mu(F_n + 2F_c) \quad F_n \geq -0.3F_c \tag{9.19b}$$

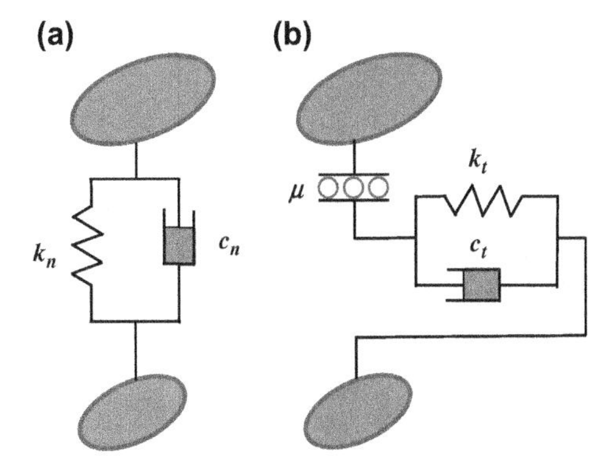

FIGURE 9.4

The typical spring-dashpot model for describing the contact interactions between particles: (a) normal contact and (b) tangential interaction.

9.2.3.1 Phenomenological Contact Models

In order to determine the contact forces in DEM calculations, various phenomenological models are proposed, in which contact stiffness and energy dissipation mechanisms are incorporated to describe the interaction between particles. The phenomenological contact models are usually composed of (1) a spring to approximate elastic deformation and/or (2) a dashpot to account for the energy dissipation due to viscoelastic deformation. The spring and dashpot can be considered simultaneously in both normal and tangential directions (see Fig. 9.4). This is the so-called spring and dashpot model, also known as the Kelvin–Voigt model (Hunt and Crossley, 1975) and is widely used in non-DEM contact problems in mechanical engineering.

For systems consisting only of elastic particles, energy is primarily dissipated by friction and the particle deformation can be described using the linear stress-strain relationship shown in Figure 8.2(a). The dashpots in the normal and tangential directions shown in Fig. 9.4 can be omitted as there is no energy dissipation due to viscoelastic deformation. The contact can then be modeled using the so-called linear spring model (Thornton et al., 2011) with the normal and tangential forces being calculated by

$$F_{\mathrm{n}} = k_{\mathrm{n}}\delta_{\mathrm{n}} \tag{9.20}$$

$$F_{\mathrm{t}}^{\mathrm{new}} = \begin{cases} F_{\mathrm{t}}^{\mathrm{old}} + k_{\mathrm{t}}\Delta\delta_{\mathrm{t}} & \left(F_{\mathrm{t}}^{\mathrm{new}} < \mu F_{\mathrm{n}}\right) \\ \mu F_{\mathrm{n}} & \left(F_{\mathrm{t}}^{\mathrm{new}} \geq \mu F_{\mathrm{n}}\right) \end{cases} \tag{9.21}$$

The normal and tangential spring stiffnesses k_{n} and k_{t} need to be selected carefully in order to appropriately represent the interaction between two elastic bodies.

Thornton et al. (2011) suggested that the normal spring stiffness k_n can be determined using the following equation:

$$k_n = 1.2024\left(m^{1/2}E^{*2}Rv_{ni}\right)^{2/5} \tag{9.22}$$

which gives the same contact duration as the Hertz-Mindlin-Deresiewicz model for the impact of two elastic spheres. According to the Mindlin theory (Mindlin, 1949), the tangential spring stiffness k_t should have a value between $2k_n/3$ and k_n (Cundall and Strack, 1979) and can be given as a function of the Poisson's ratio v as follows:

$$k_t = \frac{2(1-v)}{2-v}k_n \tag{9.23}$$

The normal and tangential spring stiffnesses k_n and k_t can also be calculated using Eqs (9.7) and (9.16), i.e., the Hertz theory and the "no-slip" theory of Mindlin (1949), respectively. This is referred to as the Hertz-Mindlin model (or the HM model, Thornton et al., 2011). Since k_n and k_t given by Eqs (9.7) and (9.16) are functions of the contact area (and the displacement), the HM model is essentially a nonlinear spring model.

The dashpots in Fig. 9.4 are introduced as contact damping to model the energy dissipation due to viscous deformation, in addition to friction. This is commonly known as the spring-dashpot model, in which the normal and tangential forces are calculated by

$$F_n = k_n\delta_n + c_n v_n \tag{9.24}$$

$$F_t^{new} = \begin{cases} F_t^{old} + k_t\Delta\delta_t + c_t v_t & \left(F_t^{new} < \mu F_n\right) \\ \mu F_n & \left(F_t^{new} \geq \mu F_n\right) \end{cases} \tag{9.25}$$

where c_n and c_t are the normal and tangential damping coefficients which are generally functions of the corresponding contact stiffness, i.e.,

$$c_n = 2\beta\sqrt{m^* k_n} \tag{9.26}$$

$$c_t = 2\beta\sqrt{m^* k_t} \tag{9.27}$$

β is the damping ratio and can be related to the normal coefficient of restitution e_n (Tsuji et al., 1992; Ting et al., 1993; Thornton et al., 2011):

$$\beta = -\frac{\ln e_n}{\sqrt{\pi^2 + \ln^2 e_n}} \tag{9.28}$$

In Eqs (9.24)–(9.27), the normal and tangential spring stiffnesses k_n and k_t can be given either by Eqs (9.7) and (9.16) or Eqs (9.22) and (9.23).

Bilinear models (see also Box 8.2) are also proposed in order to account for the effect of plastic deformation on energy dissipation. As discussed in Chapter 8, for elastoplastic materials, plastic deformation dominates the energy dissipation. It is hence reasonable that, in these hysteretic models, contact damping could be omitted.

Instead, contact stiffnesses with different values are introduced for loading and unloading. This is known as the "partially latching spring" model (Walton and Braun, 1986; Walton, 1993; Thornton et al., 2013), in which the normal contact force during loading is given as

$$F_n = k_1 \delta_n \tag{9.29}$$

and during unloading,

$$F_n = k_2(\delta_n - \delta_0) \tag{9.30}$$

where δ_0 is the relative approach when the normal force returns to zero (see Fig. 9.5),

$$\frac{k_1}{k_2} = e_n^2 \tag{9.31}$$

$$\delta_0 = \delta_{max}\left(1 - e_n^2\right) \tag{9.32}$$

where δ_{max} is the maximum relative approach.

The tangential force is calculated using Eq. (9.21) with the tangential contact stiffness (Thornton et al., 2013)

$$k_t = k_1 \frac{2(1-\upsilon)}{2-\upsilon}\left(1 - \frac{F_t - F_t^*}{\mu F_n - -F_t^*}\right)^{1/3} \quad \text{for } \Delta F_t > 0 \tag{9.33}$$

$$k_t = k_1 \frac{2(1-\upsilon)}{2-\upsilon}\left(1 - \frac{F_t^* - F_t}{\mu F_n - F_t^*}\right)^{1/3} \quad \text{for } \Delta F_t < 0 \tag{9.34}$$

where ΔF_t is the tangential force increment in a given time step, and F_t^* is the tangential force at the instant when the relative tangential displacement reverses direction.

9.2.4 DETERMINATION OF THE TIME STEP

For cyclic calculations in DEM, it is important to select an appropriate time step. On the one hand, the time step should be as large as possible so as to increase the

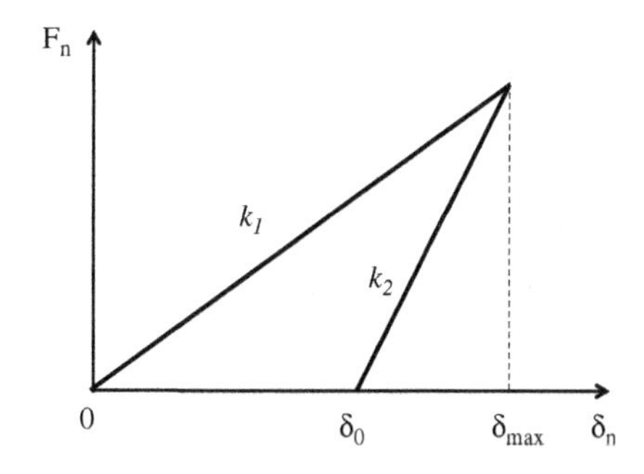

FIGURE 9.5

Schematic illustration of the bilinear contact model.

computing efficiency. On the other hand, it needs to be small enough that the calculations are stable and accurate, and a constant acceleration can be assumed at each time step for updating the kinematics of the particles (Section 9.2.1). A small time step will ensure that no new contacts occur in the current calculation cycle (i.e., current time step), so that the operational contacts are only those identified at the beginning of the current time step. The out-of-balance force at the end of the calculation cycle is then the resultant of the contact forces arising only from those identified contacts. The longest time for which this is true is the critical time step.

When a linear contact model is used (say, e.g., Eqs (9.20) and (9.21)), the critical time step can be calculated from the mass of the smallest particles in the model m_s, and the same stiffness as the normal contact stiffness k_n (Cundall, 1988; Ting et al., 1995):

$$\Delta t_c = 2\pi\sqrt{\frac{m_s}{k_n}} \tag{9.35}$$

When linear spring-dashpot models are used to describe viscous behavior, the damping ratio needs to be considered, so that the critical time step is then given as (Dziugys and Peters, 2001)

$$\Delta t_c = \frac{\pi}{\sqrt{\frac{k_n}{m_s}\left(1 - \frac{\ln^2 e_n}{\pi^2 + \ln^2 e_n}\right)}} \tag{9.36}$$

For DEM modeling with nonlinear models, a widely used approach to determination of the critical time step is based on the frequency of the Rayleigh wave, a kind of surface acoustic wave which travels on solids. In particular, the time step used must be smaller than the time for the Rayleigh wave to propagate through the smallest particle in the assembly. This leads to a critical time step given by

$$\Delta t_c = \frac{\pi d_{min}}{2\lambda}\sqrt{\frac{\rho_s}{G}} \tag{9.37}$$

where

$$\lambda = 0.1631\upsilon + 0.8766 \tag{9.38}$$

d_{min}, ρ_s, G, and υ are respectively the diameter, true density, shear modulus, and Poisson's ratio of the smallest particle in the particle assembly.

9.2.5 DETERMINATION OF DEM PARAMETERS

It can be seen from the discussion in Section 9.2.3 that several key parameters need to be defined for contact modeling in DEM, including the normal and tangential contact stiffnesses k_n and k_t and friction coefficient μ. The contact stiffnesses can be related to the material intrinsic properties, such as Young's Modulus E, Poisson's ratio υ, true density ρ, and particle size d, which can be obtained either from materials handbooks or through experimental characterization using advanced techniques such as nano-indentation and atomic force microscopy (AFM). Direct measurement of the friction coefficient μ between particles and with other surfaces is very difficult, even with advanced techniques such as AFM and use of a nano-tribometer. Alternative approaches to determining the friction coefficient include inferring it from single-particle impact experiments using the theories discussed in Chapter 8 (Lorenz et al., 1997; Foerster et al., 1994; Gorham and Kharaz, 2000) or from the friction measurement of bulk solids (Li et al., 2005). Table 9.1 lists some friction coefficients reported in the literature.

Table 9.1 Friction Coefficients for Various Materials Reported in the Literature

Materials	μ	References
Acrylic ball–acrylic ball	0.096 ± 0.006	Lorenz et al. (1997)
Acrylic ball–aluminum plate	0.140	Mullier et al. (1991)
Aluminum oxide sphere–aluminum alloy lead plate	0.180	Gorham and Kharaz (2000)
Aluminum oxide sphere–glass plate	0.092	Gorham and Kharaz (2000)
Cellulose acetate–cellulose acetate ball	0.250 ± 0.020	Foerster et al. (1994)
Cellulose acetate ball–cellulose acetate ball	$0.220 \sim 0.330$	Mullier et al. (1991)
Fresh glass ball–aluminum plate	0.131 ± 0.007	Lorenz et al. (1997)
Fresh glass ball–fresh glass ball	0.048 ± 0.006	Lorenz et al. (1997)
Glass ball–glass plate	0.155	Li et al. (2005)
Glass ball–perspex plate	0.133	Li et al. (2005)
Nylon ball–nylon ball	0.175 ± 0.100	Labous et al. (1997)
Polystyrene ball–polystyrene ball	0.189 ± 0.009	Lorenz et al. (1997)
Radish seeds–aluminum plate	0.190	Mullier et al. (1991)
Soda lime glass ball–soda lime glass ball	0.092 ± 0.006	Foerster et al. (1994)
Spent glass ball–spent glass ball	0.177 ± 0.020	Lorenz et al. (1997)
Spent glass ball–spent glass ball (stationary)	0.126 ± 0.014	Lorenz et al. (1997)
Spent glass ball–aluminum plate	0.126 ± 0.009	Lorenz et al. (1997)
Stainless steel ball–stainless steel ball	0.099 ± 0.008	Lorenz et al. (1997)
Steel ball–steel plate	0.214	Li et al. (2005)

9.3 DATA ANALYSIS

From cyclic calculations, detailed microscopic information on each particle at each time step can be obtained, including positions, velocities, accelerations, and contact forces. Normally this is just the beginning of the analysis process, and much more insight into particulate processes can be gained by appropriate choices of further data analysis techniques. The ultimate goal is to relate the macroscopic performance to the intrinsic particle properties and to the microscopic behavior at the particle level. In this section, some characteristics of particle systems that can be obtained from DEM data are introduced.

9.3.1 MACROSCOPIC DEFORMATION PATTERNS

Knowing the position of each particle at each time step, the evolution of macroscopic deformation patterns can be readily obtained by displaying each particle. To enhance the visualization of the deformation pattern, particles can be color coded either according to their initial position or their distinctive features (such as density or size). The former is generally employed to explore the deformation patterns of particles during shearing or flow, while the latter offers good visual demonstrations of particle mixing or segregation. Figs 9.6 and 9.7 present some typical examples in which color-coded particle bands are employed to show the deformation patterns. Some examples in which colors are used to indicate particles of

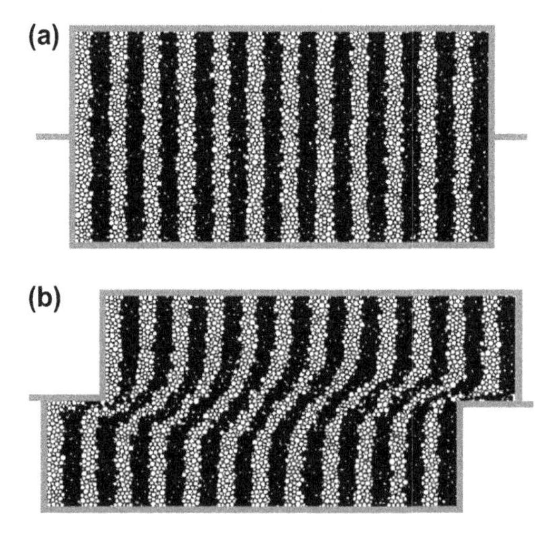

FIGURE 9.6

Deformation pattern of DEM specimen in a direct shear test observed using color-coded particle bands (Zhang, 2003). The deformation of the particle system under shear is clearly shown; (a) before shearing, and (b) end of shearing.

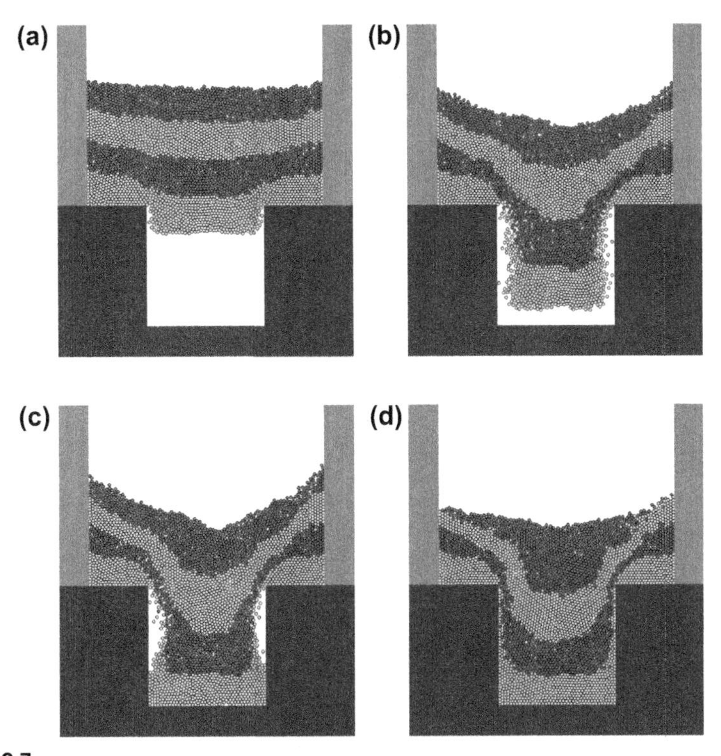

FIGURE 9.7

Powder flow patterns during die filling obtained from DEM analysis (Guo et al., 2009b), in which bands of particles are color-coded according to their initial positions. (a) $t = 8.5$ ms, (b) $t = 17.0$ ms, (c) $t - 20.5$ ms, and (d) $t - 56.2$ ms.

different properties are presented in Fig. 9.8. In addition to static representations of this kind, successive "frames" can be built up into animations, so demonstrating the system dynamics.

9.3.2 FORCE TRANSMISSION NETWORK

Contact forces between particles at each time step can be displayed to show the force transmission patterns (or contact force network) at various time instants. An example is presented in Fig. 9.9, in which the contact force is visualized as a solid line connecting the centers of the two contacting spheres (indicating the direction of the contact force), with the thickness of the line representing either the magnitude of the contact force or the ratio of the contact force to the average contact force in the system. This allows the contact force network to be classified into the strong force network formed by contact forces more than the average contact force, and the weak force network showing the contact forces less than the average. Different shades of

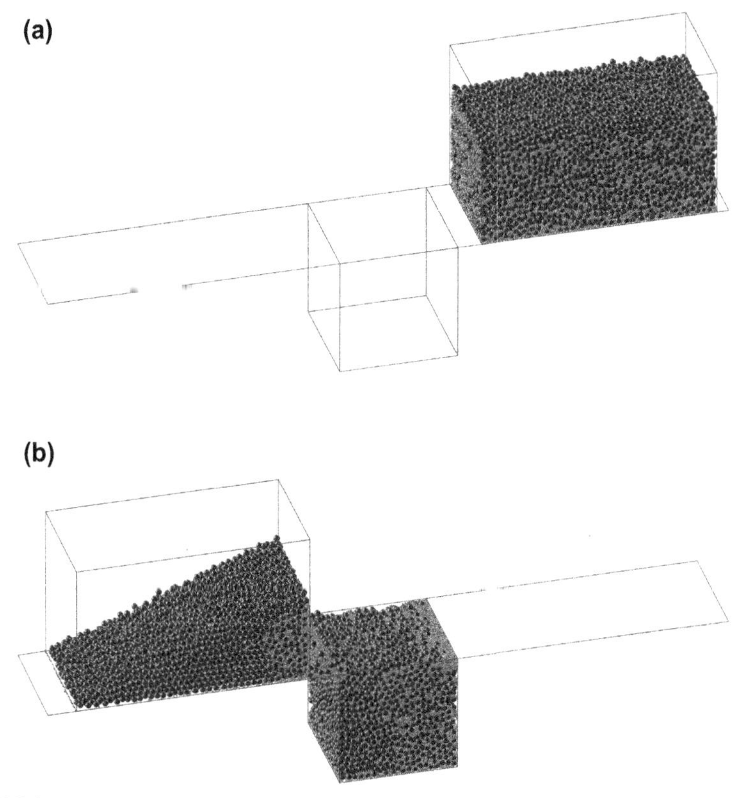

FIGURE 9.8

Packing patterns of DEM specimen during die filling observed using color-coded particles; the "shoe" containing the particles moves from right to left. Yellow (light gray in print versions) represents small particles and magenta (dark gray in print versions) represents the large particles. Segregation during die filling is clearly shown (Guo et al., 2010); (a) before die filling and (b) after die filling.

gray or different colors can then be used to visualize these two sub-networks, as shown in Fig. 9.9, in which the strong force network is shown in bold and the weak force network is shown in gray. Using this approach, the macroscopic force transmission in the system can be clearly visualized and related to forces at the single particle level.

9.3.3 PACKING DENSITY DISTRIBUTION

In a number of applications, such as packing of particles in a container, it is of practical importance to know the density distribution. This can be readily achieved from the DEM analysis, as the positions of individual particles are explicitly determined at each time step. Two approaches can be used to obtain the density

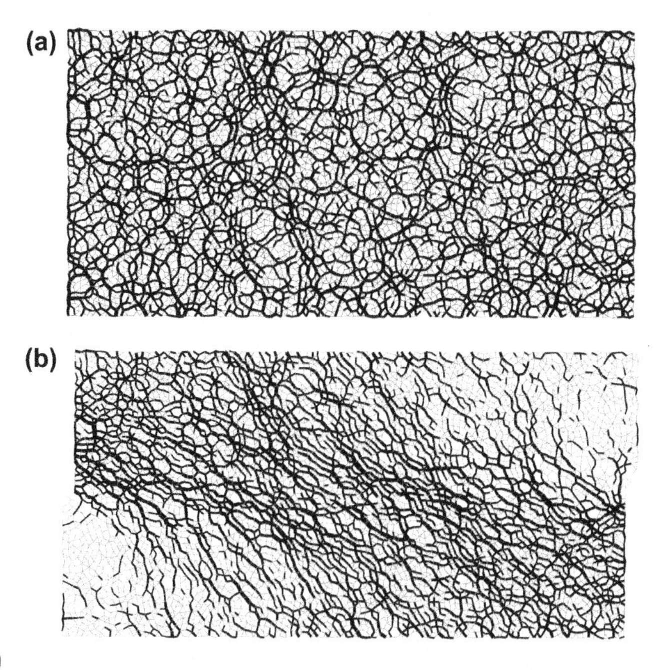

FIGURE 9.9

Force transmission patterns during the direct shear test obtained from DEM analysis (Zhang, 2003); (a) before shearing, and (b) during shearing.

distribution: (1) the box method and (2) the Voronoi cell method (after Georgy Voronoy, Ukrainian/Russian mathematician).

In the box method, the space occupied by the particle system is divided into a number of grids. The packing density in each box can be determined by treating each box as a subset of the whole particle system and using the void fraction or solid fraction defined by Eqs (2.17) and (2.18), respectively, as illustrated in Fig. 9.10. The solid fraction in the highlighted box is defined as either the ratio of the volume of the shaded particles to the volume of the box in 3D, or the ratio of the area of particles to that of the box in 2D. Knowing the packing density in each box, the spatial distribution of the packing density can then be obtained, which is generally presented in contour plots as illustrated in Fig. 9.11. It is to be expected that the calculated density (and solid fraction and void fraction) will depend to some extent on the box size, as shown in Fig. 9.10. Generally, the box size must be larger than the size of the largest particle in the system, and sensitivity studies need to be performed to select an appropriate box size.

In the Voronoi cell method, the particle system is divided into a number of cells, each of which is a polyhedron containing only one particle, such that the surface of the cell is equidistant from the surface of the particle and the surfaces of its nearest neighbors. The number of close neighbors therefore determines the number of faces, as shown in Fig. 9.12. For each Voronoi cell, the volume can be

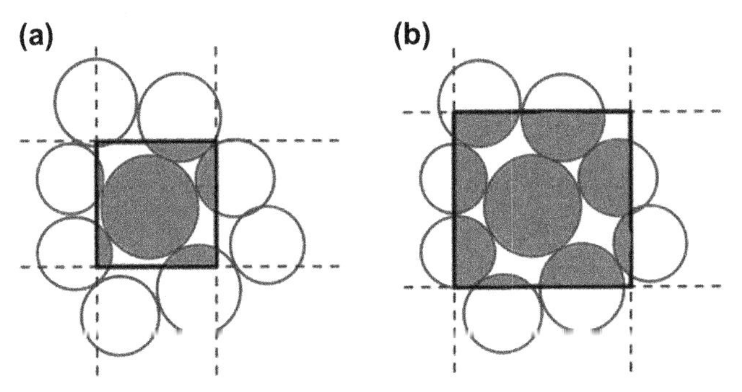

FIGURE 9.10

The box method for calculating density distributions; (a) with small box size; (b) with large box size.

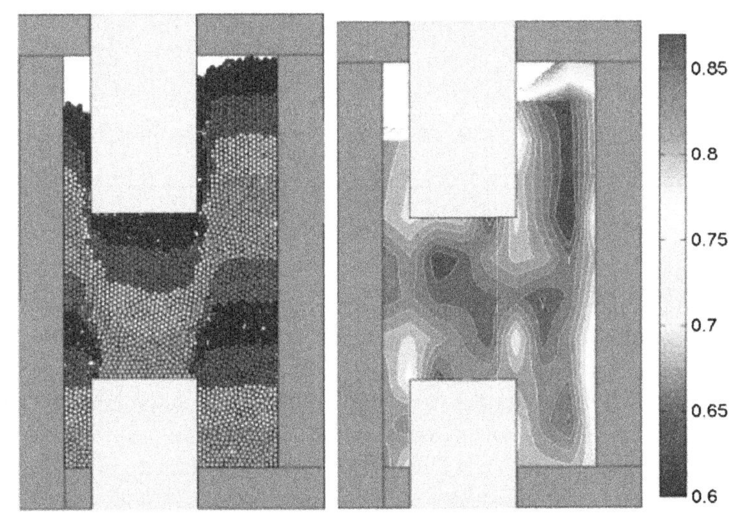

FIGURE 9.11

Particle packing pattern for a powder transfer operation and the resulting solid fraction distribution determined using the box method, from 2D DEM analysis (Coube et al., 2005).

calculated and the relative density can be determined as the ratio of the sphere volume to that of the cell, from which the density distribution can then be obtained. An example of the density distribution obtained using the Voronoi cell method is given in Fig. 9.13. The packing density using this method requires no choice about scale of scrutiny; hence, the Voronoi cell approach is in principle a superior method for determination of packing density distributions from DEM calculations.

Despite the inherent superiority of the Voronoi cell method, the box method is easier to implement and has the additional advantage that it enables distributions of other particle system characteristics to be obtained simultaneously using the same box configurations. For example, it is possible to obtain the distributions of

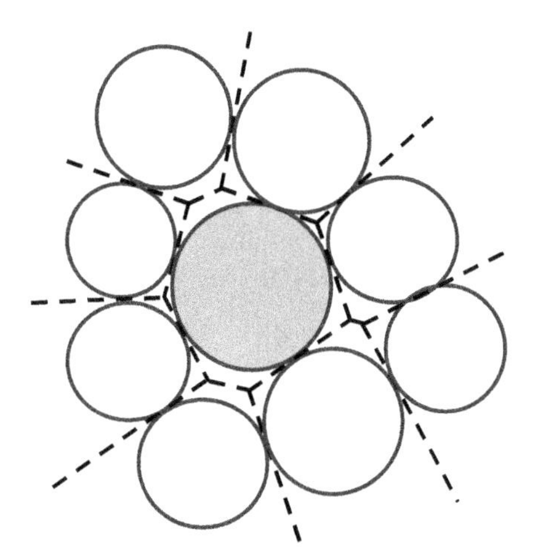

FIGURE 9.12

A Voronoi cell in 2D.

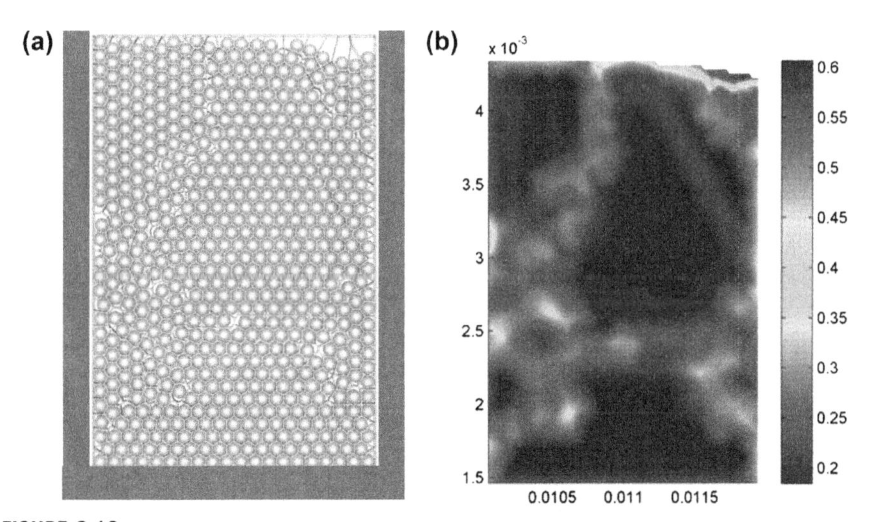

FIGURE 9.13

Packing of particles in a container: (a) the particle packing pattern and associated Voronoi cells and (b) the solid fraction distributions determined using the Voronoi cell method.

the average particle rotation, ϖ, of the sample using the same box method as used for determining the packing density distribution. Here ϖ is defined as the average rotation of all the particles in each box:

$$\langle \theta \rangle = \frac{\sum_{i=1}^{C} \omega_i}{n_\mathrm{p}} \tag{9.39}$$

where ω_i is the rotation of particle i and n_p is the total number of particles in each box.

9.3.4 RADIAL DISTRIBUTION FUNCTION

Particle packing structures can also be described statistically using the radial distribution function (RDF), $g(r)$, i.e., the probability of finding a particle at a distance r away from a reference particle. The RDF can be evaluated by counting how many particles have their centres within a distance of r to $r + \Delta r$ away from the reference particle (see Fig. 9.14), i.e., the number of particles with their centres in a spherical shell of radius r and thickness Δr. The RDF for a 2D inhomogeneous system is defined as:

$$g(r) = \frac{n(r)}{2\pi r \bar{n} \Delta r} \tag{9.40}$$

and, for a 3D system

$$g(r) = \frac{n(r)}{4\pi r^2 \bar{n} \Delta r} \tag{9.41}$$

where

r is the distance from the reference particle;
$n(r)$ is the mean number of particles within a ring (2D) or a spherical shell (3D) of thickness Δr at a distance of r;
$\bar{n}$ is the mean number density of particles.

From DEM data, the RDF can be determined by first calculating the distances between all particle pairs and then determining the distribution of that distance. Figure 9.15 shows an example of the RDF for a simple packing of particles with the same particle size and the same electrostatic charge, indicating the variation of the number density of particles with separation distance.

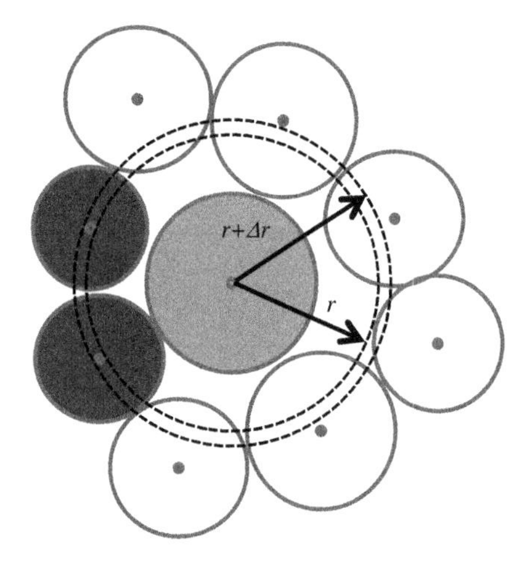

FIGURE 9.14

Definition of radial distribution function.

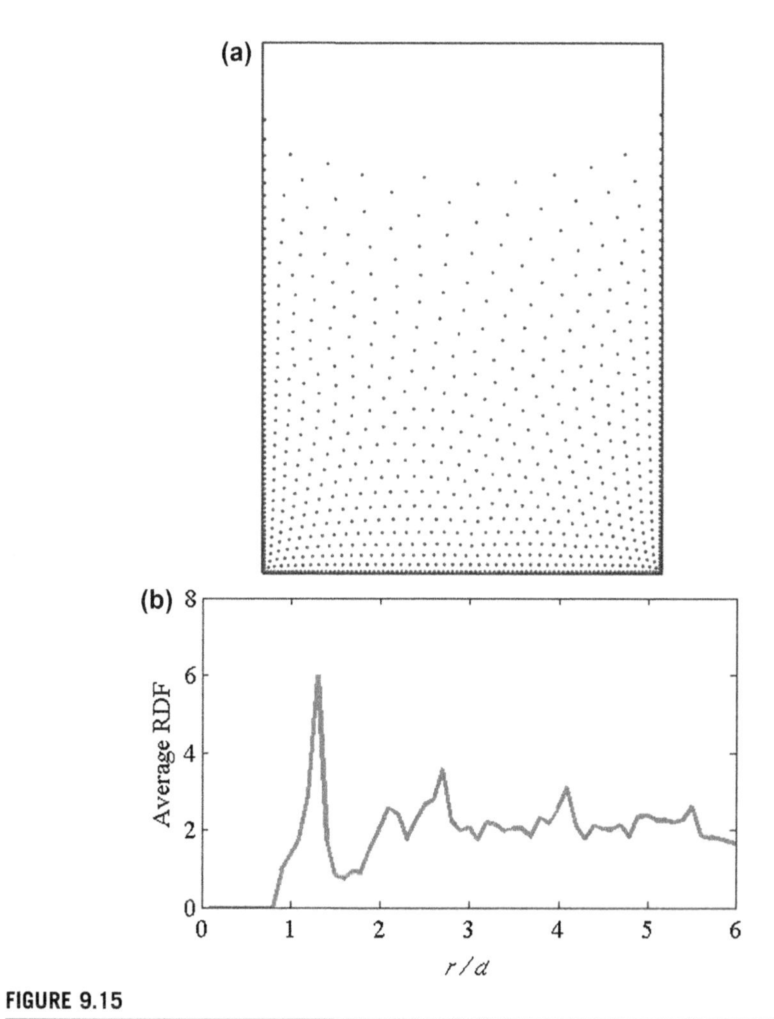

FIGURE 9.15

The particle packing pattern (a) and corresponding RDF (b) of charged particles obtained using DEM (Pei et al., 2015a).

9.3.5 COORDINATION NUMBER

The coordination number is another important statistical parameter that can be used to characterize the packing structure of a particle system. It indicates the number of contacts of a reference particle. As illustrated in Fig. 9.16, Particle 1 has no contact, so its coordination number is 0; particles 2, 5, 6, and 7 have two contacts each, so their coordination number is 2; particles 3, 4, 8, 9, and 10 have a coordination number of 3 as they have three contacts each; particle 11 has only one contact, so its coordination number is 1.

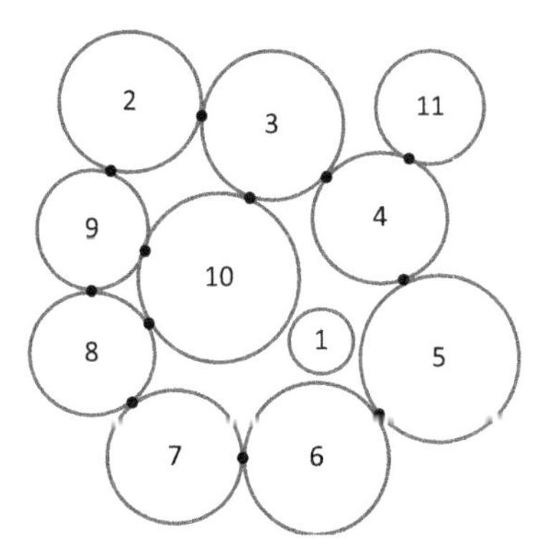

FIGURE 9.16

Coordination number.

For a particle system, the average coordination number Z is defined as:

$$Z = \frac{2N_C}{n_p} \tag{9.42}$$

where N_C is the total number of contacts and n_p is the total number of particles. For the example illustrated in Fig. 9.16, the total number of contacts (shown as black dots), N_C, is 12 and the average coordination number is therefore 2.18.

Zhang (2003) argued that particles which have no contacts with others (such as particle 1 in Fig. 9.16), and particles which have only one contact with their neighboring particles (e.g., particle 11 in Fig. 9.16) cannot make any contribution to the mechanical properties of the particle system (i.e., the removal of particles 1 and 11 in the system illustrated in Fig. 9.16 will not significantly affect the mechanical behavior of the whole). Hence, a mechanical coordination number was introduced and defined as:

$$Z_m = \frac{2N_C - n_1}{(n_p - n_1 - n_0)}, \quad Z_m \geq 2 \tag{9.43}$$

where n_1 and n_0 are the numbers of particles with one and no contacts, respectively. For the example given in Fig. 9.16, $N_C = 12$, $n_1 = n_0 = 1$, $n_p = 11$, so $Z_m = 2.56$.

The coordination number reflects the degree of connection between particles so it can be used to evaluate the packing of the systems, in applications such as force transmission and tensile strength (Cundall and Strack, 1979). Not surprisingly, published data show that the average coordination number has a strong correlation with measurements of particle packing density such as the solid fraction or void fraction (Oda, 1977), increasing for higher packing density (solid fraction). For example, for ordered packing of monosized particles, the correlation between the average coordination number and the solid fraction D can be approximated using one of the following empirical expressions, among many others (German and Park, 2008):

$$Z = \frac{\pi}{1 - D} \tag{9.44}$$

$$Z = 14 - 10.4(1 - D)^{0.38} \tag{9.45}$$

9.3.6 MICROSTRUCTURAL CHARACTERISTICS

The macroscopic properties of a particulate system depend on its microstructure, which can be described in different ways. Some microstructural properties are scalars, i.e. they have magnitude but no direction. An example is solid fraction. Many characteristics of a particle system, however, are vectors, i.e., they have both magnitude and direction, examples including contact force, displacement and velocity of particles. Two particle systems of the same material and the same solid fraction may show very different mechanical responses if they have different microstructures (Oda, 1977; Oda et al., 1980). A further complication is *isotropy.* A particle system is *isotropic* with respect to a certain property if the values of that property are identical in all directions. When a property has different values in different directions, the system is said to be *anisotropic* with respect to that property. Using DEM data, two parameters can be determined which characterize the microstructure of a particle system: (1) the *fabric tensor* and (2) the contact normal orientations. Both are based upon the distribution of contact normal vectors that determine the structural anisotropy of a particle system. If the distribution of the contact normal vectors is random and can be approximated using a uniform distribution, the particle system in effect possesses an isotropic structure. If the distribution is non-uniform, the structure is then anisotropic.

For assemblies of discs or spheres, the fabric tensor ϕ_{ij} characterizes the structural anisotropy using the contact normals n_i and n_j (Satake, 1982):

$$\varphi_{ij} = \langle n_i n_j \rangle = \frac{1}{2N_C} \sum_{1}^{2N_C} n_i n_j \tag{9.46}$$

In order to obtain a complete characterization of the microstructure, a second order fabric tensor was proposed by Oda et al. (1980),

$$F_{ij} = \frac{2N_C \overline{R}}{V} \varphi_{ij} \tag{9.47}$$

where $\overline{R}$ is the mean particle radius and V is the volume of the particle system.

Contact normal orientations can be used to describe load-induced anisotropy. From the DEM data, the distribution of the orientations of the contact normals can be obtained by plotting a histogram of the proportion of contact normals in a

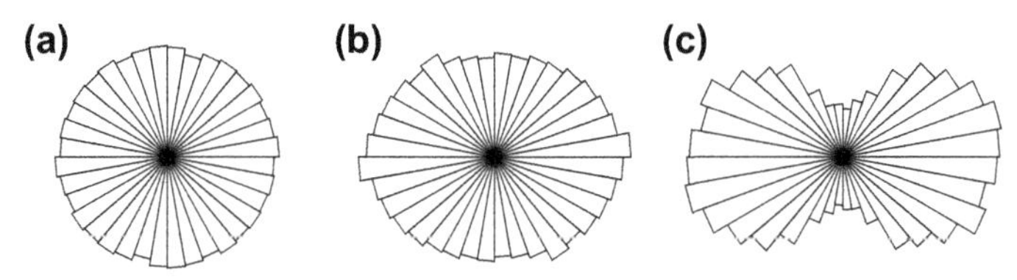

FIGURE 9.17

Typical patterns of the contact normal orientation distribution (Zhang, 2003); (a) Isotropic distribution, (b) Weak anisotropy, and (c) Strong anisotropy.

series of adjacent orientation classes that partition the full orientation space (i.e., a unit circle), as shown in Fig. 9.17. In order to construct such a diagram, each contact is interrogated to find out which of the partitioned bands (i.e., contact normal directions) its inclination belongs to. If it falls into band *i*, the total contact number for the band *i* is increased by one. After all contacts are examined, the total contact number of each band is divided by the total number of contacts in the whole system to obtain the radial coordinate as follows:

$$r_i = \frac{n_{Ci}}{N_C} \tag{9.48}$$

where n_{Ci} is the total number of contacts mapped into band i.

The same method can also be used to obtain the distribution of contact normals weighted by the magnitude of the contact normal force, in which the total normal force accumulated for each band is normalized by the total normal force of the whole system. Hence, the radial coordinate p_i for band i is given as:

$$p_i = \frac{\sum_1^{n_{Ci}} F_n^i}{\sum F_n} \tag{9.49}$$

where F_n^i is the contact normal force mapped into band i.

Some typical patterns of the contact normal orientation distribution are illustrated in Fig. 9.17. If the contact normal orientation distribution is isotropic, its shape is close to a circle (Fig. 9.17(a)). With increasing anisotropy (i.e., from weak anisotropy shown in Fig. 9.17(b) to strong anisotropy shown in Fig. 9.17(c)), the shape mutates to a peanut shape.

9.4 APPLICATIONS

Since the first publication on DEM in 1979, the method has gained in popularity in a wide range of disciplines and fields, including chemical engineering, mechanical engineering, civil engineering, physics, agriculture, astrophysics and mathematics. It has become a powerful way of simulating particle systems: not only dispersed systems in which the particle–particle interactions are collisional, but also compact systems of particles with multiple enduring contacts. It can be used to obtain data that are normally inaccessible to physical experimentation and to perform rigorous and revealing parametric studies, including those of the "what if?" kind. There is a large collection of publications on the applications of DEM, including a review article of Zhu et al. (2008) and a series of special issues on DEM modeling (Thornton, 2000, 2008, 2009; Wu, 2012a, 2012b).

DEM is a powerful tool for providing dynamic information at a microscopic level, e.g., individual particle trajectories and transient interaction forces. Such information is essential to understanding of the underlying physics of particulate materials. For example, DEM has been widely employed to investigate processes including:

- ***Heap formation*** (Luding, 1997; Baxter et al., 1997; Matuttis et al., 2000; Smith et al., 2001; Zhou et al., 2003; Tüzün et al., 2004; Fazekas et al., 2005), for which DEM was used to explore how particle properties (shape, size, size distribution, friction, density, and mechanical properties) affect the angle of repose; to analyze force transmission in heap formation and to calculate the pressure distribution at the bottom of the heap; and to examine the effects of particle properties on segregation during heap formation.
- ***Particle packing***. DEM can be used to analyze the packing of particles of different sizes and shapes, such as fine particles, cohesive particles, and wet particles (e.g., Yang et al., 2003, 2008; Li et al., 2005). It can be used to explore the microstructure (connectivity, coordination member, RDF, porosity, and density distributions) and analyze the force and stress transmission in a packed particle bed. It can also be used to explore how single particle properties (size, shape, mechanical properties) and interfacial properties (surface energy and friction) affect the microstructure and packing behavior.
- ***Powder compaction***. DEM can be used to simulate bulk deformation of packed particle systems under consolidation and to explore the microstructural characteristics, stress transmission and density variations during powder compaction (see, e.g., Ng, 1999; Martin et al., 2006).
- ***Shear tests***. DEM can be used to model deformation behavior of particle systems under different shearing conditions, such as direct shear, simple shear, and biaxial shear, and to explore the mechanics of granular materials under shear deformation at microscopic and macroscopic levels, such as stress transmission, shear band formation, and dilation (Thornton and Zhang, 2003, 2006; Zhang and Thornton, 2007; Rock et al., 2008).
- ***Hopper flow***. DEM can be used to simulate the dynamics of particle flow in and from containers such as hoppers, from which the force transmission to the hopper walls, particle velocities inside the hopper, and mass flow rate of the particles discharging from the hopper can all be determined. DEM can also be used to analyze the flow pattern (mass flow, core flow) and to guide the hopper design for a given material (see, e.g., Kohring et al., 1995; Langston et al., 1997; Ketterhagen et al., 2007, 2008).

- ***Powder flow***. Similarly to modeling hopper flow, DEM has been proved to be useful for modeling powder flow in processing devices such as the V-mixer (Kuo et al., 2002; Lemieux et al., 2008), rotating drums (Dury et al., 1998; Mishra et al., 2002), and die filling and powder transfer (Wu et al., 2003; Coube et al., 2005; Wu and Cocks, 2006; Wu, 2008; Guo et al., 2009a, 2009b; Bierwisch et al., 2009). From DEM analysis, not only the flow patterns but also quantitative information on flow rate, flow velocity, solids fluxes and stress and force distributions can be obtained. These can be used to guide process design and optimization.
- ***Mixing and segregation***. DEM is a robust method for analyzing mixing and segregation of powder mixtures as the motion of each particle in the system is explicitly determined. DEM has been used extensively to explore mixing of particles and to analyze segregation in processes such as in a rotating drum (Dury et al., 1998), during die filling (Guo et al., 2009a, 2010), and in vibrating beds (Zeilstra et al., 2006, 2008), induced by air flow, size and density differences.
- ***Electrostatics in particle systems***. With the implementation of a contact electrification model (Pei et al., 2013, 2014) and incorporation of electrostatic interactions (with proper consideration of contact detection for particle systems with long-range interaction (Pei et al., 2015a)), DEM can be used to analyze how particles become charged during handling and processing and how the electrostatic charge affects their subsequent behavior. It can also be used to investigate the uniformity of charge distribution on irregularly shaped particles (Pei et al., 2015b) with potential to model electrostatics in particle systems for real applications.

REFERENCES

Baxter, J., Tüzün, U., Burnell, J., Heyes, D.M., 1997. Granular dynamics simulations of two-dimensional heap formation. Physical Review E 55, 3546–3554.

Bierwisch, C., Kraft, T., Riedel, H., Moseler, M., 2009. Three-dimensional discrete element models for the granular statics and dynamics of powders in cavity filling. Journal of the Mechanics and Physics of Solids 57, 10–31.

Boon, C.W., Houlsby, G.T., Utili, S., November 2013. A new contact detection algorithm for three-dimensional non-spherical particles. Powder Technology 248, 94–102. ISSN:0032-5910.

Coube, O., Cocks, A.C.F., Wu, C.Y., 2005. Experimental and numerical study of die filling, powder transfer and die compaction. Powder Metallurgy 48 (1), 68–76.

Cundall, P.A., Hart, R.P., 1992. Numerical modelling of discontinua. Engineering Computation 9, 101–113.

Cundall, P.A., Strack, O.D.L., 1979. A discrete numerical model for granular assemblies. Géotechnique 29, 47–65.

Cundall, P.A., 1988. In: Satake, M., Jenkins, J.T. (Eds.), Micromechanics of Granular Materials. Elsevier, Amsterdam, pp. 113–123.

Dury, C.M., Knecht, R., Ristow, G.H., 1998. Size segregation of granular materials in a 3D rotating drum. High Perform. Computer Networks 1401, 860–862.

Dziugys, A., Peters, B., 2001. An approach to simulate the motion of spherical and nonspherical fuel particles in combustion chambers. Granular Matter 3, 231–265.

Fazekas, S., Kertesz, J., Wolf, D.E., 2005. Piling and avalanches of magnetized particles. Physical Review E 71, 061303.

Foerster, S.F., Louge, M.Y., Chang, H., Allia, K., 1994. Measurements of the collision properties of small spheres. Physics of Fluids 6 (3), 1108–1115.

German, R.M., Park, S.J., 2008. Handbook of Mathematical Relations in Particulate Materials Processing. Wiley, London.

Gorham, D.A., Kharaz, A.H., 2000. Measurement of particle rebound characteristics. Powder Technology 112, 193–202.

Guo, Y., Wu, C.-Y., Kafui, K.D., Thornton, C., 2009a. Numerical analysis of density-induced segregation during die filling. Powder Technology 197, 111–119.

Guo, Y., Kafui, K.D., Wu, C.-Y., Thornton, C., Seville, J.P.K., 2009b. A coupled DEM/CFD analysis of the effect of air on powder flow during die filling. AIChE Journal 55 (1), 49–62.

Guo, Y., Wu, C.Y., Kafui, K.D., Thornton, C., 2010. Numerical analysis of density-induced segregation during die filling in the presence of air. Powder Technology 197 (1–2), 111–119.

Hunt, K.H., Crossley, F.R.E., 1975. Coefficient of restitution interpreted as damping in vibroimpact. Journal of Applied Mechanics 42 (2), 440–445.

Johnson, K.L., Kendall, K., Roberts, A.D., 1971. Surface energy and the contact of elastic solids. Proceedings of the Royal Society of London A 324, 301–313.

Ketterhagen, W.R., Curtis, J.S., Wassgren, C.R., Kong, A., Narayan, P.J., Hancock, B.C., 2007. Granular segregation in discharging cylindrical hoppers: a discrete element and experimental study. Chemical Engineering Science 62, 6423–6439.

Ketterhagen, W.R., Curtis, J.S., Wassgren, C.R., Hancock, B.C., 2008. Modeling granular segregation in flow from quasi-three-dimensional, wedge-shaped hoppers. Powder Technology 179, 126–143.

Kohring, G.A., Melin, S., Puhl, H., Tillemans, H.J., Vermohlen, W., 1995. Computer- simulations of critical, nonstationary granular flow—through a hopper. Computer Methods in Applied Mechanics and Engineering 124, 273–281.

Kuo, H.P., Knight, P.C., Parker, D.J., Tsuji, Y., Adams, M.J., Seville, J.P.K., 2002. The influence of DEM simulation parameters on the particle behaviour in a V-mixer. Chemical Engineering Science 57, 3621–3638.

Labous, L., Rosato, A.D., Dave, R.N., 1997. Measurements of collisional properties of spheres using high-speed video analysis. Physical Review E 56, 5717–5725.

Langston, P.A., Nikitidis, M.S., Tuzun, U., Heyes, D.M., Spyrou, N.M., 1997. Microstructural simulation and imaging of granular flows in two- and three- dimensional hoppers. Powder Technology 94, 59–72.

Lemieux, A., Leonard, G., Doucet, J., Leclaire, L.A., Viens, F., Chaouki, J., Bertrand, F., 2008. Large-scale numerical investigation of solids mixing in a V-blender using the discrete element method. Powder Technology 181, 205–216.

Li, Y., Xu, Y., Thornton, C., 2005. A comparison of discrete element simulations and experiments for 'sandpiles' composed of spherical particles. Powder Technology 160, 219–228.

Lin, X., Ng, T.T., 1995. Short communication; contact detection algorithms for three- dimensional ellipsoids in discrete element modelling. International Journal for Numerical and Analytical Methods in Geomechanics 19, 653–659.

Lorenz, A., Tuozzolo, C., Louge, M.Y., 1997. Measurements of impact properties of small, nearly spherical particles. Experimental Mechanics 37, 292–298.

Luding, S., 1997. Stress distribution in static two-dimensional granular model media in the absence of friction. Physical Review E 55, 4720–4729.

Martin, C.L., Bouvard, D., Delette, G., 2006. Discrete element simulations of the compaction of aggregated ceramic powders. Journal of the American Ceramic Society 89, 3379–3387.

Matuttis, H.G., Luding, S., Herrmann, H.J., 2000. Discrete element simulations of dense packings and heaps made of spherical and non-spherical particles. Powder Technology 109, 278–292.

Mindlin, R.D., Deresiewicz, H., 1953. Elastic spheres in contact under varying oblique forces. Journal of Applied Mechanics 20, 327–344.

Mindlin, R.D., 1949. Compliance of elastic bodies in contact. Journal of Applied Mechanics 16, 259–266.

Mishra, B.K., Thornton, C., Bhimji, D., January 2002. A preliminary numerical investigation of agglomeration in a rotary drum. Minerals Engineering 15 (1–2), 27–33.

Mullier, M.A., Seville, J.P.K., Adams, M.J., 1991. The effect of agglomerate strength on attrition during processing. Powder Technol. 65, 321.

Ng, T.T., 1999. Fabric study of granular materials after compaction. Journal of Engineering Mechanics 125, 1390–1394.

Oda, M., Konishi, J., Nemat-Nasser, S., 1980. Some experimentally based fundamental results on the mechanical behaviour of granular materials. Géotechnique 30 (4), 479–495.

Oda, M., 1977. Co-ordination number and its relation to shear strength of granular materials. Soils and Foundations 17 (2), 29–42.

Ouadfel, H., Rothenburg, L., 1999. An algorithm for detecting inter-ellipsoid contacts. Computers and Geotechnics 24, 245–263.

Pei, C., Wu, C.-Y., England, D., Byard, S., Berchtold, H., Adams, M., 2013. Numerical analysis of contact electrification using DEM–CFD. Powder Technology 248, 34–43.

Pei, C., Wu, C.-Y., Adams, M., England, D., Byard, S., Berchtold, H., 2014. Contact electrification and charge distribution on elongated particles in a vibrating container. Chemical Engineering Science 125, 238–247.

Pei, C., Wu, C.-Y., Adams, M., England, D., Byard, S., Berchtold, H., 2015a. DEM-CFD modeling of particle systems with long-range electrostatic interactions. AIChE Journal 61, 1792–1803.

Pei, C., Wu, C.-Y., Adams, M., 2015b. Numerical analysis of contact electrification of non-spherical particles in a rotating drum. Powder Technology 285, 110–122.

Rock, M., Morgeneyer, M., Schwedes, J., Brendel, L., Wolf, D.E., Kadau, D., 2008. Visualization of shear motions of cohesive powders in the true biaxial shear tester. Particulate Science and Technology 26, 43–54.

Satake, M., 1982. Fabric tensor in granular materials. In: IUTAM Conference on Deformation and Failure of Granular Materials, Delft, 31 August–3 September, pp. 63–68.

Savkoor, A.R., Briggs, G.A.D., 1977. The effect of tangential force on the contact of elastic solids in adhesion. Proceedings of the Royal Society of London A 356, 103–114.

Smith, L., Baxter, J., Tüzün, U., Heyes, D.M., 2001. Granular dynamics simulations of heap formation: effects of feed rate on segregation patterns in binary granular heap. Journal of Engineering Mechanics 127, 1000–1006.

Thornton, C., Ning, Z., 1997. A theoretical model for the stick/bounce behaviour of adhesive, elastic-plastic spheres. Powder Technology 99, 154–162.

Thornton, C., Randall, W., 1988. In: Satake, M., Jenkins, J.T. (Eds.), Micromechanics of Granular Materials. Elsevier, Amsterdam, pp. 133–142.

Thornton, C., Yin, K.K., 1991. Impact of elastic spheres with and without adhesion. Powder Technology 65, 153–165.

Thornton, C., Zhang, L., 2003. Numerical simulations of the direct shear test. Chemical Engineering & Technology 26, 153–156.

Thornton, C., Zhang, L., 2006. A numerical examination of shear banding and simple shear non-coaxial flow rules. Philosophical Magazine 86, 3425–3452.

Thornton, C., Cummins, S.J., Cleary, P.W., 2011. An investigation of the comparative behavior of alternative contact force models during elastic collisions. Powder Technology 210, 189–197.

Thornton, C., Cummins, S.J., Cleary, P.W., 2013. An investigation of the comparative behaviour of alternative contact force models during inelastic collisions. Powder Technology 233, 30–46.

Thornton, C., 1991. Interparticle sliding in the presence of adhesion. Journal of Physics D Applied Physics 24, 1942–1946.

Thornton, C., 2000. Special issue on numerical simulations of discrete particle systems. Powder Technology 109, 3–265.

Thornton, C., 2008. Special issue on discrete element modelling of fluidised beds. Powder Technology 184, 132–265.

Thornton, C., 2009. Special issue on the 4th Int. Conf. on Discrete Element Methods. Powder Technology 193, 216–336.

Ting, J., Khwaja, M., Meachum, L.R., Rowell, J.D., 1993. An ellipse-based discrete element model for granular materials. International Journal for Numerical and Analytical Methods 17, 603–623.

Ting, J.M., Meachum, L., Rowell, J.D., 1995. Effect of particle shape on the strength and deformation mechanisms of ellipse-shaped granular assemblages. Engineering Computations 12, 99–108.

Tsuji, Y., Tanaka, T., Ishida, T., 1992. Lagrangian numerical simulation of plug flow of cohesionless particles in a horizontal pipe. Powder Technology 71, 239–250.

Tüzün, U., Baxter, J., Heyes, D.M., 2004. Analysis of the evolution of granular stress-strain and voidage states based on DEM simulations. Philosophical Transactions of the Royal Society A 362, 1931–1951.

Walton, O.R., Braun, R.L., 1986. Viscosity, granular-temperature, and stress calculations for shearing assemblies of inelastic, frictional disks. Journal of Rheology 30, 949–980.

Walton, O.R., 1993. Numerical simulation of inclined chute flows of monodisperse inelastic, frictional spheres. Mechanics of Materials 16, 239–247.

Wu, C.Y., Cocks, A.C.F., Gillia, O.T., Thompson, D.A., 2003. Experimental and numerical investigations of powder transfer. Powder Technology 138 (2–3), 214–226.

Wu, C.Y., Cocks, A.C.F., 2006. Numerical and experimental investigations of the flow of powder into a confined space. Mechanics of Materials 38 (4), 304–324.

Wu, C.-Y., 2008. DEM simulations of die filling during pharmaceutical tableting. Particuology 6, 412–418.

Wu, C.-Y., 2012a. Discrete Element Modelling. Special issue for Powder Technology. Elsevier, London.

Wu, C.-Y., 2012b. Discrete Element Modelling of Particulate Media. RSC Publishing, ISBN 978-1-84973-503-2.

Yang, R.Y., Zou, R.P., Yu, A.B., 2003. A simulation study of the packing of wet particles. AIChE Journal 49, 1656–1666.

Yang, R.Y., Zou, R.P., Yu, A.B., Choi, S.K., 2008. Characterization of interparticle forces in the packing of cohesive fine particles. Physical Review E 78, 031302.

Zeilstra, C., van der Hoef, M.A., Kuipers, J.A.M., 2006. Simulation study of air-induced segregation of equal-sized bronze and glass particles. Physical Review E 74, 010302(R).

Zeilstra, C., van der Hoef, M.A., Kuipers, J.A.M., 2008. Simulation of density segregation in vibrated beds. Physical Review E 77, 031309.

Zhang, L., Thornton, C., 2007. A numerical examination of the direct shear test. Géotechnique 57, 343–354.

Zhang, L., 2003. The Behaviour of Granular Material in Pure Shear, Direct Shear and Simple Shear (Ph.D. thesis). Aston University, UK.

Zhou, Y.C., Xu, B.H., Zou, R.P., Yu, A.B., Zulli, P., 2003. Stress distribution in a sandpile formed on a deflected base. Advanced Powder Technology 14, 401–410.

Zhu, H.P., Zhou, Z.Y., Yang, R.Y., Yu, A.B., 2007. Discrete particle simulation of particulate systems: theoretical developments. Chemical Engineering Science 62, 3378–3396.

Zhu, H.P., Zhou, Z.Y., Yang, R.Y., Yu, A.B., 2008. Discrete particle simulation of particulate systems: a review of major applications and findings. Chemical Engineering Science 63, 5728–5770.

CHAPTER

Finite Element Modeling 10

The finite element method (FEM) is one of the computational techniques for finding approximate solutions to differential and integral equations. Although FEM was developed in the first half of the twentieth century, with its initial application in solid mechanics, civil, and structural engineering, it has since advanced significantly into modeling of physical systems in a wide variety of engineering disciplines, including structural dynamics, heat transfer, fluid dynamics, and aerodynamics. In these and other fields of engineering and applied sciences, most real problems are complex in both geometry and boundary conditions. It is generally impossible to obtain analytical solutions for such problems, so approximate methods are needed. In FEM, a computational domain is generally set up within which the underlying physics can be defined using partial differential equations or integral equations. The defined computational domain is then subdivided into interconnecting simple geometric parts or subdomains, i.e., so-called finite elements, so that a set of element equations can be defined to approximate the original problem. An approximate solution for the whole domain can be obtained by solving these simple element equations. In outline, therefore, the practical application of FEM, i.e., finite element modeling, typically involves the following steps:

1. Defining a computational domain for the given problem.
2. Subdividing (i.e., discretizing) the computational domain into a set of elements, which is also referred to as mesh generation.
3. Developing integral or differential equations for a typical element (i.e., sub-domain), and a set of algebraic equations (i.e., element equations) among the unknown parameters (degrees of freedom) of the elements.
4. Assembling a global system of algebraic equations from the element equations through transforming the local coordinates of the elements to the global coordinates (i.e., spatial transformation).
5. Applying essential boundary and initial conditions.
6. Solving the system of algebraic equations to obtain approximate values for the unknown parameters in the global coordinate system (i.e., using the global degrees of freedom).

7. Postprocessing to compute solutions and extract data of interest from the solutions obtained in Step 6.

A detailed discussion on the theory of FEM is outside the scope of this book; for this we refer readers to Zienkiewicz et al. (2013) and other FEM-related books. In this chapter, the application of FEM in modeling particle systems is introduced, which includes (1) modeling of particle–particle interactions; (2) multiple particle finite element modeling (MPFEM), and (3) continuum modeling, with an example in powder compaction.

10.1 MODELING OF PARTICLE–PARTICLE INTERACTION

For the contact or impact between a particle and another particle or a wall, which are ubiquitous in particle technology, the boundary conditions at the contact are unknown *a priori*, as neither the actual contact surface nor the stresses and displacements on the contact surface are known prior to the solution of the problem. Contact/impact problems can also involve geometrical and material nonlinearity (i.e., complex material deformation behavior) so that it is very difficult if not impossible to obtain rigorous analytical solutions. However, with the advances in FEM, the contact/impact problems can now be effectively analyzed with desired accuracy. A detailed description of the principle of FEM for analyzing contact/impact problems can be found in Zienkiewicz et al. (1989), Hughes (1987), and Bathe (1982). Some distinct aspects involved in modeling of such problems using FEM are introduced in this section.

10.1.1 CONTACT MODELING TECHNIQUES

In FEM, contact boundaries are approximated by collections of polygons for three-dimensional (3D) problems and lines for two-dimensional (2D) applications. The polygons are normally either three-vertex triangular facets or four-vertex quadrilaterals. The polygons and lines composing the contact boundaries are referred to as contact segments. Each edge of the contact segment is known as a contact edge. A node on the contact edge or the contact segments is a contact node. A contact surface is then defined as the collection of all the contact segments that approximate to a complete physical boundary. Note that a contact object may have more than one contact surface.

In the contact between two different contact surfaces of either a single body or two contacting bodies, one of the contact surfaces is specified as a "slave" surface, and the other a "master" surface. The contact algorithms based on this specification are therefore known as master-slave algorithms (Zhong, 1993). Between two contact surfaces, the contact pressure results in an external load exerted on the contact body. In the FEM model, the contact load is replaced by a set of forces exerted at the nodes. Hence, the problem can be simplified by taking the nodal forces instead of the contact pressures as primary unknowns. The contact nodal forces can then be obtained by solving the governing equations for each contact segment (see, for example, Hughes et al., 1976 and Zhong, 1993), from which the contact pressure can in turn be calculated directly from the nodal forces.

The solution for the contact condition requires that the total number of the contacting slave nodes and the contact force at each be known *a priori*, implying that algorithms are needed to find the total number of contacting slave nodes and to determine the contact nodal force at each. The former is the task of contact searching procedures, while the latter requires certain contact constraint methods. Here contact search algorithms and the commonly used contact constraint methods (the kinematic constraint method and the penalty method) are introduced in Boxes 10.1 and 10.2, respectively.

BOX 10.1 CONTACT SEARCHING PROCEDURE

In the master-slave method, the objective of the contact searching procedure is to determine the total number of contacting slave nodes n_c at time instant t and for each contacting slave node to find the corresponding contact point in the master segment. To determine n_c, a trial-and-error procedure is widely used. Assuming that the solution at time $t - \Delta t$ is known, and that the total number of contacting slave nodes n_c and the position of a contact node in relation to all the contact segments are also known, it is possible to select new potential contact nodes at time t by evaluating the distance of a new potential contact node from any master segment/edge/node. It is then necessary to introduce a prescribed control distance, which can be specified either to be zero or on the basis of the relative velocities $\dot{u}_c$ of the contact boundaries as

$$l_c = \Delta t \dot{u}_c(t - \Delta t) \cdot \mathbf{N}_1(t - \Delta t) \tag{10.1}$$

where $\mathbf{N}_1$ is a unit normal vector at a potential target point. The selection of the potential contacting slave nodes is performed in three phases (Zhong, 1993): (1) a master node that is closest to the slave node being considered is found; (2) a master segment that contains the master node and is closest to the slave node is found as a target segment, and (3) the shortest distance between the slave node and its target master segment is calculated. If this distance is smaller than l_c, the slave node is then considered to be in contact with the master segment and regarded as a new potential contact slave node. Based on the selected potential contact slave nodes, an approximate solution is obtained. By enforcing the contact conditions at all the contact slave nodes, it may be found that the contact forces at some of the selected slave nodes are inconsistent with the physical conditions, implying that the mechanical contact condition is violated. A selected slave node that does not satisfy the mechanical contact condition is identified as an error node. To obtain a correct solution, all the error nodes must be removed from the selection of contacting nodes, and then the updated selection is obtained. The updated selection is used to find the solution. If no error nodes are found, the updated selection is regarded as the true selection, and the contact searching procedure can be terminated. For each identified slave node, the corresponding contact target point, defined as the closest point on the master segment to the slave node, can then be identified.

BOX 10.2 CONTACT CONSTRAINT METHODS: KINEMATIC VERSUS PENALTY

Once a potential contacting slave node and the corresponding target point on the master segment are found, contact constraints must be imposed to model the contact boundary conditions. Two approaches are commonly used to enforce the contact constraints: the kinematic constraint method and the penalty method.

In the kinematic constraint method, also known as the Lagrange multiplier method, constraints are enforced on the system of equations in the global coordinate system. With the FE discretization of the contact surface, impenetrability of contact (i.e., no node can penetrate its contacting surface) needs to be satisfied for all contact constraints, which can be established explicitly. With the kinematic constraint method, a Lagrange multiplier, having the contact forces as its elements, is introduced into the element equations. By treating the Lagrange multiplier vector as an unknown and solving the element equations using the given boundary conditions simultaneously, the Lagrange multiplier vector (i.e., the set of contact forces) can be determined.

The contact constraint condition is enforced exactly in the kinematic method. It should be noted that the enforcement of contact constraint is normally applied only on the slave nodes. Problems may hence arise when the master surface is discretized more finely than the slave surface, so that some master nodes may penetrate into the slave surface. Therefore, the master surface should be discretized with coarser elements than the slave surface.

The penalty method uses an explicit approach to enforce the contact constraints. As illustrated in Fig. 10.1, a slave node is about to contact the master surface. At the time instant t, the slave node makes contact with the master surface and penetrates a distance $\tilde{p}$ into the it. This is permitted so no action is taken to avoid the penetration at time t. However, at the next time instant $t + \Delta t$, an interface "spring" is added between the slave node and the master surface. The force generated from the interface spring is equal to the product of spring stiffness and the penetration distance $\tilde{p}$. The interface spring stiffness is normally determined as the multiplication of a penalty parameter $\tilde{\alpha}$ with the stiffness of the contacting bodies. Because the penalty method resolves any contact penetration that exists at the beginning of each increment and the contact force is directly computed by introducing the penalty parameters, it is therefore an explicit method.

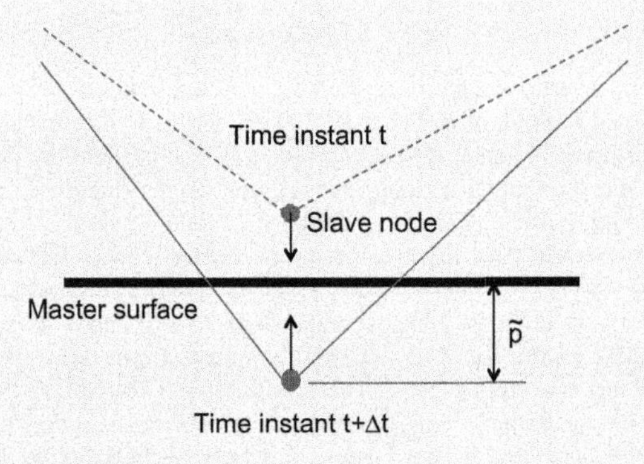

FIGURE 10.1

Illustration of the penalty method.

BOX 10.2 CONTACT CONSTRAINT METHODS: KINEMATIC VERSUS PENALTY—cont'd

In the penalty method, the contact constraints are enforced explicitly without any additional unknown, so the computation can be performed efficiently. Since the penetrations are controlled by the penalty parameters, the accuracy of the solution depends on the choice of these parameters. If the penalty parameters are too small, the resulted penetration may be too large, which may cause unacceptable calculation errors. If these parameters are too large, severe numerical problems may be caused and it may become impossible to obtain a solution. Thus the choice of the penalty parameters is critical.

10.1.2 APPLICATIONS

As an example, consider the normal impact of a spherical particle with a planar surface as illustrated in Fig. 10.2. The sphere has a radius $R = 10\ \mu m$ and collides with the substrate at an incident velocity $v_{ni} = 5.0$ m/s. The sphere and the substrate are assumed to be elastic and have the same material properties: Young's modulus: 208.0 GPa; Poisson's ratio: 0.3; density: $7.85 \times 10^3\ kg/m^3$. The normal impact shown in Fig. 10.2 is geometrically axisymmetric along the vertical axis passing through the center of the sphere; it can therefore be analyzed using a 2D FE model. Because of this symmetry, only half of the sphere and the substrate are considered (see Fig. 10.3).

The effects of impact are not confined to the contact area; a mechanical stress wave travels through both the particle and the substrate and carries energy away from the contact. The substrate in the FEM should be sufficiently large that the influence of the boundary constraints can be ignored. To choose the size of the substrate appropriately, it is necessary to estimate the maximum deformation by taking the loading condition and material properties into account. A systematic study on the

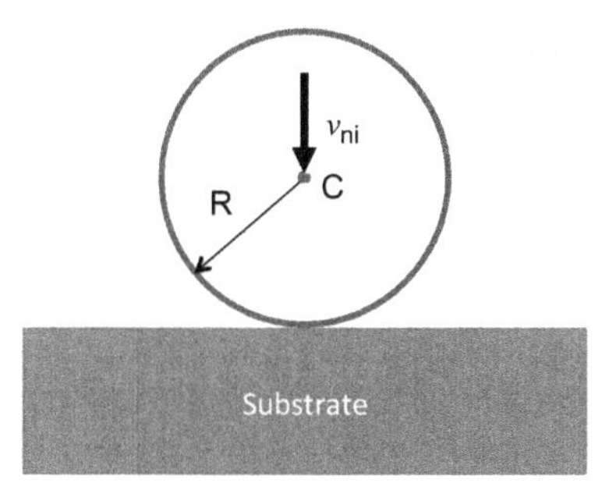

FIGURE 10.2

Illustration of the normal impact of a sphere with a planar surface.

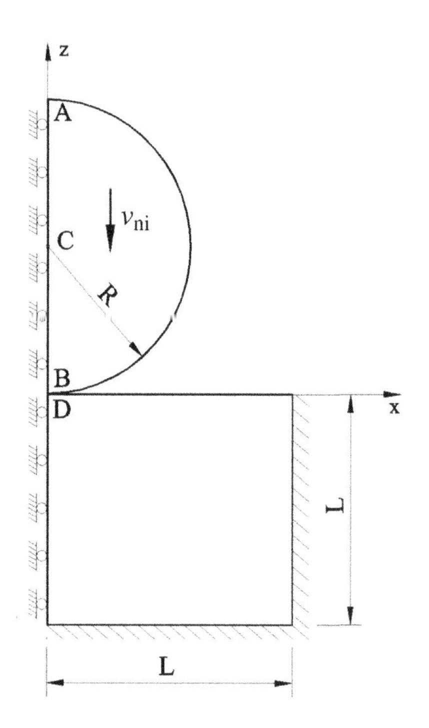

FIGURE 10.3

The finite element model for the impact of a sphere with a substrate (Wu, 2002).

effect of the substrate size ($L/R = 1 \sim 20$) reported in Wu (2002) showed that a substrate of dimension $L/R = 1$ is too thin for modeling the impact accurately, even at low impact velocities. For impact with a substrate of dimension $L/R \geq 2$, the effects of the boundary constraints and the propagation of the stress wave are sufficiently small, so that dimension $L/R = 2$ is adopted here.

The corresponding FE model is shown in Fig. 10.4. Finer mesh is employed near the contact zone (Fig. 10.4(b)) in order to model the contact deformation more accurately, as the deformation is mainly localized to the region of the initial contact point (see Chapter 8).

The interaction between the sphere and the substrate is modeled with a master-slave slideline, consisting of two contact lines: one on the sphere side and the other on the substrate side. This type of slideline is based on the contact searching algorithm discussed in Box 10.1, and the kinematic constraint method (Box 10.2) is used to enforce the contact constraints. As specified in the kinematic constraint method, the master surface on the sphere side is discretized with coarse meshes, while finer mesh is used for the slave surface on the substrate side.

The collision process is modeled by specifying an initial velocity v_{ni} (i.e., 5.0 m/s) to every node inside the sphere. The boundary condition is illustrated in Fig. 10.3.

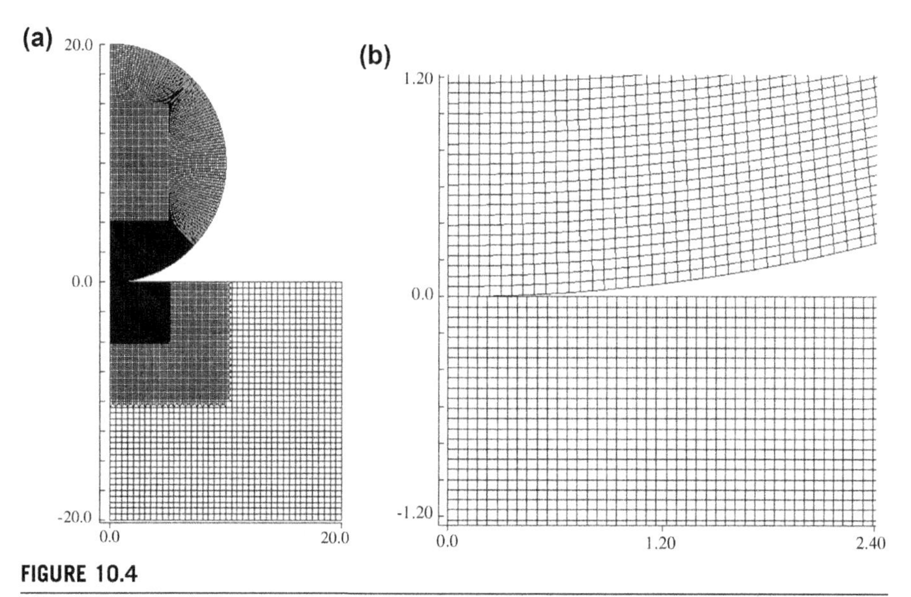

FIGURE 10.4

(a) Finite element model for the impact of a sphere with a half-space; (b) finer mesh used in the contact region (Wu, 2002).

Owing to the symmetry, all nodes along the z axis are constrained to move only in the z-direction (i.e., the displacement $u_x = 0$), while the nodes along the bottom line and the side edge of the substrate are fixed (i.e., $u_x = \mu_z = 0$).

Using the model presented in Fig. 10.4, and the given initial and boundary conditions, the response of each node in this model can then be solved using an FE solver. Here the solver DYNA2D (Whirley and Engelmann, 1992) is chosen but this type of problem can also be analyzed using other FE solvers, such as ABAQUS and ANSYS. From the response of each node, much information of interest can be obtained; some typical results are discussed in this section.

During the impact, both bodies deform and develop internal stresses. A typical pattern for the stress distribution at maximum compression is presented in Fig. 10.5, in which the effective stress is shown in Fig. 10.5(a), with the corresponding maximum shear stress in Fig. 10.5(b). It is clear that the pattern for the stresses inside the sphere is quite similar to that inside the substrate. The "+" signs in Fig. 10.5 indicate the location of the maximum value of the corresponding stresses. It can be seen that the maximum values of both the effective stress and shear stress are located inside the substrate at a depth of about half of the contact radius, implying that plastic deformation will initiate inside the substrate if the contacting bodies are elastic-plastic in nature.

From the FE analysis, the impact response, such as displacement and velocity at each node, the kinetic energy of each body and the contact force at the interface can also be obtained, as presented in Figs 10.6 and 10.7. Figure 10.6(a) shows the time evolutions of displacements at four distinct nodes: node A (the top); B (the bottom);

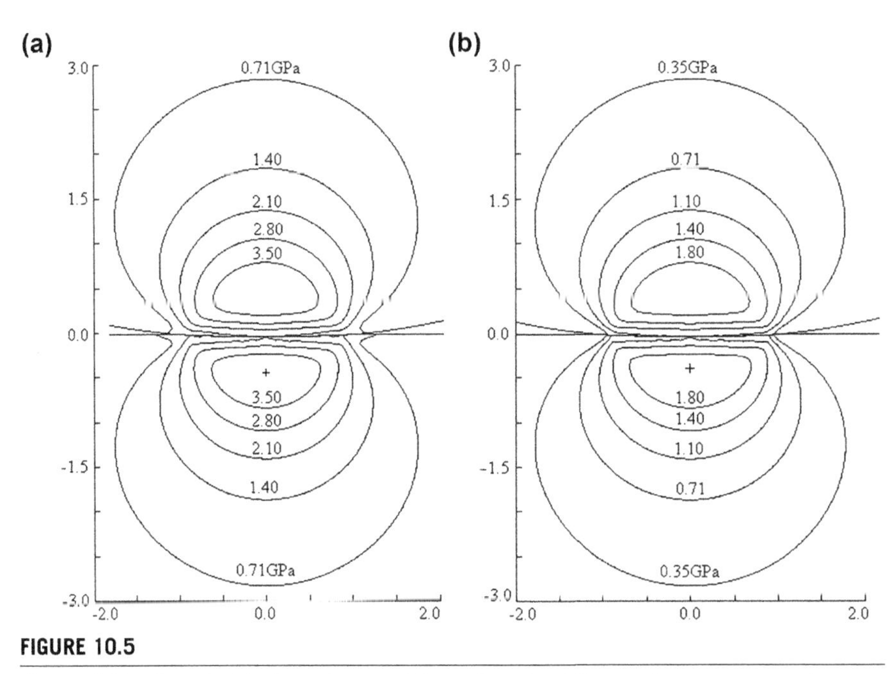

FIGURE 10.5

Stress distributions inside the sphere and the substrate at the maximum compression: (a) the effective stress; (b) the maximum shear stress (Wu, 2002).

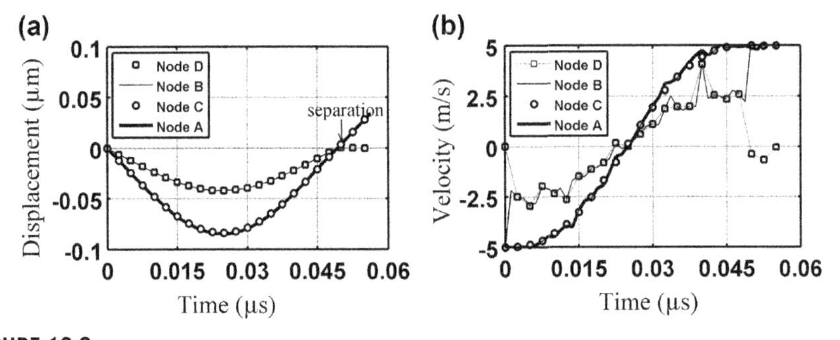

FIGURE 10.6

Time histories of (a) nodal displacements and (b) velocities during the impact of an elastic sphere with an elastic substrate. Nodes A, B, C, and D correspond to positions marked in Fig. 10.3 (Wu, 2002).

C (the center), and D (the initial contact point of the substrate), as marked in Fig. 10.3. The corresponding velocity evolutions are given in Fig. 10.6(b). Figure 10.7(a) shows the evolution of the kinetic energy, and the evolution of the contact force is presented in Fig. 10.7(b). It can be seen that, for the whole impact process, there are two distinct phases: compression and restitution. The compression

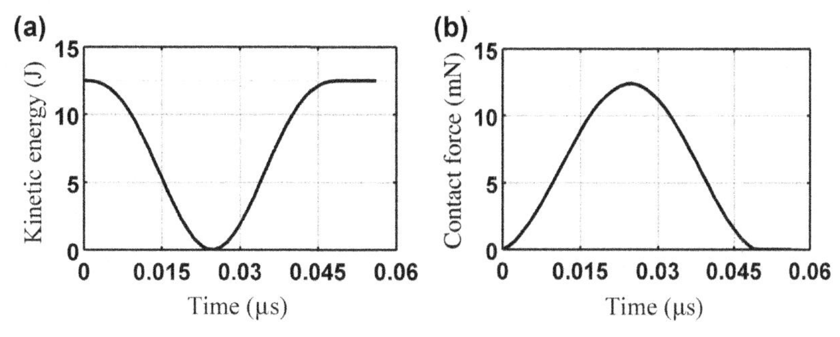

FIGURE 10.7

Time histories of (a) kinetic energy and (b) impact force during the impact of an elastic sphere with an elastic substrate (Wu, 2002).

phase terminates when the nodal displacements and the contact force reach their maximum values, and the nodal velocities and the kinetic energy approach zero. Thereafter the restitution phase starts. The restitution phase ends when the nodal displacements and the contact force reduce to zero. After the separation of the two contacting bodies, the kinetic energy remains unchanged at a constant value close to the initial one, which implies that the energy loss due to stress wave propagation is negligible, and the coefficient of restitution is close to unity. The results shown in Fig. 10.6 also indicate that the deformation of the sphere is mainly concentrated in the bottom hemisphere as the nodal displacements at the top (A) and the center (C) of the sphere are essentially identical, while those at nodes C and B are very different.

From these impact responses, further information can be obtained, including the maximum pressure, the contact area (Fig. 8.4), the force–displacement relationship (see Figs 8.5 and 8.18), the duration of impact (Fig. 8.6), and the coefficient of restitution (Fig. 8.19), which provide physical insight into the contact/impact problems of particles.

Similar FE analysis can be readily applied to other more complex impact problems, including contact of irregularly shaped particles, interaction in the presence of surface energy (i.e., adhesive impacts), and interaction with liquid bridges, provided that an appropriate method can be established to model the liquid at the interface.

For modeling adhesive contact/impact using FEM, two techniques have been developed: (1) A body force model, in which adhesion is represented by intermolecular interactions, so that the van der Waals force between two bodies is modeled in the FE formulation as a body force derived from the Lennard-Jones potential (Zhang et al., 2011; Cho and Park, 2004; Sauser and Li, 2007; Sauer and Wriggers, 2009); (2) An interfacial adhesion model (Feng et al., 2009; Wang et al., 2012), in which the adhesive force is approximated by introducing spring elements. The deformation of the spring elements is modeled with a specified constitutive equation describing the relationship between the adhesion stress and the separation distance of the two surfaces.

Modeling particle interaction due to liquid bridges using FEM was attempted by Xu and Fan (2004), who applied the Laplace pressure and surface tension on the solids at the interface between the liquid and solid, and at the three-phase contact line respectively. The induced force and pressure were applied on the surfaces of the contacting bodies as the boundary conditions. In addition, the interaction between the liquid bridge and the deformation of the contacting bodies was considered. However, the liquid bridge was not modeled explicitly so its deformation during the contact was not considered, which deserves further investigation.

10.2 MULTIPLE PARTICLE FINITE ELEMENT MODELING

In many applications, such as tableting and roll compaction, particle systems are subjected to high consolidation pressures, resulting in large deformations of particles. For these systems, the discrete element method (DEM) discussed in Chapter 9 is currently not applicable since large deformation (i.e., gross change of the geometry) of individual particles is not considered. For consolidation of particle systems, two computational approaches can be used:

1. Continuum modeling using FEM, in which a particle system is treated as a continuum instead of a collection of individual entities. This approach will be discussed in detail in Section 10.3;
2. MPFEM, which can be regarded as an extension of the approach described in Section 10.1 to model interactions of many particles simultaneously using FEM. This is sometimes referred to as the combined finite-discrete element method. In principle, it uses the FE procedure to model the deformation of a system consisting of discrete particles, each of which is considered explicitly.

In MPFEM, each particle is discretized with its own FEM mesh and its deformation individually modeled. As the particles interact with each other, their deformation depends on their loading conditions (including body forces, external forces, and contact forces) and the stress state. One of the key issues is then how to model the interaction between particles. The contact modeling techniques discussed in Section 10.1.1 can be used directly in MPFEM so that the contact can be modeled at each contact node. This approach was adopted by Procopio and Zavalinagos (2005), Zhang (2009), Frenning (2008, 2010), Harthong et al. (2009, 2012), Guner et al. (2015) and Gustafsson et al. (2013). It requires careful design of FE meshes so that the contact can be modeled accurately (generally requiring fine meshes) and with computational efficiency. An alternative approach was proposed by Ransing et al. (2000) and Lewis et al. (2005), who discretized the boundary of each particle with a layer of interface elements. The interface layer has a finite thickness Δ, a stiffness k, and a damping coefficient c.

The contact force F is then calculated from the overlap δ and impact velocity v as follows

$$F = \frac{k\delta}{\Delta - \delta} + \frac{cv}{\delta} \tag{10.2}$$

By introducing this interface layer, the kinematic behavior can be computed. Ransing et al. (2000) found that the choice of the stiffness and damping coefficient had negligible effect on the force–displacement relationship for the compression of particle systems, but could significantly affect the convergence stability and speed of the numerical calculation.

MPFEM is a useful approach as it combines the strength of conventional DEM, in which particles are modeled explicitly as individuals, and FEM, in which the deformation of individual particles and the interaction between them can be modeled. The combination can be used to model deformation of particles in a granular assembly under high compression pressures. It can also model large deformations of particles (see Fig. 10.8), arbitrary particle shapes (Fig. 10.9), and complex tooling geometries (Fig. 10.10). A further advantage is that the stress and strain distributions at the particle level can be determined. At the macroscopic level, both yield surfaces (Harthong et al., 2012) and overall compression behavior can be determined (Procopio and Zavalinagos, 2005; Frenning, 2008, 2010; Harthong et al., 2012).

However, MPFEM is very computationally intensive since each particle needs to be discretized into hundreds (even thousands) of elements, and the contact search needs to be carried out at the nodal level; this severely limits the number of particles that can be modeled. So far, the maximum number of particles in MPFEM analysis reported in literature is 1680 (Gustafsson et al., 2013), which is many orders of magnitude below the number of particles handled in powder handling and processing applications. Furthermore, using MPFEM, the FE mesh needs to be designed carefully so that both computational convergence and good accuracy can be obtained.

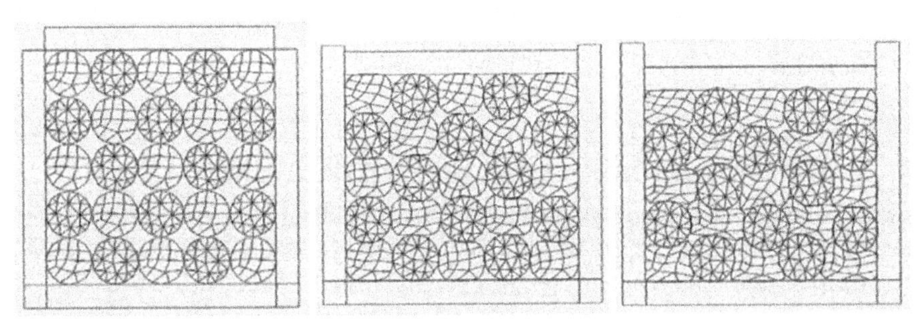

FIGURE 10.8

Modeling of compression of a mixture of ductile and brittle particles using FEM (Ransing et al., 2000).

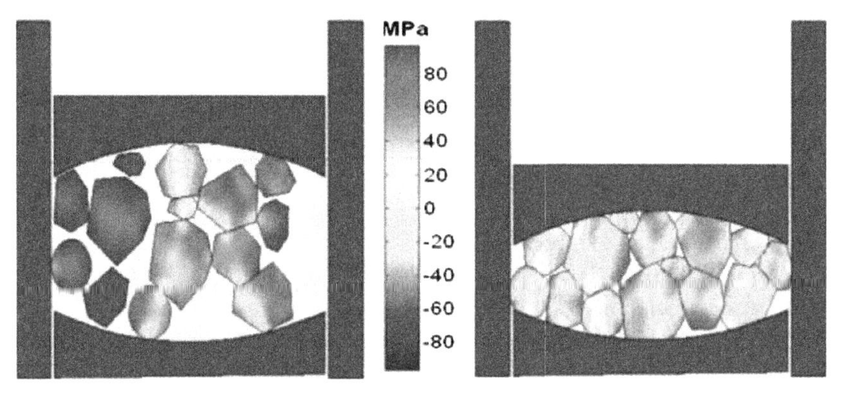

FIGURE 10.9

Modeling of compression of arbitrarily shaped particles using FEM: distribution of shear stresses (Lewis et al., 2005).

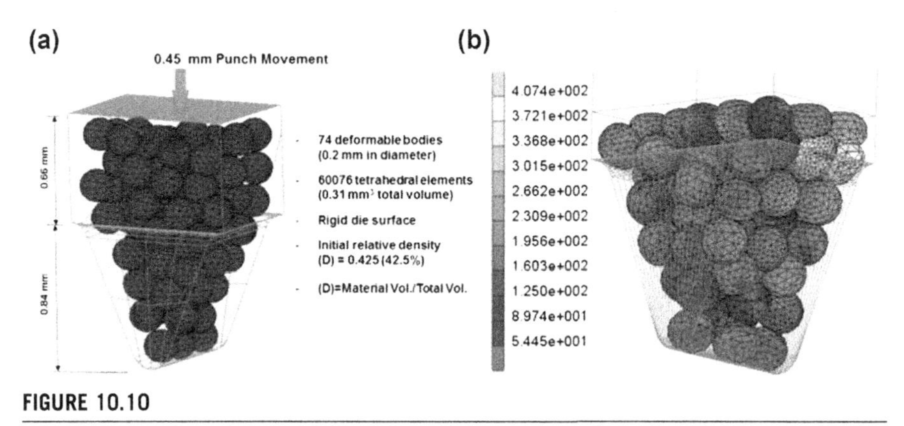

FIGURE 10.10

Modeling of compression of deformable spheres in a complex shaped die using MPFEM: (a) model set up; (b) distribution of effective stresses (Guner et al., 2015).

10.3 CONTINUUM MODELING OF POWDER COMPACTION

Powder compaction is a process used in the pharmaceutical, fine chemicals, powder metallurgy, ceramic, and agrochemical industries to manufacture particulate products, such as tablets, mechanical components, and pellets. During this process, powders are compressed in a mold (i.e., a die) under high pressure, and undergo a complicated response, ranging from particle rearrangement and particle deformation to fragmentation. Powder compaction can be adequately modeled using the DEM approach discussed in Chapter 9 if the applied pressure is very low so that the process is dominated by particle rearrangement and deformation of particles is negligible. Powder compaction involving high pressures and significant particle deformation can also be modeled using MPFEM, as discussed in Section 10.2, if only a limited number of particles or granules is involved. However, most powder compaction processes involve many millions of particles and high pressure so that significant particle deformation takes place. For these processes, neither DEM nor MPFEM is applicable.

A feasible alternative is the continuum approach using FEM, and it has been widely employed for modeling the compaction of metallic, ceramic, and pharmaceutical powders. Using this approach, powders are treated as continuous media instead of assemblies of individual particles, and the compaction processes are then described as boundary value problems, for which partial differential equations are developed to represent mass, energy, and momentum balances, and constitutive laws are introduced to describe stress-strain relationships and die-wall friction. The powders are generally modeled as elastic—plastic materials for which the yield behavior is approximated using appropriate yield models, widely known as the yield surfaces. (Here yield implies that the deformed material cannot recover to its initial state, such as after onset of plastic deformation.) The choice of material model and calibration of the associated material parameters are very important aspects in continuum modeling of powder compaction, which will be discussed in detail here.

10.3.1 MATERIAL MODELS

During powder compaction, a powder can be modeled as an elastic—plastic continuum and the deformation of the powder can be described using the incremental plasticity theory. In this approach, the total strain increment $d\varepsilon_{ij}$ can be decomposed into an elastic strain increment $d\varepsilon_{ij}^{e}$ and a plastic one $d\varepsilon_{ij}^{p}$:

$$d\varepsilon_{ij} = d\varepsilon_{ij}^{e} + d\varepsilon_{ij}^{p} \quad (i = 1, 3 \text{ and } j = 1, 3) \tag{10.3}$$

where subscript i and j indicate the coordinates.

The elastic behavior of isotropic materials (i.e., those for which the material properties are independent of direction) can be either linear or nonlinear. For elastic deformation, the relationship between macroscopic stress increment $d\sigma_{ij}$ and strain increment $d\varepsilon_{ij}^{e}$ can be given as (Timoshenko and Goodier, 1951):

$$d\varepsilon_{ij}^{e} = \frac{ds_{ij}}{2G} + \frac{d\sigma_m}{3K}\delta_{ij} \quad (i = 1, 3 \text{ and } j = 1, 3) \tag{10.4}$$

where δ_{ij} is the Kronecker delta function (i.e., $\delta_{ij} = 1$ if $i = j$, $\delta_{ij} = 0$ if $i \neq j$), σ_m is the mean stress and given as

$$\sigma_m = \frac{1}{3}(\sigma_{11} + \sigma_{22} + \sigma_{33}) \tag{10.5}$$

s_{ij} is the deviatoric stress and given as

$$s_{ij} = \sigma_{ij} - \sigma_m\delta_{ij} \tag{10.6}$$

G and K are the shear and bulk moduli, respectively, which are material properties. For continuous solid materials, they are normally constants, but for powders, they are functions of the relative density. G and K are also related to the Young's modulus E and Poisson's ratio υ as follows

$$E = \frac{9GK}{3K + G} \tag{10.7}$$

$$\upsilon = \frac{3K - 2G}{2(3K + G)} \tag{10.8}$$

For elastic—plastic materials there is an elastic limit, above which plastic deformation initiates. This limit is defined by a yield criterion. For an unconfined uniaxial test, the yield criterion is simply that the yield stress Y is reached (see Fig. 8.2). However, when several stress components act on a material simultaneously, defining the yield criterion becomes complicated. Generally, the yield criterion is defined using a scalar yield function

$$F[\sigma_{ij}, \varsigma_n] = 0 \tag{10.9}$$

where ς_n denotes n specific material parameters, such as the relative density. In the stress space, Eq. (10.9) defines a surface representing the elastic limit of the stress state, i.e., the set of maximal permissible stresses. In the plasticity theory, this surface is the so-called yield surface. An exemplar yield surface in 2D is illustrated in Fig. 10.11. The yield function is generally defined in such a way that the stress state is located inside the yield surface, i.e., $F[\sigma_{ij}, \varsigma_n] < 0$, if the deformation is elastic; while the stress state is positioned on the yield surface, i.e., $F[\sigma_{ij}, \varsigma_n] = 0$, if plastic deformation occurs. Note that it is inadmissible to have $F[\sigma_{ij}, \varsigma_n] > 0$, i.e., it is impossible to have a stress state represented by a point outside the yield surface. This implies that in the plastic deformation range, the stress state can only be redistributed between different stress components so that the stress point is still on

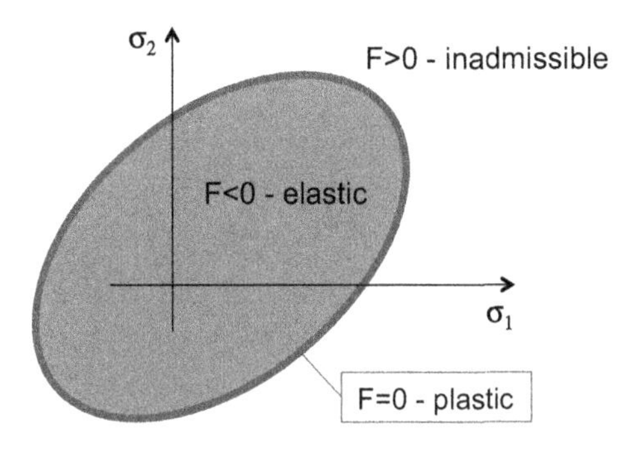

FIGURE 10.11

An example of yield surface in 2D.

the yield surface but it may "slide" into a new position. The change from plastic deformation to elastic deformation (e.g., during unloading after powder compaction) leads to the stress point moving from the boundary of the yield surface to the inside, and *vice versa*.

Once plastic deformation takes place, the material may be subjected to further loading so that it may deform further. To describe the material response when plastic deformation prevails, a flow rule (or flow potential) needs to be defined to relate the plastic strain increment $d\varepsilon_{ij}^{p}$ at yield to the stress state. The plastic strain increment $d\varepsilon_{ij}^{p}$ can be given as (Green, 1972)

$$d\varepsilon_{ij}^{p} = \lambda \frac{\partial \Psi}{\partial \sigma_{ij}} = \lambda \nabla \Psi \qquad (10.10)$$

where λ is a plastic multiplier (i.e., a positive scalar of proportionality) and Ψ is the plastic flow potential, which is also a function of the stresses and material parameters, similarly to the yield function $F[\sigma_{ij}, \varsigma_n]$. In the stress space, $\Psi[\sigma_{ij}, \varsigma_m] = 0$ can also be represented as a surface, termed the flow potential surface. Because $\nabla\Psi$ is a vector normal to the surface described by $\Psi[\sigma_{ij}, \varsigma_m] = 0$, Eq. (10.10) implies that the plastic strain increment $d\varepsilon_{ij}^{p}$ can be plotted as a vector normal to the surface with a length determined by the plastic multiplier λ, as illustrated in Fig. 10.12. The plastic flow potential Ψ needs to be defined for a given elastic–plastic material in a similar way to the yield function. For some materials, such as metals, the plastic flow potential can be approximated using exactly the same expression as the yield function, i.e., $\Psi[\sigma_{ij}, \varsigma_m] = F[\sigma_{ij}, \varsigma_m]$. In this case, Eq. (10.10) becomes

$$d\varepsilon_{ij}^{p} = \lambda \frac{\partial F}{\partial \sigma_{ij}} = \lambda \nabla F \qquad (10.11)$$

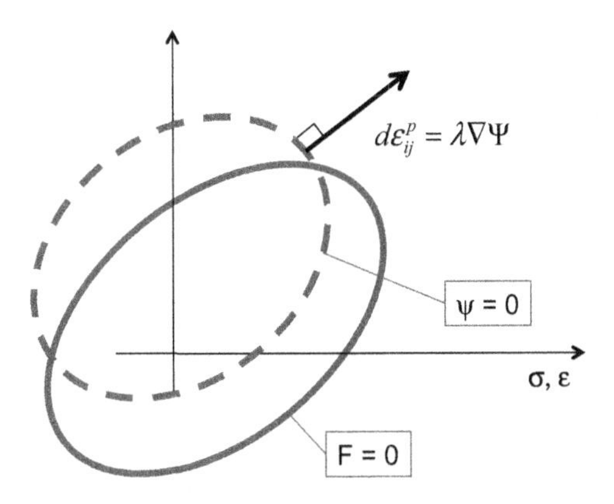

FIGURE 10.12

Illustration of the plastic strain increment in 2D.

Since the flow rule is associated with the yield function (Eq. (10.11)), this is so-called *associated* flow. If $\Psi[\sigma_{ij}, \varsigma_m]$ has a different expression from $F[\sigma_{ij}, \varsigma_m]$, the flow rule is *nonassociated*, because the plastic strain increment is not associated with the yield function. The nonassociated flow rule can be applied to most elastic–plastic materials.

Various yield surfaces were proposed to describe the plastic behavior of particulate materials, including the Drucker–Prager–Cap (DPC) model (Drucker and Prager, 1952), the Cam-Clay model (Schofield and Wroth 1968), and the Gurson model (Dimaggio and Sandler, 1971). Originally developed for soil mechanics, these have been used for analyzing consolidation of soils, the compaction of metallic, ceramic, and pharmaceutical powders (Aydin et al., 1996; Wu et al., 2005; Coube and Riedel, 2000; Sinka et al., 2003; Michrafy et al., 2002). Among these models, the DPC model is the most widely used yield surface for modeling powder compaction, because it can describe both the shear failure and the plastic yield of particulate materials; in addition, it can be readily calibrated experimentally for most powders. It is therefore discussed in the next section.

10.3.2 DPC MODEL

The DPC model was initially developed as an extension of the Mohr–Coulomb model (see Chapter 7) with a compaction surface introduced as the yield criterion, so it originally consisted of a shear failure segment F_s that is essentially the Mohr–Coulomb failure line and a cap segment F_c (Dracker and Prager, 1952). In order to ensure computational stability, a modified DPC model was developed and implemented in some FE solvers, such as ABAQUS, in which a transition segment F_t was introduced to provide a smooth surface bridging the shear failure segment and the cap segment (ABAQUS, 2013). These three segments are generally defined using two stress invariants: the hydrostatic pressure p and the von Mises equivalent stress q in the (p, q) plane (see Fig. 10.13),

$$p = \frac{1}{3}(\sigma_1 + \sigma_2 + \sigma_3) \tag{10.12}$$

$$q = \sqrt{\frac{1}{6}\left[(\sigma_1 - \sigma_2)^2 + (\sigma_2 - \sigma_3)^2 + (\sigma_3 - \sigma_1)^2\right]} \tag{10.13}$$

where σ_i ($i = 1, 2, 3$) are the principal stresses.

The shear failure surface F_s is a straight line determined by the cohesion d and the angle of friction β:

$$F_s[p, q] = q - p\tan\beta - d = 0 \tag{10.14}$$

This defines a criterion for the occurrence of shear flow. For a given material, if the stress state defined by the hydrostatic pressure p and the von Mises equivalent stress q satisfies Eq. (10.14), i.e., $F_s[p, q] = 0$, shear failure is induced.

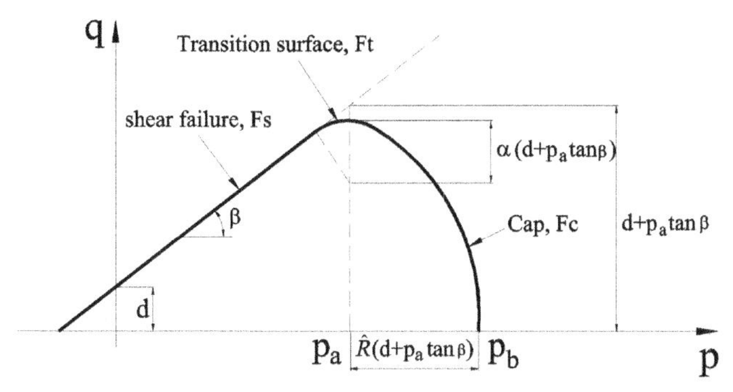

FIGURE 10.13

The Drucker–Prager–Cap model (Wu et al., 2005).

The cap surface F_c describes compaction and hardening of a material. If the stress state positions on the cap surface, densification and hardening occurs. The cap surface is an elliptical curve of constant eccentricity in the p–q plane and intersects the hydrostatic pressure axis (i.e., p axis). It is defined by:

$$F_c[p, q] = \sqrt{(p - p_a)^2 + \left(\frac{\hat{R}q}{1 + \alpha - \alpha/\cos\beta}\right)^2} - \hat{R}(d + p_a \tan\beta) = 0 \quad (10.15)$$

where $\hat{R}$ and α are parameters defining the shape of the cap surface and the transition surface, respectively. p_a is an evolution parameter characterizing the hardening or softening behavior driven by the volumetric plastic strain, ε_v^p, and given as:

$$p_a = \frac{p_b - \hat{R}d}{1 + \hat{R}\tan\beta} \quad (10.16)$$

where p_b is the hydrostatic pressure that defines the position of the cap. p_b is generally defined as a function of the volumetric plastic strain ε_v^p, i.e.,

$$p_b = f\left[\varepsilon_v^p\right] \quad (10.17)$$

Equation (10.17) determines hardening or softening of the cap surface: hardening is caused by volumetric plastic compaction, while softening is induced by volumetric plastic dilation.

The transition surface is always relatively small and controlled by the parameter α with a typical value of 0.01-0.05. The transition surface is given by:

$$F_t[p, q] = \sqrt{(p - p_a)^2 + \left[q - \left(1 - \frac{\alpha}{\cos\beta}\right)(d + p_a \tan\beta)\right]^2} - \alpha(d + p_a \tan\beta) = 0 \quad (10.18)$$

Two flow potentials are defined as the plastic flow rules: an associated flow potential Ψ_c on the cap surface and a nonassociated flow potential Ψ_s for the shear failure surface and the transition surface. The associated flow potential Ψ_c and the nonassociated flow potential Ψ_s are defined as

$$\Psi_c[p, q] = \sqrt{(p - p_a)^2 + \left(\frac{\hat{R}q}{1 + \alpha - \alpha/\cos\beta}\right)^2} \tag{10.19}$$

$$\Psi_s[p, q] = \sqrt{[(p - p_a)\tan\beta]^2 + \left(\frac{q}{1 + \alpha - \alpha/\cos\beta}\right)^2} \tag{10.20}$$

In the (p, q) plane, Eqs (10.19) and (10.20) represent two elliptic segments of a continuous and smooth potential surface, as shown in Fig. 10.14.

The DPC model introduces six parameters: d, β, $\hat{R}$, p_a, α, and p_b. They are not necessarily constant for powders but it was shown that d, β, $\hat{R}$, p_a are functions of the relative density, and p_b varies with the volumetric plastic strain (Eq. (10.17)). The cohesion d and internal frictional angle β define the shear failure surface; $\hat{R}$ and p_a determine the cap surface; α is one of the parameters defining the transition surface; and p_b specifies the hardening or softening of the cap surface. All six parameters are required to define the constitutive model for a given powder, in addition to the elasticity parameters, such as Young's modulus E and Poisson's ratio υ.

10.3.3 DETERMINATION OF CONSTITUTIVE PROPERTIES

In order to use the models discussed above, all parameters have to be determined from experimental measurements on the material under consideration. This exercise is called experimental calibration of the constitutive models (or simply "model calibration"). Full calibration of the models requires triaxial compression test data. In the triaxial test, the stresses acting on a test sample can be controlled

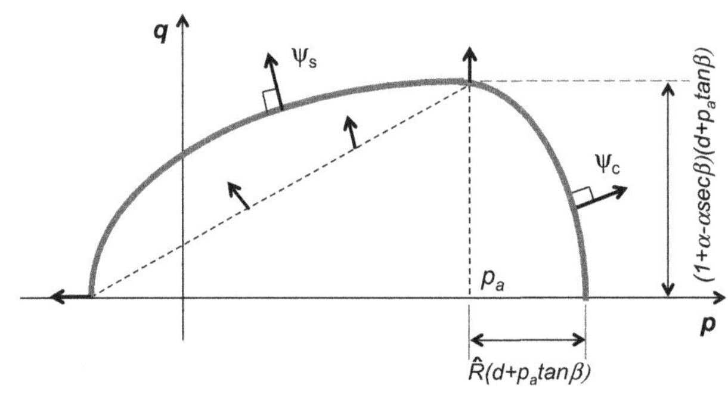

FIGURE 10.14

Illustration of the plastic flow rules in the Drucker–Prager–Cap model (Aydin et al., 1996).

separately in three directions so that a wide range of stress states can be achieved, and stress probing can also be performed to identify the yield, loading, hardening, and softening at different stress levels. However, the triaxial test requires a sophisticated testing system and complicated data analysis, and its application is limited to scientific research. Furthermore, powders can deform significantly during triaxial testing, which makes the interpretation of the test data extremely difficult.

Alternatively, model calibration using simple and standard test systems has also been developed, which generally involves uniaxial compression in a die instrumented with radial pressure sensors; in addition to measuring the radial stresses, diametrical compression and unconfined uniaxial compression can also be measured, as illustrated in Fig. 10.15.

10.3.3.1 Young's modulus E *and Poisson's ratio* υ

Determination of elastic properties (i.e., Young's modulus E and Poisson's ratio υ) can be achieved using the uniaxial compression test with an instrumented cylindrical die (Aydin et al., 1996; Wu et al., 2005; Michrafy et al., 2002). During the test, a powder is first compressed to a specified maximum compression force and then unloaded. The typical relationship between the axial stress σ_z and axial strain ε_z (i.e., the stress and strain in the z-direction) is illustrated in Fig. 10.16. During loading, the stress increases with the strain at an increasing rate. In the early stage of unloading, the stress decreases linearly with the strain, implying that elastic deformation dominates.

For uniaxial compression:

$$\sigma_1 = \sigma_z \tag{10.21}$$

$$\sigma_2 = \sigma_3 = \sigma_r \tag{10.22}$$

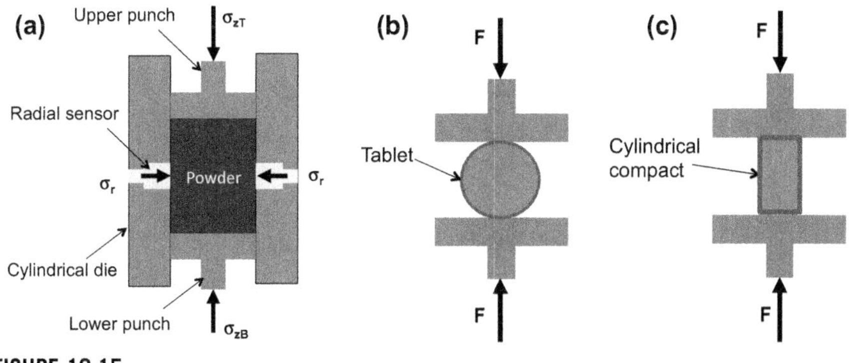

FIGURE 10.15

Some tests involved in model calibration: (a) uniaxial compression, (b) diametrical compression, and (c) unconfined uniaxial compression.

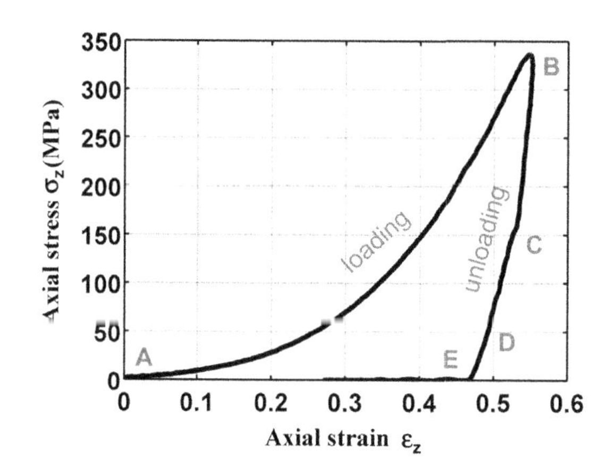

FIGURE 10.16

The variation of axial stress with axial strain during compaction (Wu et al., 2005).

Using Eqs (10.12) and (10.13), we have

$$p = \frac{1}{3}(\sigma_z + 2\sigma_r) \tag{10.23}$$

$$q = \frac{1}{\sqrt{3}}|\sigma_z - \sigma_r| \tag{10.24}$$

From the measurement of the radial stress, the evolution of stress states (p, q) during the whole compaction process can be obtained. Figure 10.17 shows the

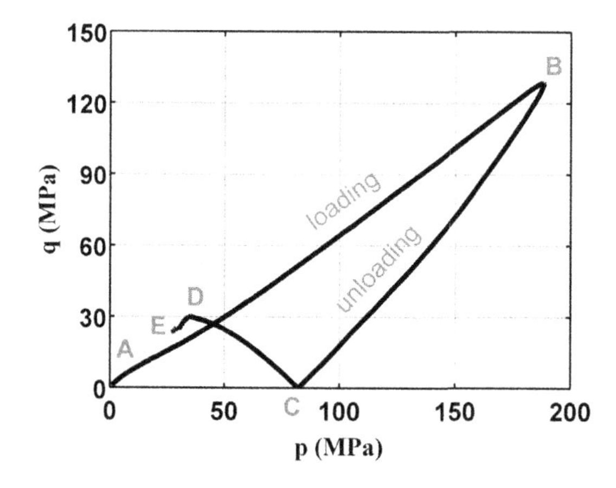

FIGURE 10.17

The variation of deviatoric stress with mean stress during compaction (Wu et al., 2005).

corresponding evolution of stress states in the p–q plane. During loading, the stress increases from (0, 0) (point A, in Fig. 10.17) until the maximum compression is reached (point B), at which point the powder has been plastically deformed (i.e., irreversible deformation has taken place). Therefore, from the incremental plasticity theory, point B should be on the yield surface or, in other words, it should be positioned on the cap surface in the DPC model.

During unloading, elastic deformation takes place and the stress state moves away from the yield surface. As the axial stress σ_z decreases, at a certain point it will have the same value as the radial stress σ_r, i.e., $\sigma_z = \sigma_r$. From Eq. (10.24), $q = 0$. This is the hydrostatic state, denoted as point C in Figs 10.16 and 10.17. Once the unloading progresses beyond the hydrostatic state, the axial stress becomes smaller than the radial stress (i.e., $\sigma_z < \sigma_r$). Since the von Mises equivalent stress q is nonnegative, as shown in Eq. (10.24), q increases as the unloading continues, until it reaches the shear failure surface $F_s\ [p, q] = 0$ (point D). The magnitudes of the slopes of BC and CD in the (p, q) plane are very close, provided that the change in relative density of the powder system is not significant. Further unloading will be accompanied by dilation (DE).

For elastic unloading from uniaxial compaction in a cylindrical die,

$$d\varepsilon_{11} = d\varepsilon_z \tag{10.25a}$$

$$d\varepsilon_{22} = d\varepsilon_r = 0 \tag{10.25b}$$

$$d\varepsilon_{33} = d\varepsilon_\theta = 0 \tag{10.25c}$$

Substituting Eqs (10.21)–(10.25) into Eq. (10.4), we obtain

$$\frac{d\sigma_z}{d\varepsilon_z} = K + \frac{4}{3}G \tag{10.26a}$$

$$\frac{dq}{dp} = \frac{2G}{\sqrt{3}K} \tag{10.26b}$$

Equation (10.26) indicates that the slope of the unloading curve (BC or CD) in Fig. 10.16 can be described using Eq. (10.26a), while the slopes of BC and CD in Fig. 10.17 can be approximated using Eq. (10.26b). Therefore, we have

$$K + \frac{4}{3}G = \frac{\sigma_{zB} - \sigma_{zC}}{\varepsilon_{zB} - \varepsilon_{zC}} \tag{10.27a}$$

$$\frac{2G}{\sqrt{3}K} = \frac{q_B - q_C}{p_B - p_C} = \frac{q_B}{p_B - p_C} \tag{10.27b}$$

where the subscripts B and C denote the values at points B and C in Figs 10.16 and 10.17, respectively.

From Eq. (10.27), the bulk modulus K and shear modulus G can be calculated. Consequently, the Young's modulus E and Poisson's ratio υ can be determined using Eqs (10.7) and (10.8). It should be noted that the Young's modulus E and Poisson's ratio υ obtained using Eq. (10.27) are for the powder with a specific relative density, which was achieved with a maximum compression pressure of σ_{zB}, and therefore assumes that dilation during unloading is insignificant. To obtain Young's modulus E and Poisson's ratio υ at different relative densities, the powder needs to be compressed at appropriate maximum pressures, i.e., σ_{zB}, and for each compression, one set of values for Young's modulus E and Poisson's ratio υ can be determined. As a result, the variation of E and υ with relative density can be determined. The use of Eq. (10.27) to obtain the elastic properties for powders undergoing significant dilation during unloading needs caution, as the variation in the relative density during unloading can be very large and thus significant errors may be introduced. For powders undergoing significant dilation during unloading, Eq. (10.26) can be used to determine the E and υ at different stress states during unloading, and the relative density can also be calculated from the increment axial strain so that the relationship between the elastic properties (E and υ) and the relative density can be determined.

10.3.3.2 Cohesion d and angle of friction β

The cohesion d and the angle of friction β can be determined from either (1) uniaxial compression in an instrumented die or (2) a combination of diametrical compression and unconfined uniaxial compression tests.

From uniaxial compression in an instrumented die, the nonlinear part of the stress–strain curve at the very end of unloading (DE in Fig. 10.16) is attributed to shear failure, and in the p–q plane, the corresponding stress state falls on the shear failure surface. Therefore, by fitting the regime DE with a straight line in the p–q plane (see Fig. 10.17), the cohesion d and the friction angle β can be determined as follows

$$d = \frac{p_E q_D - p_D q_E}{p_E - p_D} \tag{10.28a}$$

$$\beta = \tan^{-1}\left(\frac{q_D - q_E}{p_D - p_E}\right) \tag{10.28b}$$

This approach only involves the uniaxial compression test and can only be used for powders with insignificant dilation during unloading. In practice, some degree of dilation will inevitably occur, which will introduce a degree of error. A further practical problem in these measurements is that the stress level during unloading is very small by comparison with the maximum compression level, and so cannot be determined with high accuracy.

Alternatively, a combination of diametrical compression (Fig. 10.15(b)) and unconfined uniaxial compression tests (Fig. 10.15(c)) can be performed to determine the cohesion d and the angle of friction β. For these tests, it is critical to prepare some compacts with the same relative densities so that two different tests can be performed with samples of the same relative density. The maximum tensile stress σ_t in the diametrical compression test (Fig. 10.15(b)) can be calculated from the maximum break force using Eq. (2.36), while the maximum compression stress σ_c in the unconfined uniaxial compression test (Fig. 10.15(c)) can be determined from the break force F and the cross-sectional area of the cylindrical compact, i.e.,

$$\sigma_c = \frac{4F}{\pi D^2} \tag{10.29}$$

where D is the diameter of the compact. When the compact breaks during diametrical compression, the stress state at the center can be given as (Cunningham et al., 2004)

$$\sigma_1 = \sigma_t \tag{10.30a}$$

$$\sigma_2 = -3\sigma_t \tag{10.30b}$$

$$\sigma_3 = 0 \tag{10.30c}$$

Using Eqs (10.12) and (10.13), we have

$$p = \frac{2}{3}\sigma_t \tag{10.31a}$$

$$q = \sqrt{13}\sigma_t \tag{10.31b}$$

Similarly, for the unconfined uniaxial compression test, we have

$$\sigma_1 = -\sigma_c \tag{10.32a}$$

$$\sigma_2 = \sigma_3 = 0 \tag{10.32b}$$

and

$$p = \frac{1}{3}\sigma_c \tag{10.33a}$$

$$q = \sigma_c \tag{10.33b}$$

Substituting Eqs (10.31) and (10.33) into (10.14), and solving the system of equations, we obtain

$$d = \frac{(\sqrt{13}-2)\sigma_t\sigma_c}{\sigma_c - 2\sigma_t} = \frac{2.33\sigma_t\sigma_c}{\sigma_c - 2\sigma_t} \tag{10.34a}$$

$$\beta = \tan^{-1}\left[3\left(\frac{\sigma_c - d}{\sigma_c}\right)\right] \tag{10.34b}$$

Once the strength of a powder compact in diametrical compression and unconfined uniaxial compression is known (i.e., σ_t and σ_c), the cohesion d and the angle of friction β can be calculated using Eq. (10.34). This is a very robust approach for determining the cohesion d and the angle of friction β when the relative density is high, i.e., when strong compacts can be produced for testing. It may become problematic for compacts of low relative densities, as they tend to be too fragile to be used in these tests.

The two approaches introduced above are complementary so that they could be used in combination for determining the cohesion d and the angle of friction β over a wide range of relative densities. The first approach can be used for powders at low relative densities (i.e., under low compression pressure in uniaxial compression), for which the dilation during unloading is generally very small so that the measurement error can be minimized, while the second approach can be used for powder compacts with high relative densities as they are generally strong enough to be tested under diametrical compression and unconfined uniaxial compression.

10.3.3.3 Cap parameters: $\hat{R}$, p_a, α, *and* p_b

Four additional parameters, $\hat{R}$, p_a, α, and p_b are required to define the cap surface and the transition surface. Among these parameters, α is generally very small with a value of 0.01–0.05. Within this range, the variation of α has little effect on the yield behavior, as it is introduced primarily to ensure computational stability. It may therefore be chosen arbitrarily within this range.

The values of $\hat{R}$ and p_a can be determined from the analysis of the stress state on the cap surface. As discussed above, point B in Fig. 10.17 is located on the cap surface. Hence from Eq. (10.15), we have

$$F_c[p_B, q_B] = \sqrt{(p_B - p_a)^2 + \left(\frac{\hat{R}q_B}{1+\alpha-\alpha/\cos\beta}\right)^2} - \hat{R}(d + p_a \tan\beta) = 0 \tag{10.35}$$

At point B, the plastic strain increment in the radial direction $d\varepsilon_r^p$ can be assumed to be zero, as the deformation of the powder in the radial direction is constrained by the die wall. Thus, using Eq. (10.10), the radial plastic strain increment can be given as

$$d\varepsilon_r^p = \lambda \frac{\partial \Psi_c(p_B, q_B)}{\partial \sigma_r} = 0 \tag{10.36}$$

Since λ is a positive scalar, to satisfy Eq. (10.36), we have

$$\left.\frac{\partial \Psi_c}{\partial \sigma_r}\right|_{(p_B, q_B)} = 0 \tag{10.37}$$

Substituting Eq. (10.19) into Eq. (10.37) gives

$$2(p - p_a)\left.\frac{\partial p}{\partial \sigma_r}\right|_{(p_B,q_B)} + \frac{2\hat{R}^2 q}{(1+\alpha-\alpha/\cos\beta)^2}\left.\frac{\partial q}{\partial \sigma_r}\right|_{(p_B,q_B)} = 0 \tag{10.38}$$

Using Eqs (10.23), (10.24), and (10.38), we obtain

$$-\frac{2}{3}(p_B - p_a) + \frac{\hat{R}^2 q_B}{(1+\alpha-\alpha/\cos\beta)^2} = 0 \tag{10.39}$$

Hence, the cap parameter $\hat{R}$ can be solved from Eq. (10.39) as

$$\hat{R} = \sqrt{\frac{2\hat{\alpha}^2(p_B - p_a)}{3q_B}} \tag{10.40}$$

where $\hat{\alpha} = 1 + \alpha - \alpha/\cos\beta$.

Substituting Eq. (10.40) into Eq. (10.39), p_a can then be solved in terms of the known parameters p_B, q_B, α, d, and β and is given as (Han et al., 2008)

$$p_a = \frac{-3q_B - 4d\hat{\alpha}^2 \tan\beta + \sqrt{9q_B^2 + 24dq_B\hat{\alpha}^2 \tan\beta + 8\hat{\alpha}^2(3p_Bq_B + 2q_B^2)\tan^2\beta}}{4\hat{\alpha}^2 \tan^2\beta} \tag{10.41}$$

Once p_a is determined, the cap parameter $\hat{R}$ can be calculated from Eq. (10.40). Hence, using Eq. (10.16), p_b can be determined, i.e.,

$$p_b = \hat{R}d + p_a(1 + \hat{R}\tan\beta) \tag{10.42}$$

The corresponding volumetric plastic strain can be calculated from the relative densities at point B (i.e., the relative density at the maximum compression), ρ_B, and at point A, ρ_A (i.e., the initial relative density prior to compaction) as follows

$$\varepsilon_v^p = \varepsilon_{vB}^p = \ln\left(\frac{\rho_B}{\rho_A}\right) \tag{10.43}$$

Strictly speaking, the measured cap parameters $\hat{R}$, p_a, and p_b are related to the material state at maximum compression (i.e., point B in Figs 10.16 and 10.17); hence, the corresponding relative density and volumetric plastic strain should be ρ_B and ε_{vB}^p, respectively. To obtain the dependence of these parameters on relative density, the powder needs to be compressed to different maximum pressures. For each compaction, a specific value for each of these parameters can be obtained, together with the corresponding volumetric plastic strain ε_v^p. Hence the cap parameters $\hat{R}$ and p_a can be obtained as functions of relative density, as can p_b as a function of the volumetric plastic strain ε_v^p. For some applications, a density-independent calibration can be performed, which is discussed in Box 10.3.

BOX 10.3 DENSITY-INDEPENDENT CALIBRATION

If the effect of relative density on the constitutive parameters is insignificant, i.e., the elastic properties (Young's modulus and Poisson's ratio), the cohesion d, the angle of friction β, and the cap parameters $\hat{R}$ and p_a are independent of the relative density, the relationship between p_b and ε_v^p may also be determined from a single set of compaction data, as adopted by Aydin et al. (1996), Michrafy et al. (2002), and Wu et al. (2005). In this case, the Young's modulus and Poisson's ratio are constants, and the unloading curves in the stress-strain diagram will be parallel to each other if the powder is compressed to various maximum compression pressures, as the slopes of the unloading curves are identical as defined by Eq. (10.26a). The volumetric plastic strain can be determined as:

$$\varepsilon_v^p = \varepsilon_z - \frac{\sigma_z}{K + 4G/3} \tag{10.44}$$

Using the corresponding stress state (p, q) on the loading curve, p_a can be determined from Eq. (10.41) by replacing (p_B, q_B) with (p, q), and p_b can then be determined using Eq. (10.42). Therefore, the variation of p_b with ε_v^p can be determined.

10.3.3.4 DPC model calibration procedure

The calibration procedure for the DPC model using uniaxial compression and diametrical compression described above can be summarized as follows:

Step 1:
Produce cylindrical samples of a certain relative density D_1 and perform diametrical compression and unconfined uniaxial compression tests to obtain the tensile stress σ_t and the compression strength σ_c, respectively. Use Eq. (10.34) to calculate the cohesion d_1 and the internal frictional angle β_1.
Step 2:
Perform uniaxial compression (loading and unloading) of the sample powder in an instrumented die to a maximum compression pressure that results in a relative density of the same value as D_1 at the maximum compression. Plot σ_z versus ε_z, and p versus q from the uniaxial data in the similar way to that shown in Figs 10.16 and 10.17. Identify the critical stress states pointed out as A, B, C, D, and E. Use Eq. (10.27) to calculate the bulk modulus K_1 and shear modulus G_1, then calculate the Young's modulus E_1 and Poisson's ratio υ_1 using Eqs (10.7) and (10.8).
Step 3:
If it is impossible to get the cohesion d_1 and the internal frictional angle β_1 in Step 1, especially for powders in a certain relative density range, fit the regime DE in the $p-q$ plot using a straight line, and calculate the cohesion d_1 and the internal frictional angle β_1 using Eq. (10.28).
Step 4:
Choose a value for α ($= 0.01 \sim 0.05$), calculate $\hat{R}$ using Eq. (10.40), p_a from Eq. (10.41) and p_b and ε_v^p from Eqs (10.42) and (10.43).

Repeating the above steps for different relative densities, the variations of the DPC parameters with relative densities can then be obtained. Tables 10.1−10.6 list a collection of DPC-based constitutive parameters for various powders reported in the literature.

10.3.4 MODELING PROCEDURE AND APPLICATIONS

The theoretical approaches discussed in this chapter together with other algorithms can be implemented in a computer program, which forms the FEM solver. A variety of FE packages are available as FEM solvers. Using these solvers to model the compaction of a powder, a range of input parameters are needed, which can be classified into two categories: material parameters and process parameters. For a given powder, as illustrated in Fig. 10.18, it is necessary not only to choose a material model that can represent the deformation behavior of the powder, but also to perform model calibration (see Section 10.3.3) so that the input material parameters are specific to the powder considered. Process parameters are also needed so that the geometry and process conditions can be specified; these are generally defined in the FE model as boundary and loading conditions. Using these input parameters, FE modeling can be performed to explore the responses of the system, which include

Table 10.1 Density-Independent DPC Parameters Reported in the Literature

Powder	E (GPa)	v	d (MPa)	β(°)	α	$\hat{R}$	p_b (MPa)
α-Al_2O_3 (AKP-30), an agglomerated alumina powder (Aydin et al., 1996)	9.03	0.28	5.50	16.50	0.03	0.56	$p_b = c_1 + c_2 \times 10^{C_3 \cdot \varepsilon_v^p}$ $c_1 = 3.07$, $c_2 = 0.42$, $c_3 = 7.34$
Lactose monohydrate (Wu et al., 2005)	3.57	0.12	9.10×10^{-4}	41.02	0.03	0.60	$p_b = c_1 e^{C_2 \cdot \varepsilon_v^p}$ $c_1 = 8.12 \times 10^{-4}$, $c_2 = 9.05$
Lactose (Michrafy et al., 2002)	4.60	0.17	0.46	29.30	0.03	0.058	Column A in Table 10.2
Distaloy AE, a steel powder (Bejarano et al., 2003)	$E = k \exp\left(\frac{\rho}{\rho_0}\right)^n$ $k = 8.757$ GPa $n = 5$	0.28	3.254	70.84	0.01	0.68	4.14

Table 10.2 The Variation of p_b with ε_v^p Reported in Michrafy et al. (2002)

p_b (MPa)	ε_v^p
5.00×10^{-4}	0.20
5.90	0.26
21.51	0.32
19.58	0.37
29.02	0.41
56.64	0.48
97.93	0.55
145.84	0.62
227.72	0.64
239.28	0.65

Table 10.3 DPC Parameters for MCC PH 102 from Cunningham et al. (2004): $\alpha = 0.03$

Relative Density	*E* (GPa)	v	*d* (MPa)	β (°)	$\hat{R}$
0.30	0.50	0.02	0.10	41.00	0.05
0.35	0.55	0.03	0.20	46.00	0.08
0.40	0.60	0.04	0.30	48.00	0.11
0.45	0.70	0.05	0.50	52.00	0.13
0.50	0.80	0.07	0.70	54.00	0.16
0.55	0.90	0.08	1.10	57.00	0.20
0.60	1.20	0.10	1.50	58.00	0.25
0.65	2.00	0.12	2.00	61.00	0.32
0.70	2.50	0.14	2.60	63.00	0.39
0.75	3.00	0.17	3.30	65.00	0.45
0.80	4.00	0.20	4.20	66.00	0.51
0.85	4.60	0.23	5.30	68.00	0.58
0.90	6.50	0.26	6.60	71.00	0.65
0.95	8.50	0.29	8.20	72.00	0.75

the response of the powder during the process, e.g., compaction behavior, properties of final products, such as density and stress distributions, and the response of the tooling if it is of interest. In this section, typical output from FE modeling will be discussed in order to illustrate the FE capability.

As an example, consider the compaction of a lactose powder in a die of 8 mm diameter. The calibrated material properties are given in Wu et al. (2005) and are presented in Table 10.1. The powder is compressed with a moving upper punch while the lower punch is stationary. As it is an axisymmetrical problem, it can be modeled with

Table 10.4 The Variation of p_b with ε_v^p for MCC PH 102 from Cunningham et al. (2004): $\alpha = 0.03$

p_b (MPa)	ε_v
0.043	0
0.62	0.12
1.61	0.26
2.59	0.37
4.08	0.48
6.87	0.57
12.00	0.66
16.80	0.74
23.70	0.81
31.80	0.88
42.30	0.95
56.60	1.01
79.60	1.07
119.00	1.12
143.00	1.14

Table 10.5 DPC Parameters for a steel Powder (Distaloy AE) from Shang et al. (2012)

Relative Density	E (GPa)	v	d (MPa)	β (°)	$\hat{R}$
0.57	9.64	0.19	0.59	71.31	0.48
0.61	13.50	0.16	0.98	70.97	0.48
0.65	15.81	0.16	1.64	70.63	0.50
0.69	19.86	0.15	2.70	70.82	0.52
0.71	22.56	0.15	3.63	70.83	0.53
0.75	30.08	0.14	6.55	70.67	0.56
0.78	37.22	0.12	8.41	70.50	0.58
0.80	39.72	0.12	10.20	70.33	0.60
0.82	45.90	0.12	12.93	70.51	0.62
0.83	51.68	0.11	15.12	69.99	0.63
0.85	55.73	0.11	17.78	70.52	0.65
0.86	60.36	0.11	20.24	70.17	0.67

Table 10.6 The Variation of p_b with ε_v^p for a steel Powder (Distaloy AE) from Shang et al. (2012)

p_b (MPa)	ε_v
2.63	0.00
9.33	0.33
18.88	0.42
31.07	0.49
41.52	0.54
54.61	0.59
71.19	0.64
93.89	0.70
111.36	0.74
134.07	0.78
150.68	0.80
172.52	0.83
204.87	0.86
229.35	0.88
250.34	0.90
278.32	0.92
298.44	0.93

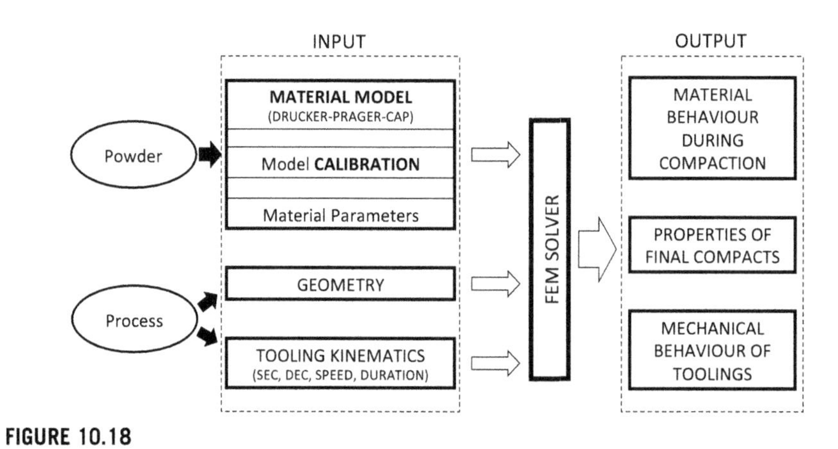

FIGURE 10.18

Illustration of typical powder compaction modeling process using FEM.

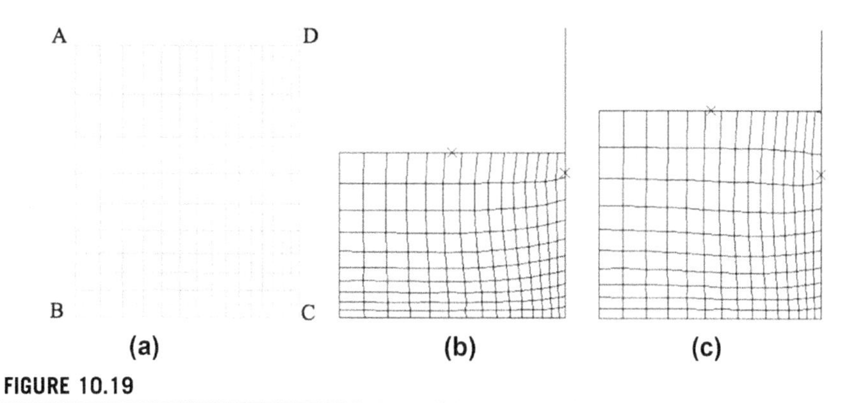

FIGURE 10.19

Finite element meshes: (a) before compaction, (b) at the end of compression, and (c) during decompression (Wu et al., 2005).

a 2D FEM: only half of the evolving section is meshed using axisymmetric continuum elements. The initial FE mesh is shown in Fig. 10.19(a). The die wall and the upper punch are modeled as rigid bodies since their deformation is negligible compared to that of the powder. The master-slave contact with finite sliding discussed in Section 10.1.1 is employed to model the interaction between the powder and tooling surface. The friction in the contact was determined experimentally using the method described in Section 7.3 and was given in Table 7.1. For the boundary condition, the nodes on the symmetrical axis (AB) are constrained to move in the horizontal direction and the nodes at boundary BC are constrained to move in the vertical direction. The upper punch only moves vertically at a specified compression speed of 3 mm/s. A uniform distribution of the initial relative density is assumed.

During powder compression, the volume of the powder bed reduces, which manifests itself in the deformation and distortion of the FE meshes as illustrated in Fig. 10.19(b) which shows the FE meshes at the maximum compression. Fig. 10.19(c) shows the meshes at the end of compaction. By examining the FE meshes at different time instants, the deformation of the powder bed is clearly represented: the powder bed is compressed during loading and dilated during unloading (i.e., relaxation). Furthermore, the packing density and stress distributions under the applied pressure can also be determined, which are shown as contour plots in Figs 10.20 and 10.21, respectively.

The density distribution is produced by mapping the relative density of every element. Fig. 10.20 shows the relative density distribution at the maximum compression. It can be seen that the density distribution is not uniform: a high density is induced around the top edge, with a low density at the bottom edge, which is in good agreement with the experimental observations of many others (Train, 1957; Kim, 2003). The nonuniform density distribution is primarily due to the presence of friction at the die wall, which constrains the motion of the powder in that region. Consequently the stress transmitted to the region near the bottom edge is limited and the powder is less compacted there. Near the top edge the

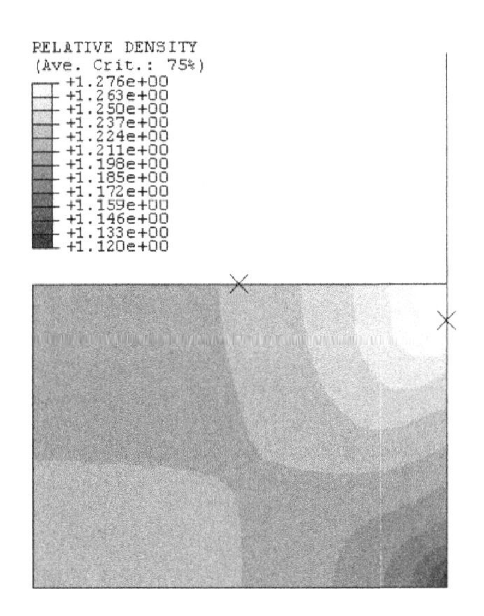

FIGURE 10.20

Density distribution at maximum compression (Wu et al., 2005).

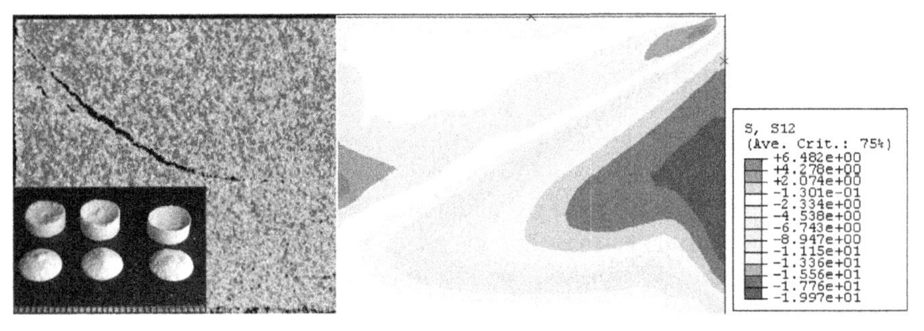

FIGURE 10.21

The distribution of shear stress obtained using FEM (left), and X-ray computed tomographic images (right) and photographs of ejected lactose tablets (insert) (Wu et al., 2005).

powder is subject to a larger stress induced by the combined effect of the compression pressure and the shear stress along the wall, leading to a higher degree of compression and a higher relative density.

The typical distribution of shear stress at the early stage of the unloading is presented on the right-hand side of Fig. 10.21. It is apparent that there is an intensive localized shear band from the top edge towards the mid-center of the compressed powder. Dilation is generally caused by shear deformation in powders. Consequently, the powder in the shear zone is less compacted and the bonding strength is relatively weak. This is further confirmed by the tablet failure pattern during powder compaction observed from the X-ray tomographic images of the ejected tablet and the photograph of three broken tablets after ejection in the same tests. This experimental evidence shows that cracks are developed from the top edge toward the mid-center, similarly to the pattern of shear banding obtained from the FE analysis. This demonstrates the utility of FE analysis in indicating possible failure patterns during powder compaction. Nevertheless, it remains a challenging task to model and predict the propagation of cracks in compacted products, a task which deserves further study.

REFERENCES

ABAQUS Theory Manual, Version 6.2, 2013. HKS Inc.

Aydin, I., Briscoe, B.J., Sanlitürk, K.Y., 1996. The internal form of compacted ceramic components: a comparison of a finite element modelling with experiment. Powder Technology 89, 239–254.

Bathe, K.J., 1982. Finite Element Procedures in Engineering Analysis. Prentice-Hall, Englewood Cliffs, New Jersey.

Bejarano, A., et al., 2003. Journal of Materials Processing Technology 143–144, 34–40.

Cho, S.S., Park, S., 2004. Finite element modeling of adhesive contact using molecular potential. Tribology International 37 (9), 763–769.

Coube, O., Riedel, H., 2000. Numerical simulation of metal powder die compaction with special consideration of cracking. Powder Metallurgy 43, 123–131.

Cunningham, J.C., Sinka, I.C., Zavaliangos, A., 2004. Analysis of tablet compaction. I. Characterization of mechanical behavior of powder and powder/tooling friction. Journal of Pharmaceutical Sciences 93 (8), 2022–2039.

DiMaggio, F.L., Sandler, I.S., 1971. Material model for granular soils,. Journal of Engineering Mechanics ASCE 96, 935–950.

Drucker, D.C., Prager, W., 1952. Soil mechanics and plastic analysis or limit design. Quarterly of Applied Mathematics 10, 157–165.

Feng, X.Q., Li, H., Zhao, H.P., Yu, S.W., 2009. Numerical simulations of the normal impact of adhesive microparticles with a rigid substrate. Powder Technology 189 (1), 34–41.

Frenning, G., 2008. An efficient finite/discrete element procedure for simulating compression of 3D particle assemblies. Computer Methods in Applied Mechanics and Engineering 197 (49–50), 4266–4272.

Frenning, G., 2010. Compression mechanics of granule beds: a combined finite/discrete element study. Chemical Engineering Science 65 (8), 2464. ISSN 0009-2509.

Green, R.J., 1972. A plasticity theory for porous solids. International Journal of Mechanical Sciences 14, 215–222.

Güner, F., Cora, Ö.N., Sofuoğlu, H., 2015. Numerical modeling of cold powder compaction using multi particle and continuum media approaches. Powder Technology 271, 238–247.

Gustafsson, G., Häggblad, H.-Å., Jonsén, P., 2013. Multi-particle finite element modelling of the compression of iron ore pellets with statistically distributed geometric and material data. Powder Technology 239, 231–238.

Han, L.H., Elliott, J.A., Bentham, A.C., Mills, A., Amidon, G.E., Hancock, B.C., 2008. A modified Drucker-Prager Cap model for die compaction simulation of pharmaceutical powders. International Journal of Solids and Structures 45 (10), 3088–3106.

Harthong, B., Jérier, J.-F., Dorémus, P., Imbault, D., Donzé, F.-V., 2009. Modeling of high-density compaction of granular materials by the discrete element method. International Journal of Solids and Structures 46, 3357–3364.

Harthong, B., Imbault, D., Doremus, P., 2012. The study of relations between loading history and yield surfaces in powder materials using discrete finite element simulations. Journal of the Mechanics and Physics of Solids 60 (4), 784–801.

Hughes, T.J.R., Taylor, R.L., Sackman, J.L., Curnier, A., Kanoknukulchai, W., 1976. Finite element method for a class of contact-impact problems. Computer Methods in Applied Mechanics and Engineering 8, 249–276.

Hughes, T.J.R., 1987. The Finite Element Method-linear Static and Dynamic Finite Element Analysis. Prentice-Hall, Englewood Cliffs, New Jersey.

Kim, H.S., 2003. Densification modelling for nanocrystalline metallic powders. Journal of Materials Processing Technology 140, 401–406.

Lewis, R.W., Gethin, D.T., Yang, X.S., Rowe, R.C., 2005. A combined finite-discrete element method for simulating pharmaceutical powder tableting. International Journal for Numerical Methods in Engineering 62 (7), 853–869.

Michrafy, A., Ringenbacher, D., Techoreloff, P., 2002. Modelling the compaction behaviour of powders: application to pharmaceutical powders. Powder Technology 127, 257–266.

Procopio, A.T., Zavaliangos, A., 2005. Simulation of multi-axial compaction of granular media from loose to high relative densities. Journal of the Mechanics and Physics of Solids 53 (7), 1523–1551.

Ransing, R.S., Gethin, D.T., Khoei, A.R., Mosbah, P., Lewis, R.W., 2000. Powder compaction modelling via the discrete and finite element method. Materials and Design 21 (4), 263–269.

Sauer, R.A., Li, S., 2007. An atomic interaction-based continuum model for adhesive contact mechanics. Finite Elements in Analysis and Design 43 (5), 384–396.

Sauer, R.A., Wriggers, P., 2009. Formulation and analysis of a three-dimensional finite element implementation for adhesive contact at the nanoscale. Computer Methods in Applied Mechanics and Engineering 198 (49–52), 3871–3883.

Schofield, A.N., Wroth, C.P., 1968. Critical State Soild Mechanics. McGraw-Hill, London.

Shang, C., Sinka, I.C., Pan, J., 2012. Constitutive model calibration for powder compaction using instrumented die testing. Experimental Mechanics 52, 903–916.

Sinka, I.C., Cunningham, J.C., Zavaliangos, A., 2003. The effect of wall friction in the compaction of pharmaceutical tablets with curved faces: a validation study of the Drucker-Prager Cap model. Powder Technology 133, 33–43.

Train, D., 1957. Transmission of forces through a powder mass during the process of pelleting. Transactions of the Institution of Chemical Engineers 35, 258–266.

Timoshenko, S., Goodier, J.N., 1951. Theory of Elasticity, third ed. McGraw-Hill, New York.

Wang, Z.-Z., Xu, Y., Gu, P., 2012. Adhesive behaviour of gecko-inspired nanofibrillar arrays: combination of experiments and finite element modelling. Journal of Physics D: Applied Physics 45 (14), 142001.

Whirley, R.G., Engelmann, B.E., 1992. DYNA2D, a Nonlinear, Explicit, Two-dimensional Finite Element Code for Solid Mechanics: User Manual. Lawrence Livermore National Laboratory, University of California, USA.

Wu, C.Y., 2002. Finite Element Analysis of Particle Impact Problems. Aston University. PhD Thesis.

Wu, C.Y., Ruddy, O., Bentham, A.C., Hancock, B.C., Best, S.M., Elliott, J.A., 2005. Modelling the mechanical behaviour of pharmaceutical powders during compaction. Powder Technology 152 (1–3), 107–117.

Xu, J.G., Fan, H., June 2004. Elastic analysis for liquid-bridging induced contact. Finite Elements in Analysis and Design 40 (9–10), 1071–1082.

Zhang, J., October 2009. A study of compaction of composite particles by multi-particle finite element method. Composites Science and Technology 69 (13), 2048. ISSN 0266-3538.

Zhang, X.J., Zhang, X.H., Wen, S.Z., 2011. Finite element modeling of the nano-scale adhesive contact and the geometry-based pull-off force. Tribology Letters 41 (1), 65–72.

Zhong, Z.H., 1993. Finite Element Procedures for Contact-Impact Problems. Oxford University Press, Oxford.

Zienkiewicz, O.C., Taylor, R.L., Zhu, J.Z., 1989. The Finite Element Method: Its Basis and Fundamentals, seventh ed. Elsevier, London.

Zienkiewicz, O.C., Taylor, R.L., Zhu, J.Z., 2013. The Finite Element Method: Its Basis and Fundamentals. Elsevier, London.

Index

'*Note*: Page numbers followed by "f" indicate figures, "t" indicate tables, and "b" indicate boxes.'

K

L

M

N

P

R

S

T

U

V

W

Y

Printed in the United States
By Bookmasters